Advances in Understanding Rock Mass Structural-Dependent Cyclic and Fatigue Behaviors

Advances in Understanding Rock Mass Structural-Dependent Cyclic and Fatigue Behaviors

Topic Editors

Yu Wang
Yingjie Xia

MDPI Basel • Beijing • Wuhan • Barcelona • Belgrade • Novi Sad • Cluj • Manchester

Topic Editors
Yu Wang
School of Resource and Safety Engineering
University of Science and Technology Beijing
Beijing
China

Yingjie Xia
School of Civil Engineering
Dalian University of Technology
Dalian
China

Editorial Office
MDPI AG
Grosspeteranlage 5
4052 Basel, Switzerland

This is a reprint of the Topic, published open access by the journals *Applied Sciences* (ISSN 2076-3417), *Geosciences* (ISSN 2076-3263), *Materials* (ISSN 1996-1944) and *Minerals* (ISSN 2075-163X), freely accessible at: https://www.mdpi.com/topics/91T57D0166.

For citation purposes, cite each article independently as indicated on the article page online and as indicated below:

Lastname, A.A.; Lastname, B.B. Article Title. *Journal Name* **Year**, *Volume Number*, Page Range.

ISBN 978-3-7258-7258-9 (Hbk)
ISBN 978-3-7258-7259-6 (PDF)
https://doi.org/10.3390/books978-3-7258-7259-6

Contents

Hongzhong Zhang, Linqi Huang, Xibing Li, Xingmiao Hu and Yangchun Wu
Study on the Evolution of Physical Parameters and Dynamic Compression Mechanical Properties of Granite after Different Heating and Cooling Cycles
Reprinted from: *Materials* **2023**, *16*, 2300, https://doi.org/10.3390/ma16062300 . . . . **1**

Qi Wu, Bowen Li and Xuehai Jiang
Triaxial Test Study on Energy Evolution of Marble after Thermal Cycle
Reprinted from: *Minerals* **2023**, *13*, 428, https://doi.org/10.3390/min13030428 . . . . **21**

Yunfeng Wu, Yu Wang, Changhong Li, Baokun Zhou, Zicheng Tian, Changkun Sun and Youdong Zhu
Damping and Stiffness Responses of Silica Rock under Constant Amplitude and Variable Rate Cyclic Loading
Reprinted from: *Appl. Sci.* **2024**, *14*, 4713, https://doi.org/10.3390/app14114713 . . . . **42**

Shiqi Li, Yuan Li, Dongjue Fan, Liang Zhao and Litian Zhang
Piecewise Linear Strength Models for Analyzing Multiple Failure Mechanisms in Rocks Materials
Reprinted from: *Materials* **2024**, *17*, 4102, https://doi.org/10.3390/ma17164102 . . . . **59**

Buchu Zhang, Shichuan Zhang, Baotang Shen, Yangyang Li, Shilong Song and Xuexian Han
Conducting Research to Identify Key Features and Critical Nodes in the Coalescence and Instability of Pre-Fabricated Jointed Rock
Reprinted from: *Appl. Sci.* **2024**, *14*, 7905, https://doi.org/10.3390/app14177905 . . . . **78**

Yongchun Yu, Yu Wang, Xuefeng Yi and Zhenzhen Chen
Mechanical Responses and Fracture Evolution of Marble Samples Containing Stepped Fissures under Increasing-Amplitude Cyclic Loading
Reprinted from: *Appl. Sci.* **2024**, *14*, 7919, https://doi.org/10.3390/app14177919 . . . . **97**

Yonghong Liu, Fujun Zhao, Qiuhong Wu and Zhouyuan Ye
Mechanical Properties of Rock-like Materials Under Disturbance Loads at Different Lateral Pressures
Reprinted from: *Materials* **2024**, *17*, 5439, https://doi.org/10.3390/ma17225439 . . . . **113**

Meimei Feng, Xiaoxiao Cao, Taifeng Wu and Kangsheng Yuan
Mechanical Properties and Damage Constitutive Model of Saturated Sandstone Under Freeze–Thaw Action
Reprinted from: *Materials* **2024**, *17*, 5905, https://doi.org/10.3390/ma17235905 . . . . **127**

Peng Li, Tianqi Chen, Yan Liu, Meifeng Cai, Liang Sun, Peitao Wang, et al.
Automatic Identification of Rock Discontinuity Sets by a Fuzzy C-Means Clustering Method Based on Artificial Bee Colony Algorithm
Reprinted from: *Appl. Sci.* **2025**, *15*, 1497, https://doi.org/10.3390/app15031497 . . . . **146**

Zhongzhong Xu, Jiulong Cheng and Hongpeng Zhao
The High-Precision Monitoring of Mining-Induced Overburden Fractures Based on the Full-Space Inversion of the Borehole Resistivity Method: A Case Study
Reprinted from: *Geosciences* **2025**, *15*, 320, https://doi.org/10.3390/geosciences15080320 . . . . . **165**

Xianda Yang, Peng Zeng, Kui Zhao, Dong Zhang, Hepeng Zhang, Nan Liang and Lihui Sun
Experimental Study on Damage Evolution Characteristics of Granite Under Short-Term Freeze–Thaw Cycles
Reprinted from: *Materials* **2026**, *19*, 853, https://doi.org/10.3390/ma19050853 . . . . **185**

materials

MDPI

Article

Study on the Evolution of Physical Parameters and Dynamic Compression Mechanical Properties of Granite after Different Heating and Cooling Cycles

Hongzhong Zhang, Linqi Huang *, Xibing Li, Xingmiao Hu and Yangchun Wu

School of Resources and Safety Engineering, Central South University, Changsha 410083, China; zhz0804@csu.edu.cn (H.Z.); xbli@csu.edu.cn (X.L.); huxingmiao@csu.edu.cn (X.H.); wuyangchun1995@csu.edu.cn (Y.W.)

* Correspondence: huanglinqi@csu.edu.cn

Abstract: The study of the evolution law of basic physical parameters and dynamic compression performance of deep granite under the environment of the heating-cooling cycle is of great significance for the stability evaluation of deep underground engineering and the development of deep resources. In this study, heating-cooling cycle tests and dynamic compression tests were conducted on a large number of fine-grained granite specimens with heating temperatures from 200 to 600 °C and times from one to twenty times using a box-type high-temperature muffle furnace and Hopkinson pressure bar (SHPB) test system, and the evolution law of basic physical parameters and dynamic compression mechanical properties of fine-grained granite were studied using theoretical and fitting analysis. The test results showed that: the changes of the basic physical parameters of granite have obvious temperature effect; 600 °C is a threshold value for the changes of each physical parameter of granite; the sensitivity of each physical parameter to the number of heating and cooling cycles is small before 600 °C; and the sensitivity of each physical parameter to the number of heating and cooling cycles significantly increases at 600 °C. The dynamic compressive strength and elastic modulus of granite decreased with the increase in heating and cooling cycles, and the maximum decrease rate was 89.1% and 85.9%, respectively, and the strain rate linearly increased with the increase in heating and cooling cycles, and the maximum strain rate was 123 s^{-1}. The temperature, the number of heating and cooling cycles, and the impact air pressure, all had significant effects on the damage mode and crushing degree of granite.

Keywords: heating-cooling cycle; cooling mode; physical parameters; microscopic damage; dynamic compression properties

1. Introduction

With the construction of deep underground engineering (nuclear waste storage), the mining of dry hot rock (Figure 1) and the development of deep mining of metal mineral resources, the mechanical properties and stability of rock masses are significantly influenced by the deep water-heat alternating action environment, and the evolution law of basic physical parameters and dynamic properties of deep rock masses under the heating-cooling cycle action environment have become a hot research topic in recent years in rock mechanics. The study of the evolution of basic physical parameters and dynamic compression properties of rocks after the heating-cooling cycle is important for the drilling and mining of dry hot rock, which can effectively improve the mining efficiency of dry hot rock; on the other hand, it can provide a theoretical basis for the development of the composite rock-breaking method of the heating-cooling cycle and mechanical impact, as well as high-temperature solution leaching mining.

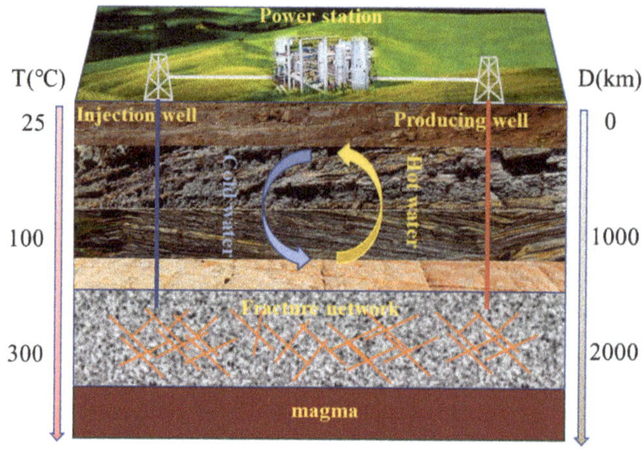

Figure 1. Schematic diagram of hot dry rock mining.

Many scholars at home and abroad have studied the effect of temperature on granite from various directions and have achieved a lot of results [1,2]. Li et al. [3], Wang et al. [4], and Zhao et al. [5] studied the effect of different heat treatments on the fracture characteristics of granite. Yin et al. [6,7] and Li et al. [8] studied the effect of temperature on the dynamic mechanical properties of granite. Sun et al. [9] studied the effect of temperature on the physical properties of granite such as quality and surface color. Xu et al. [10] and Zhou et al. [11] investigated the thermal microcrack extension of granite using optical microscopy, thin section observation and scanning electron microscopy (SEM). Zhang et al. [12] and Hao et al. [13] investigated the effect of temperature on the physical and mechanical properties, as well as the intrinsic structure relationship, of granite. Ding et al. [14] studied the effect of temperature on the permeability of fractured granite. Wang et al. [15,16] investigated the effect of two different heating methods, slow heating and fast heating, on the thermal cracking of granite.

Many others have studied the effect of different cooling methods on the mechanical properties of granite. Xi et al. [17] studied the effects of natural cooling, water cooling and liquid nitrogen (LN2) cooling on the dynamic mechanical properties of rocks and their fracture characteristics. Zhao et al. [18] studied the effect of different temperatures and cooling methods on parameters such as surface hardness and the uniaxial compressive strength of granite. Tang et al. [19] investigated the effect of temperature on the basic friction angle of granite cracks under water and liquid nitrogen cooling methods, different specimen sizes and tilt rates. Shao et al. [20] and Chen et al. [21] investigated the physical and mechanical damage characteristics of granite at different high temperatures under the effect of liquid nitrogen cooling. Guan et al. [22] studied the damage characteristics of granite specimens and their mechanisms after different heating-cooling cycles using air-cooling and liquid nitrogen cooling.

The physical and mechanical properties of granite under the action of heating and cooling cycles have also been studied. Yin et al. [23] and Li et al. [24] studied the effects of cyclic heating and cooling treatments on the microscopic pore structure of granite. Chen et al. [25] studied the effects of combined cooling on the mechanical properties of granite at different temperatures and numbers of hot and cold cycles. Zhu et al. [26] studied the effects of longitudinal wave velocity, uniaxial compressive strength, and elastic modulus of granite specimens to different cyclic heating and water cooling treatments. Wu et al. [27] and Wang et al. [28] studied the physical and mechanical properties and crack expansion of granite under single and multiple heating-cooling cycles. Yin et al. [29] investigated the microscopic thermal damage mechanism and mechanical properties of granite under cyclic heating-cooling. Zhang et al. [30] developed a method to realize the mechanical properties of granite in deep ground and established a numerical simulation method that enables long-term mechanical analysis of reservoir rocks considering thermal damage in deep geothermal engineering. Zhu et al. [31] analyzed changes in the mechanical properties of granite after various thermal impacts. Sun et al. [32] investigated the effect of cyclic heating and cooling effects on the thermal diffusion coefficient of granite at different temperatures. Ge et al. [33] and Jiang et al. [34] studied the effect of heating, and Li et al. [35] studied the effect of different cooling methods on the dynamic mechanical properties of granite.

From the above analysis, it can be seen that there have been relatively few studies on the dynamic compressive properties and damage characteristics of granite after the cyclic action of heating and cooling at different temperatures by domestic and foreign scholars, and most of the studies on the effect of temperature on the physical and mechanical properties of granite are single cooling action or using a single cooling mode, there are few studies on the physical and mechanical properties of granite after heating and cooling cycles under multiple cooling methods. In this study, the evolution law of basic physical parameters and dynamic compression mechanical properties and damage characteristics of granite after cyclic heating and cooling are analyzed by using water cooling and natural cooling, and the physical and mechanical properties of granite under heating and cooling cycles are further studied, which is important for improving the reservoir modification and drilling efficiency of dry hot rocks.

In this study, many fine-grained granite specimens were subjected to heating-cooling cycles and dynamic impact tests from 200 to 600 °C one to twenty times using a box-type high-temperature muffle furnace and SHPB test system, and the evolution of the basic physical parameters such as mass and volume of fine-grained granite under the effect of heating-cooling cycles and dynamic compression mechanical properties were summarized.

2. Specimen Preparation and Experimental Process

2.1. Preparation of Rock Specimens

The test material selected for this test was fine-grained granite from Zhangzhou City, Fujian Province, China, and the specimens were processed from the same granite raw material, which had a dense structure and high strength, and could be better processed for heating and cooling cycles. The specimens were processed in strict accordance with the requirements of T/CSRME 001-2019 Rock Dynamic Properties Test Procedure [36]. It was required that the non-parallelism and non-perpendicularity of each specimen were less than 0.02 mm, and the error of specimen height and diameter was less than 0.3 mm, and the specimen was processed into a cylindrical rock specimen with the specification of φ50 mm × 50 mm (Figure 2).

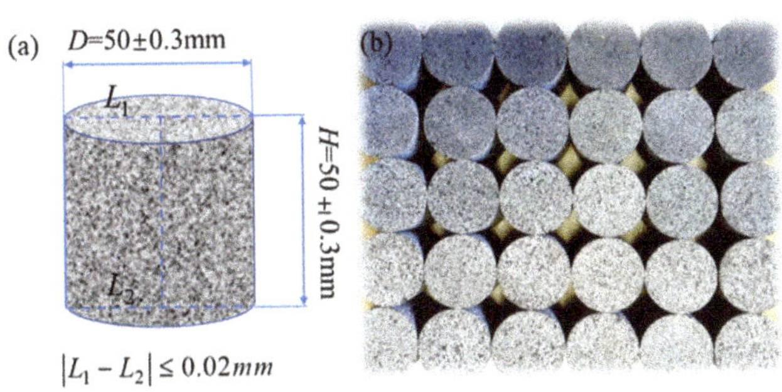

Figure 2. (**a**) Rock specimen processing requirements. (**b**) Partially processed fine-grained granite specimen.

2.2. Experimental Process

The whole test process is divided into four steps: measurement of physical parameters before treatment, heating and cooling cycle treatment, measurement of physical parameters after treatment and dynamic compression; the test process and equipment used are shown in Figure 3.

(1) Measurement of physical parameters before treatment. The physical quantities to be measured are mass, volume, density, longitudinal wave velocity and porosity of the specimen, which are measured by high precision electronic scales, electronic vernier calipers and HS-YS4A Rock Acoustic Wave Parameter Tester.

(2) Heating and cooling cycle treatment. Heating and cooling cycle treatment mainly has two kinds of fast water cooling and natural cooling; the heating temperature is 200 °C, 400 °C and 600 °C, and the number of cycles is 1, 5, 10, 20 times, respectively. Firstly, put the completed numbered specimens into the high temperature muffle furnace, heat them to the specified temperature at a heating rate of 5 °C/min, keep them warm for 2 h so that the specimens are evenly heated, and then quickly use the corresponding cooling method (fast water cooling, natural cooling) to fully cool them for 2 h after the heating is completed, and repeat the above steps until the heating and cooling treatment is completed for 5, 10 and 20 times.

(3) Measure the relevant physical parameters of the specimens treated by the heating and cooling cycle again, analyze their change characteristics and summarize the influence law of heating and cooling cycle effects on the physical parameters of the granite.

(4) The specimens after the heating and cooling cycle were subjected to uniaxial dynamic compression experiments using a Hopkinson pressure bar (SHPB) test system with an impact air pressure of 0.6, 0.7 and 0.8 MPa to analyze the effect of the heating and cooling cycle on the dynamic compression performance of the granite.

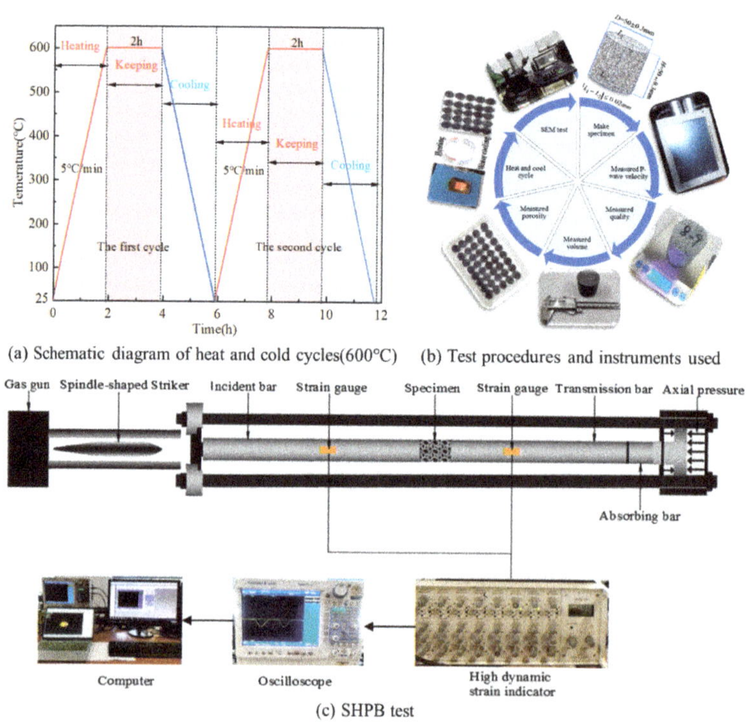

Figure 3. Schematic diagram of the test process and equipment. (**a**) Schematic diagram of heating and cooling cycles (600 °C), (**b**) Test procedures and instruments used, (**c**) SHPB test system.

2.3. Introduction of SHPB Test System

As shown in Figure 3, the SHPB test system consists of a dynamic loading system, a strain acquisition system, a data processing system, etc. The dynamic loading system consists of cylinders, spindle-shaped bullets, an incidence bar, a transmission bar, an absorption bar, etc. The high-pressure nitrogen gas is used as the power source to push the bullets by instantaneous release, so that the bullets can obtain high speed and then achieve dynamic loading on the specimen.

3. Evolutionary Patterns of Basic Physical Parameters of Granite

In order to be able to better describe the effect of the heating and cooling cycle on the basic physical parameters of the granite, the physical parameters obtained from the test are uniformly processed in the form of rate of change; the specific calculation method is shown in Equation (1), and the average value of the rate of change of each physical parameter of granite after different heating and cooling cycle treatments is shown in Table 1.

$$\delta_x = \frac{x_b - x_a}{x_b} \times 100\% \quad (1)$$

where δ_x is the change rate of a physical parameter, x_b is the value of the parameter before the test treatment and x_a is the value of the parameter after the test treatment.

Table 1. Change rate of physical parameters of granite.

T (°C)	Type of Cooling	N	Rate of Change of Physical Parameter				
			δ_m/%	δ_v/%	δ_ρ/%	δ_p/%	δ_{wv}/%
200	Water	1	0.09	0.13	0.22	108.9	17.0
		5	0.09	0.45	0.56	180.7	21.5
		10	0.10	0.80	0.91	260.9	22.2
		20	0.12	0.50	0.66	430.3	28.0
	Natural	1	0.04	0.06	0.19	54.2	12.2
		5	0.08	0.39	0.27	164.6	13.6
		10	0.10	0.48	0.53	213.5	15.5
		20	0.10	0.50	0.62	228.7	23.1
400	Water	1	0.11	0.52	0.64	338.5	41.9
		5	0.12	1.01	1.15	427.2	59.9
		10	0.11	1.14	1.27	499.4	64.8
		20	0.18	1.53	1.71	693.7	67.7
	Natural	1	0.03	0.46	0.64	66.5	32.1
		5	0.12	0.48	0.63	179.1	39.1
		10	0.13	0.78	0.71	260.4	44.6
		20	0.14	0.96	0.94	422.5	46.3
600	Water	1	0.12	1.99	2.10	773.9	78.1
		5	0.31	3.51	3.72	1518.4	86.9
		10	0.54	6.77	6.88	2976.4	-
		20	1.01	9.49	9.63	4014.5	-
	Natural	1	0.17	2.03	2.21	838.5	74.9
		5	0.20	2.34	2.53	1042.5	81.2
		10	0.20	3.21	3.33	1226.6	82.9
		20	0.21	3.49	3.62	1419.4	86.2
800	Water	1	0.23	2.51	2.61	1476.2	78.6
	Natural	1	0.21	2.48	2.66	1454.4	75.7
	Furnace	1	0.20	2.02	2.21	1277.7	74.9

3.1. Mass Variation Pattern

The variation relationship between the rate of change of granite mass and the number of heating and cooling cycles is shown in Figure 4.

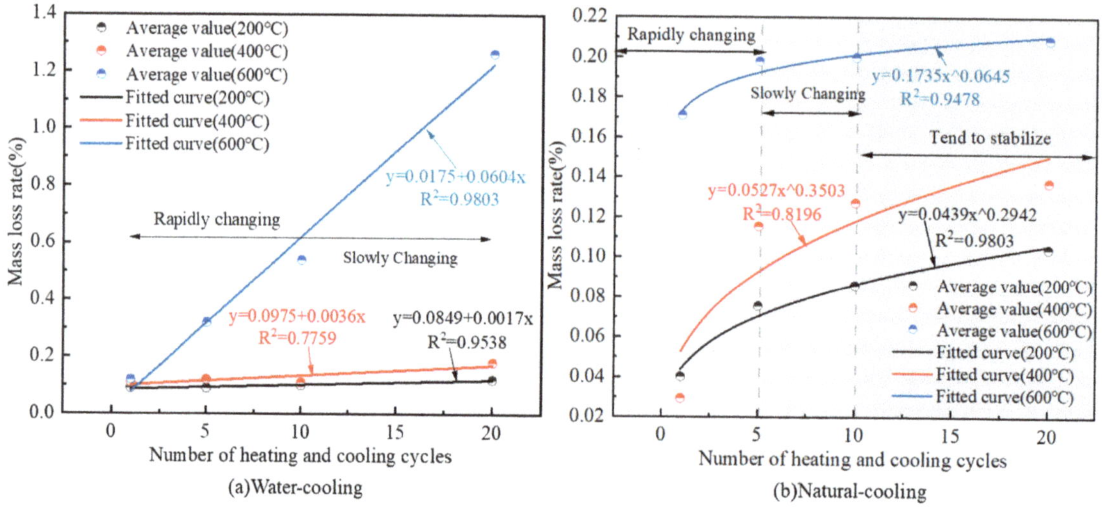

Figure 4. Variation pattern of granite mass change rate under the effect of different heating and cooling cycles.

The effect of heating and cooling cycles on the quality of the rock is mainly reflected in the loss of bound water and volatile substances inside the rock specimen at high temperature, which leads to the reduction in the quality of the rock specimen. As can be seen from Figure 4, when water cooling, the granite mass loss rate and the number of heating and cooling cycles show a linear relationship, and the slope of the straight line increases with the increase in heating temperature, indicating that the higher the temperature, the greater the rate of change in granite mass loss; when natural cooling, the granite mass loss rate and the number of heating and cooling cycles show a non-linear relationship, and the growth process with the increase of heating and cooling cycles can be divided into rapid growth (1–5), slow growth (5–10) and basic stability (10–20). From Figure 4a, it can be seen that at 200 °C and 400 °C, the change in the granite mass loss rate with the number of heating-cooling cycles is small, and the growth of the mass loss rate is only 0.03% and 0.07%, and at 600 °C, the influence of the number of heating-cooling cycles on the quality of granite is significantly enhanced, and its mass loss rate increases from 0.12% to 1.26%, and the mass loss rate increases by more than 10 times, which shows that 600 °C is an important temperature threshold for granite quality change. The same conclusion was reached by Li et al. [8], who concluded that 600 °C is the turning temperature for property degradation of granite specimens, mainly because of the α/β phase transition of quartz is around 500 °C. From Figure 5, it is also clear that the mass loss rate when water cooling under the same conditions is higher than that when natural cooling. The fitted relationship between the mass loss rate and the number of heating-cooling cycles at different temperatures is shown in Equation (2).

$$\begin{cases} \delta_{m(T=200^{\circ}\mathrm{C},W)} = 0.0849 + 0.0017N & R^2 = 0.9538 \\ \delta_{m(T=400^{\circ}\mathrm{C},W)} = 0.0975 + 0.0036N & R^2 = 0.7759 \\ \delta_{m(T=600^{\circ}\mathrm{C},W)} = 0.0175 + 0.0604N & R^2 = 0.9803 \\ \delta_{m(T=200^{\circ}\mathrm{C},NC)} = 0.0439N^{0.2942} & R^2 = 0.9803 \\ \delta_{m(T=400^{\circ}\mathrm{C},NC)} = 0.0527N^{0.3503} & R^2 = 0.8196 \\ \delta_{m(T=600^{\circ}\mathrm{C},NC)} = 0.1735N^{0.0645} & R^2 = 0.9478 \end{cases} \quad (2)$$

where: δ_m is the mass change rate, %; T is the heating temperature, °C; W is the water cooling; NC is the natural cooling; N is the number of heating and cooling cycles.

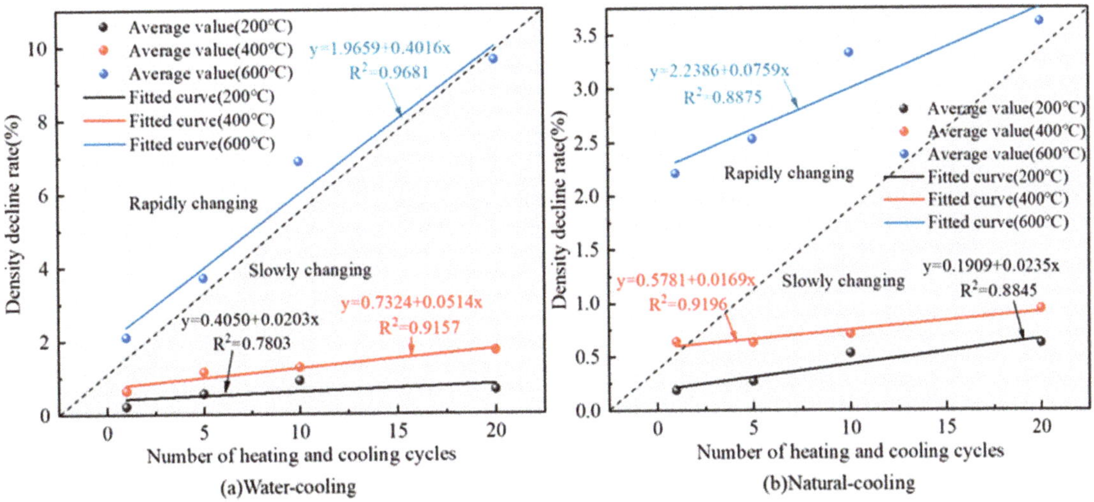

Figure 5. Variation pattern of granite density change rate under different heating and cooling cycles.

3.2. Density Variation Pattern

The variation relationship between the rate of change of granite density and the number of heating and cooling cycles is shown in Figure 5.

Under the conditions of water cooling and natural cooling, the density decrease rate of granite and the number of heating and cooling cycles both show a linear relationship, and the slope of the straight line increases with the increase in heating temperature, indicating that the higher the temperature, the greater the rate of change in the granite density decrease rate. The density decrease of granite is mainly caused by the expansion of volume and the loss of mass during the heating process. From Figure 5a,b, the granite specimens have different sensitivities to different cooling methods, and the degree of damage of granite under water cooling is greater than that under natural cooling. At 600 °C and when the number of heating and cooling cycles increased from one to twenty times, the rate of change in density when water cooling method increased from 2.10% to 9.63%, which increased 3.59 times; while the density change rate using the natural cooling method only increased from 2.21% to 3.62%, which was only 0.64 times. At the same time, the density change has a significant temperature effect. From Figure 5a, the density change rates at a heating temperature of 600 °C are 2.10%, 3.72%, 6.88% and 9.63%, which are significantly higher than those at 200 °C (0.22%, 0.56%, 0.91% and 0.66%) and 400 °C (0.64%, 1.15%, 1.27% and 1.71%), respectively. This result is in agreement with the findings of Zhao [5], Sun [9], Shao [20] and Li [24], who found that high temperatures cause large temperature stresses inside granite specimens, resulting in severe deterioration of their physical properties. The fitted relationship between the rate of change in density and the number of heating-cooling cycles at different temperatures is shown in Equation (3).

$$\begin{cases} \delta_{\rho(T=200^{\circ}C,W)} = 0.4049 + 0.0203N & R^2 = 0.7803 \\ \delta_{\rho(T=400^{\circ}C,W)} = 0.7324 + 0.0514N & R^2 = 0.9157 \\ \delta_{\rho(T=600^{\circ}C,W)} = 1.9659 + 0.4016N & R^2 = 0.9681 \\ \delta_{\rho(T=200^{\circ}C,NC)} = 0.1909 + 0.0235N & R^2 = 0.8845 \\ \delta_{\rho(T=400^{\circ}C,NC)} = 0.5781 + 0.0169N & R^2 = 0.9196 \\ \delta_{\rho(T=600^{\circ}C,NC)} = 2.2386 + 0.0759N & R^2 = 0.8875 \end{cases} \quad (3)$$

where: δ_ρ is the rate of change in density, %.

3.3. Volume Variation Pattern

The variation relationship between the volume expansion rate of granite and the number of heating and cooling cycles is shown in Figure 6.

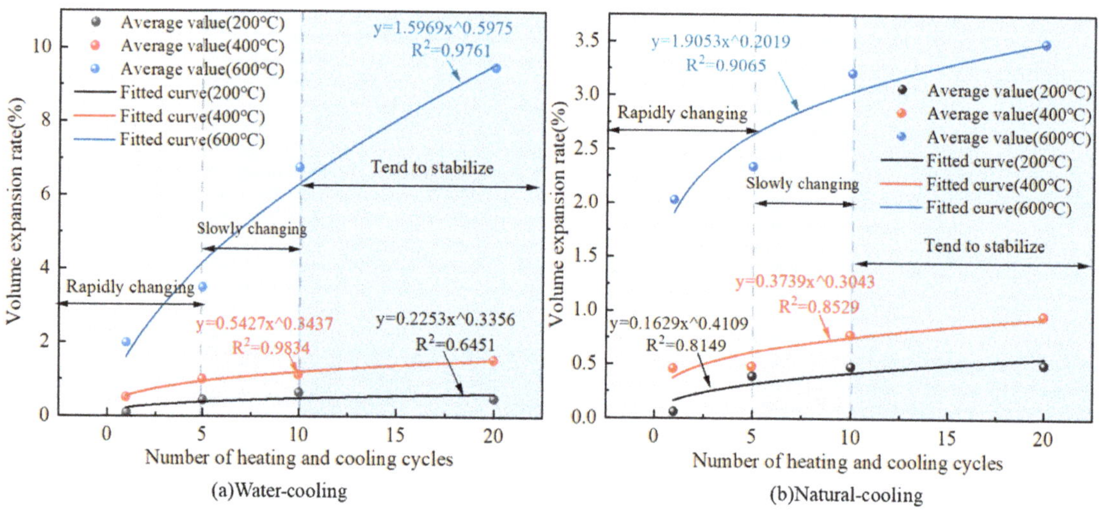

Figure 6. Variation pattern of granite volume change rate under different heating and cooling cycles.

From Figure 6, the volume expansion rate of granite and the number of heating and cooling cycles show a positive correlation under different cooling methods and heating temperature conditions. The whole process can be divided into three stages: one to five times for the fast rising stage; five to ten times for the slow rising stage; ten to twenty times for the basic stable stage. Under the water-cooling condition, the exponent of the multiplicative power function continuously increases with the increase in heating temperature, indicating that the higher the temperature, the more sensitive the granite volume is to the number of heating and cooling cycles when water cooling is used. Under natural cooling conditions, the sensitivity of the granite volume expansion rate to the number of heating and cooling cycles is much less than that under water cooling conditions. The heating temperature is 600 °C, the granite volume expansion rate increases from 1.99% to 9.49% when water cooling is used, which is an increase of 376.8%, while the granite volume expansion rate increases from 2.03% to 3.49% when natural cooling is used, which is only an increase of 71.9%, and the maximum volume expansion rate when water cooling is used is 2.7 times of that when the natural cooling method is used. The same conclusion was obtained by Li [8]. The fitted relationship between the volume expansion rate and the number of heating-cooling cycles at different temperatures is shown in Equation (4).

$$\begin{cases} \delta_{v(T=200^{\circ}C,W)} = 0.2253N^{0.3356} & R^2 = 0.6451 \\ \delta_{v(T=400^{\circ}C,W)} = 0.5427N^{0.3437} & R^2 = 0.9834 \\ \delta_{v(T=600^{\circ}C,W)} = 1.5969N^{0.5975} & R^2 = 0.9761 \\ \delta_{v(T=200^{\circ}C,NC)} = 0.1629N^{0.4109} & R^2 = 0.8149 \\ \delta_{v(T=400^{\circ}C,NC)} = 0.3739N^{0.3043} & R^2 = 0.8530 \\ \delta_{v(T=600^{\circ}C,NC)} = 1.9053N^{0.2019} & R^2 = 0.9065 \end{cases} \quad (4)$$

where: δ_v is the rate of volume change, %.

3.4. Porosity Variation Pattern

The variation relationship between the rate of change of granite porosity and the number of heating and cooling cycles is shown in Figure 7.

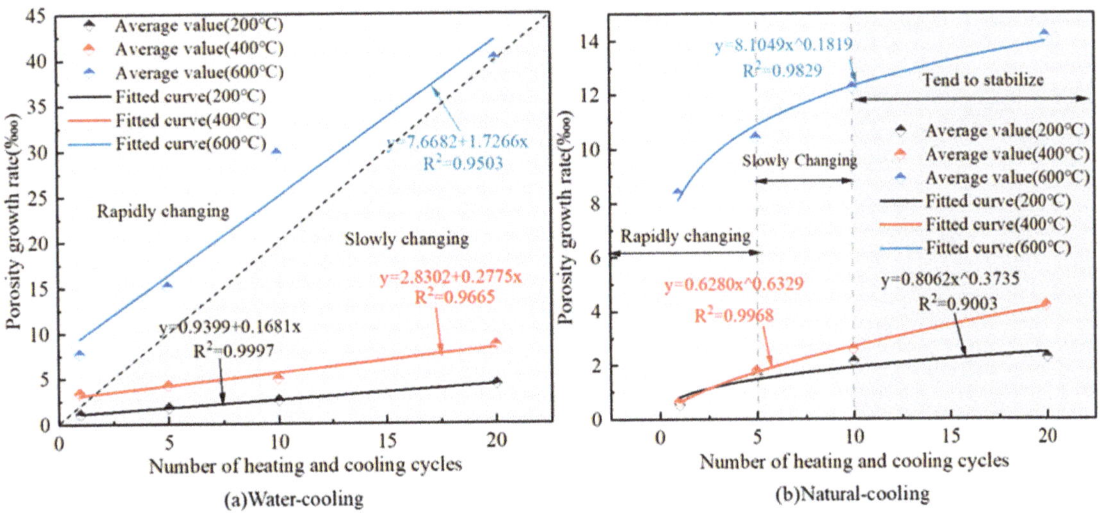

Figure 7. Variation pattern of granite porosity change rate under different heating and cooling cycles.

From Figure 7a, when water cooling is used, a good linear relationship is shown between the porosity growth rate and the number of heating and cooling cycles, and the slope of the straight line increases with the increase in heating temperature; indicating that the higher the temperature, the greater the porosity growth rate with the increase in heating and cooling cycles. When the heating temperature was 200 °C and 400 °C, the level of porosity growth rate of the specimens was low, which were below 10‱. When the heating temperature increased to 600 °C, the rate of change of the porosity growth rate of the specimens with the number of heating and cooling cycles significantly increased from 7‱ to 40‱, which was an increase of 4.7 times. From Figure 7b, the porosity growth rate increases with the number of heating and cooling cycles as a multiplicative power function, and the whole process can be divided into three stages: rapid increase stage; slow increase stage; basic stability stage. The growth rate of the porosity change rate gradually decreases with the increase in heating and cooling cycles, which is due to the large number of cracks and defects generated inside the granite specimens in the first few cycles of heating and cooling cycles, and when the number of heating and cooling cycles increases again, the mineral expansion and deformation caused by its generated thermal stress has room to move, which effectively reduces the damage of thermal stress. When the heating temperature was 200 °C and 400 °C, the porosity change rate was below 4‱. When the heating temperature increased to 600 °C, the porosity growth rate of the specimen rapidly increased from 8.39‱ to 14.19‱, an increase of 0.69 times. Thus, it seems that 600 °C is an important threshold value for the porosity change of granite. The fitted relationship between the rate of change of granite porosity and the number of heating-cooling cycles at different temperatures is shown in Equation (5).

$$\begin{cases} \delta_{p(T=200^{\circ}\mathrm{C},W)} = 0.9399 + 0.1681N & R^2 = 0.9997 \\ \delta_{p(T=400^{\circ}\mathrm{C},W)} = 2.8302 + 0.2775N & R^2 = 0.9665 \\ \delta_{p(T=600^{\circ}\mathrm{C},W)} = 7.6682 + 1.7266N & R^2 = 0.9503 \\ \delta_{p(T=200^{\circ}\mathrm{C},NC)} = 0.8062N^{0.3735} & R^2 = 0.9003 \\ \delta_{p(T=400^{\circ}\mathrm{C},NC)} = 0.6280N^{0.6329} & R^2 = 0.9968 \\ \delta_{p(T=600^{\circ}\mathrm{C},NC)} = 8.1049N^{0.1819} & R^2 = 0.9830 \end{cases} \quad (5)$$

where: δ_p is the rate of change in porosity, ‱.

3.5. P-Wave Speed Variation Pattern

The variation relationship between the rate of change of granite wave velocity and the number of heating and cooling cycles is shown in Figure 8.

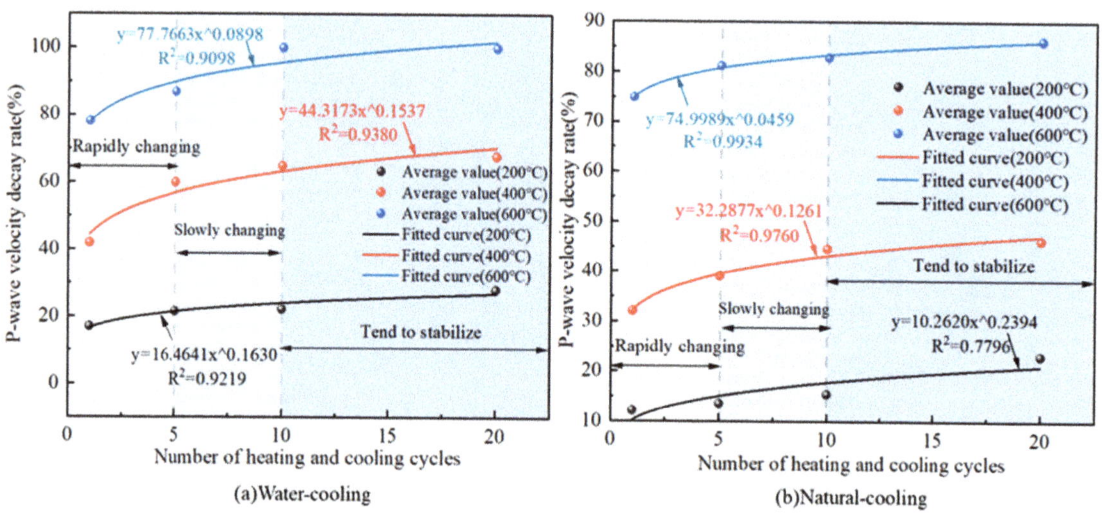

Figure 8. Variation pattern of longitudinal wave velocity change rate of granite under different heating and cooling cycles.

As can be seen from Figure 8, the P-wave velocity drop rate of the granite specimens gradually increased with the increase in the number of heating and cooling cycles, and the rise process can be divided into three stages: one to five times for the rapid rise stage; five to ten times for the slow rise stage; ten to twenty times for the basic stability stage. The P-wave velocity decreases the most after experiencing one to five heating and cooling cycles. This is because a large number of new cracks and defects are generated inside the granite specimen after one heating-cooling cycle action; at the same time, the volatilization of volatile substances, such as bonded water in the granite specimen under the high temperature action, leads to the increase in the volume of the original cracks and holes inside it, which finally makes the P-wave in most of the area unable to pass, so the P-wave velocity significantly drops. Meanwhile, it can be seen from Figure 8a that the temperature effect of granite wave velocity change is very significant. The number of water cooling and heating-cooling cycles is 20 times, and the wave velocity variation rate is 28% when the heating temperature is 200 °C; 67.7% when it is 400 °C; and 100% when it is 600 °C. Compared with the wave velocity variation rate at 200 °C, it increased 1.4 times and 2.6 times at 400 °C and 600 °C, respectively. Sun [9] also obtained similar conclusions in his study. Meanwhile, the wave velocity at 600 °C could not be measured due to the formation of macroscopic penetration cracks inside the granite specimens at this time, which resulted in the inability of P waves to pass through the specimens. The fitted relationship between the P-wave velocity drop rate and the number of heating-cooling cycles at different temperatures is shown in Equation (6).

$$\begin{cases} \delta_{wv(T=200^\circ\mathrm{C},W)} = 16.4641N^{0.1630} & R^2 = 0.9219 \\ \delta_{wv(T=400^\circ\mathrm{C},W)} = 44.3173N^{0.1537} & R^2 = 0.9380 \\ \delta_{wv(T=600^\circ\mathrm{C},W)} = 77.7663N^{0.0898} & R^2 = 0.9098 \\ \delta_{wv(T=200^\circ\mathrm{C},NC)} = 10.2619N^{0.2394} & R^2 = 0.7796 \\ \delta_{wv(T=400^\circ\mathrm{C},NC)} = 32.2877N^{0.1261} & R^2 = 0.9760 \\ \delta_{wv(T=600^\circ\mathrm{C},NC)} = 74.9989N^{0.0459} & R^2 = 0.9934 \end{cases} \quad (6)$$

where: δ_{wv} is the rate of change of wave velocity, %.

3.6. Granite Surface Color Variation Pattern

As shown in Figure 9, the surface color of the granite specimen gradually changed from dark cyan at 25 °C to beige at 600 °C as the heating temperature increased and the number of heating and cooling cycles increased. The effect of the number of heating and cooling cycles on the color of the granite specimens was very small, and the color of the specimens was only slightly brightened when the number of heating and cooling cycles increased from one to ten times. The main reason for this phenomenon is, on the one hand, the loss of potassium feldspar from the composition of granite under high temperature conditions, which gradually lightens the color of potassium feldspar, and, on the other hand, the dehydration of water-rich dark minerals such as mica under high temperature, which lightens the color [37].

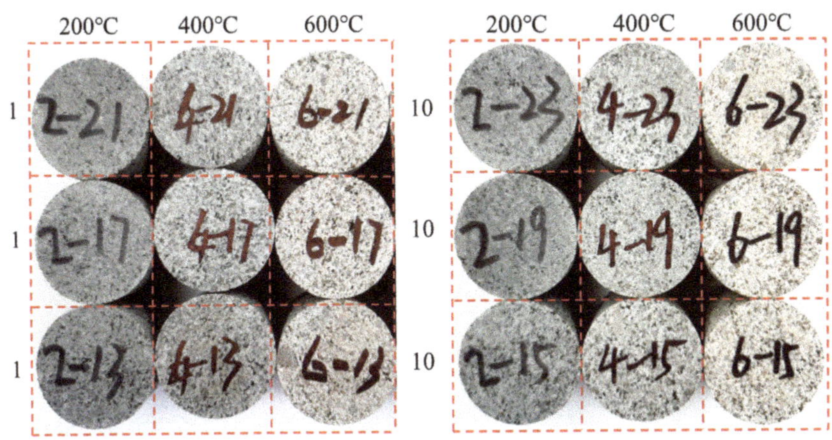

Figure 9. Changes in surface color of granite specimens under the action of different heating and cooling cycles.

3.7. Granite Microstructure Variation Pattern

The changes in granite microstructure after different heating and cooling cycles are shown in Figure 10. From the SEM images in Figure 10, at room temperature (25 °C), the granite fracture surface has smooth grains and a complete structure without obvious microcracks. With the gradual increase in temperature, more along-crystal and through-crystal damage occurred in the granite section [38], which produced many cracks, and these cracks were the main cause of the deterioration of the mechanical properties of granite. Secondly, the microcracks produced inside the granite under water cooling conditions are redundant with natural cooling conditions, mainly because of the large temperature difference between the inside and outside of the granite during water cooling, the fast-cooling rate and the large thermal stress generated inside the specimen. Finally, with the increase in the number of heating and cooling cycles and the increase in temperature, the microcracks inside the specimen gradually sprouted and expanded, and finally formed macro cracks.

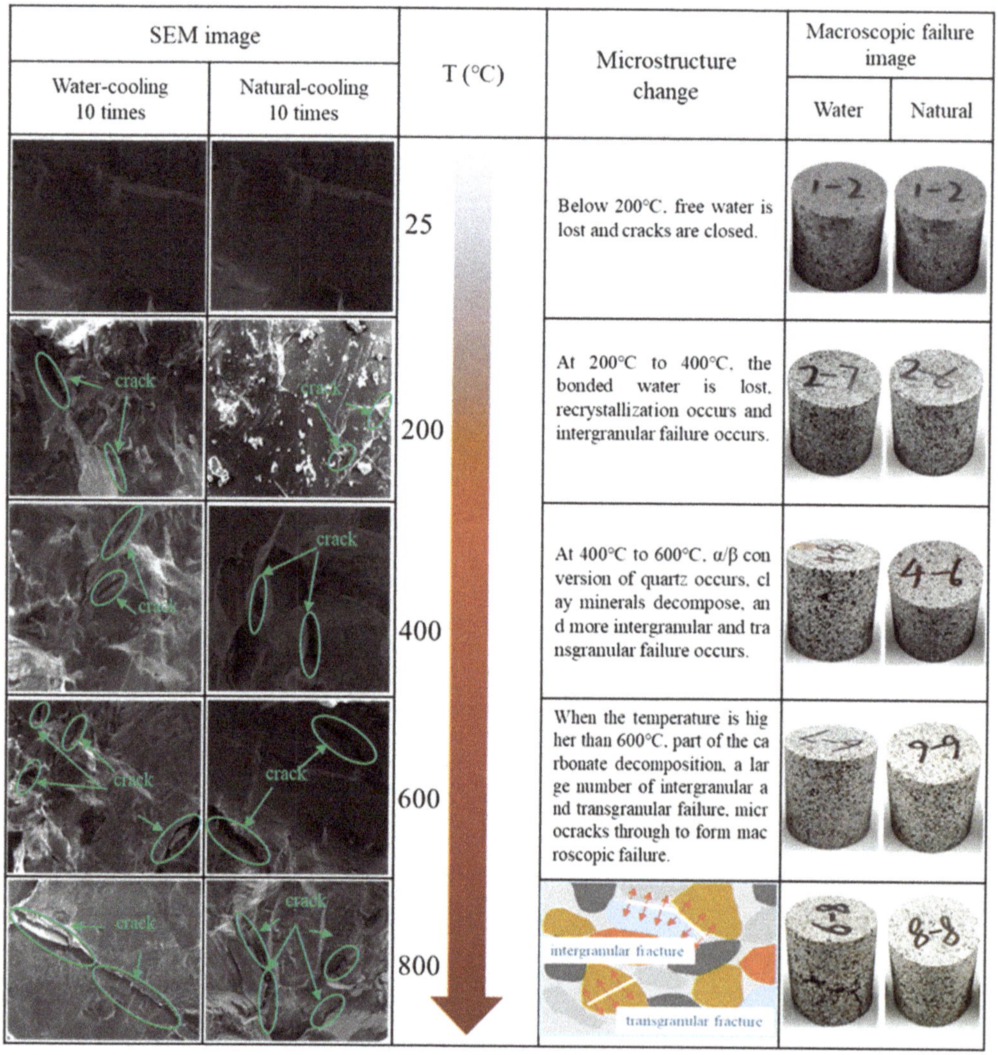

Figure 10. Changes in granite microstructure under different heating and cooling cycles.

4. Dynamic Compression Properties of Granite after Heating-Cooling Cycle

4.1. Dynamic Stress-Strain Curve of Granite

The dynamic stress-strain curves of the granite specimens after different heating and cooling cycles are shown in Figure 11, and the dynamic mechanical parameters are shown in Table 2.

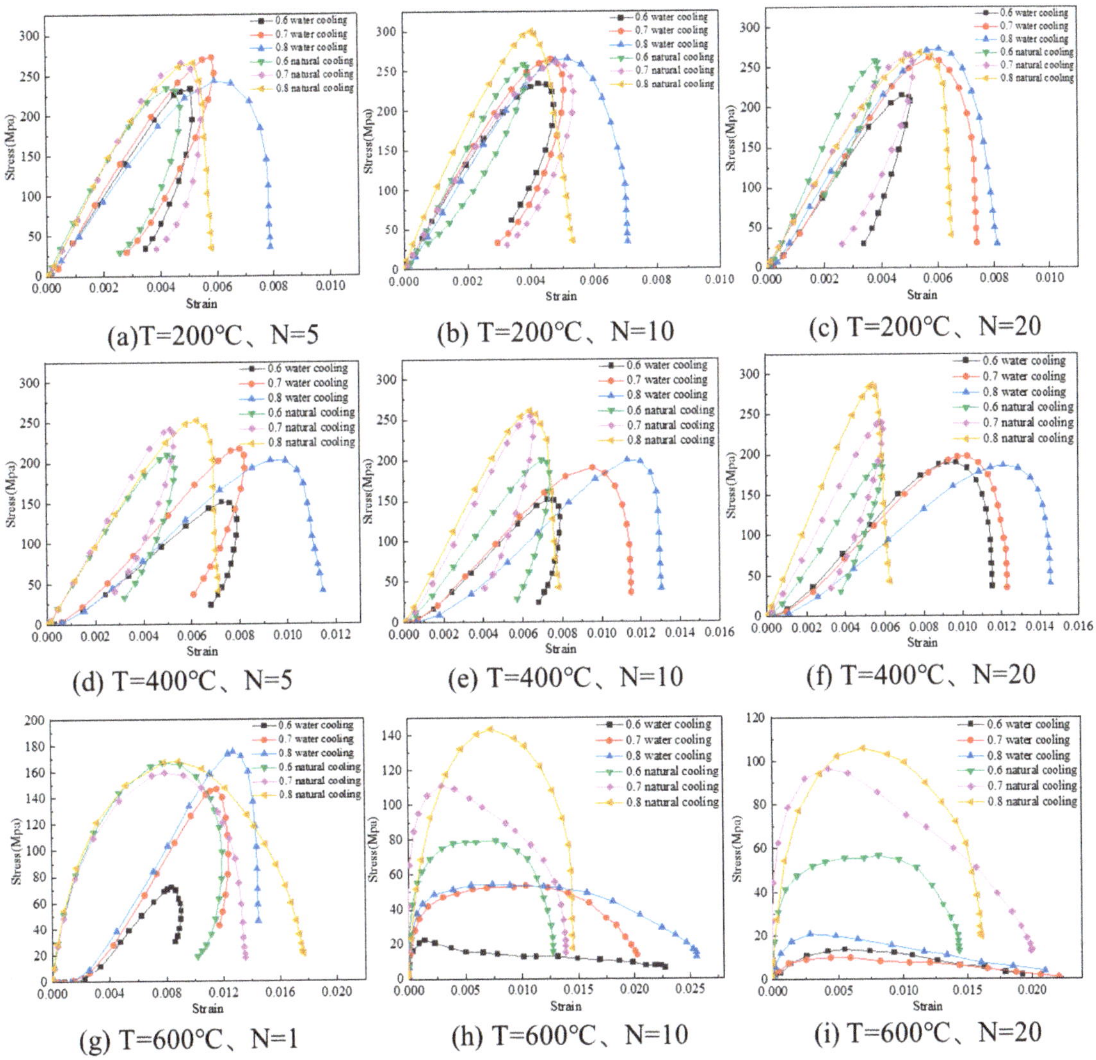

Figure 11. This is a figure. Schemes follow the same formatting.

As can be seen from Figure 11, the dynamic stress-strain curves do not have an obvious compaction phase at higher impact gas pressure, which is because the closing rate of microcracks inside the specimen is much lower than the loading rate, resulting in the specimen crossing the compaction phase and reaching the elastic change phase before it can close. From Figure 11a, the stress-strain curve gradually changes from type II to type I with the increase in impact gas pressure at the same temperature, as well as the number of heating-cooling cycles. From Figure 11a,c, the stress-strain curve will gradually change from type II to type I with the increase in heating and cooling cycles at the same temperature and impact gas pressure; meanwhile, it is easier to change the stress-strain curve from type II to type I with water cooling at the same temperature and impact gas pressure. From Figure 11b,h, the stress-strain curve will gradually change from type II to type I with the increase in heating temperature under the same number of heating and cooling cycles and impact gas pressure. The experimental results were the same as those obtained by Wu [39] and Tarasov [40]. In addition, the peak stress of the specimen tends to decrease with the increase in heating temperature and the number of heating and cooling cycles at the same impact gas pressure.

Table 2. Dynamic mechanical parameters of specimens at different conditions.

T (°C)	Type of Cooling	N	Pressure of Impact (Mpa)								
			0.6			0.7			0.8		
			σ_d	E_d	$\dot{\varepsilon}_d$	σ_d	E_d	$\dot{\varepsilon}_d$	σ_d	E_d	$\dot{\varepsilon}_d$
25		1	283	66.5	41.6	288	61.4	42.0	295	64.8	43.8
200	Water	1	255	73.0	32.5	277	77.7	34.5	289	74.9	45.6
		5	232	61.2	39.2	272	60.1	45.2	277	65.4	48.7
		10	231	54.1	43.7	265	58.5	47.4	268	58.3	49.0
		20	215	53.9	45.7	257	54.5	48.4	262	60.0	52.0
	Natural	1	260	71.8	34.7	267	68.3	38.7	301	72.4	44.1
		5	234	65.5	35.4	266	69.0	40.7	299	69.8	45.9
		10	259	63.4	37.6	261	66.0	42.3	289	65.5	46.2
		20	220	54.1	39.3	260	59.2	44.0	269	58.1	48.4
400	Water	1	178	35.5	53.6	221	38.5	55.8	206	33.0	61.3
		5	152	25.9	57.8	219	31.9	61.7	206	28.7	71.2
		10	151	24.3	60.3	191	25.1	70.7	201	23.7	82.2
		20	191	22.1	71.0	190	24.7	74.0	197	21.4	84.3
	Natural	1	174	54.1	47.8	244	58.1	54.6	265	55.4	53.0
		5	210	47.5	49.1	242	54.9	61.2	253	49.2	57.7
		10	200	43.4	54.3	255	47.1	63.1	261	47.4	62.6
		20	192	39.5	57.5	240	42.5	68.8	286	40.5	74.1
600	Water	1	71	32.1	62.9	147	27.8	70.7	175	29.4	77.3
		5	86	22.8	77.0	107	26.6	86.6	115	24.8	95.8
		10	22	14.8	89.8	54	14.0	99.5	56	10.5	108
		20	14	3.4	99.7	16	4.9	105	21	4.2	123
	Natural	1	167	39.1	57.6	159	39.6	67.2	168	40	86.3
		5	114	33.3	61.3	154	28.7	72.3	93	30.9	89.2
		10	80	28.7	74.5	111	28.4	86.3	144	29.1	94.8
		20	57	20.9	86.2	97	21.9	92.6	106	22.2	98.8
800	Water	1	90	20.7	73.3	147	21.5	85.1	140	22.7	96.6
	Natural	1	123	26.8	71.8	154	24.1	81.9	168	29.3	88.4
	Furnace	1	141	21.8	72.5	187	36.1	77.0	188	33.9	86.4

4.2. Dynamic Compression Strength Variation Law of Granite

The relationship between the dynamic compression strength of granite specimens and the number of heating and cooling cycles are shown in Figure 12.

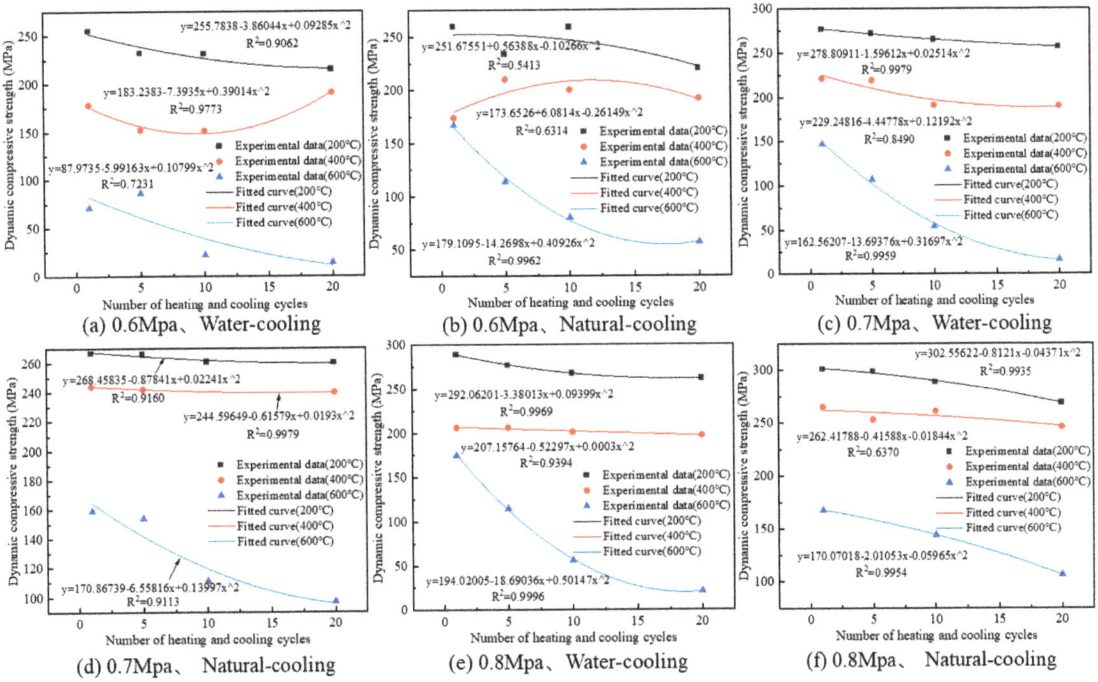

Figure 12. Dynamic compression strength variation pattern of granite specimens after different heating-cooling cycle treatments.

From Figure 12, the relationship between the dynamic compression strength of granite and the number of heating and cooling cycles satisfies the quadratic function and decreases with the increase in the number of heating and cooling cycles. When the temperature is lower than 600 °C, the dynamic compression strength of granite decreases less with the increase in heating and cooling cycles; when the temperature reaches 600 °C, the dynamic compression strength of granite sharply decreases with the increase in heating and cooling cycles, which makes the dynamic compression strength of granite rapidly decrease. The same experimental results were obtained for Li [8]. From Figure 12c, the dynamic compression strength of granite decreased by 20 MPa, 31 MPa and 131 MPa when the heating temperature was 200 °C, 400 °C and 600 °C, and the decrease rate was 7.2%, 14.0% and 89.1%, respectively. The dynamic compression strength of granite decreases 12.4 times and 6.4 times at 600 °C than at 200 °C and 400 °C, respectively, which shows that it has an obvious temperature effect. From Figure 12c,d, the dynamic compression strength of granite decays faster when the water cooling method is used. With the natural cooling method, the dynamic strength decay rate of granite is 2.6%, 1.6% and 39.0% when the heating temperature is 200 °C, 400 °C and 600 °C, respectively, and the dynamic compression strength decay rate of granite with the water-cooling method is 2.8 times, 8.8 times and 2.3 times that when using the natural cooling method, respectively.

4.3. Dynamic Modulus of Elastic Change Law of Granite

The relationship between the dynamic modulus of elastic of granite specimens and the number of heating and cooling cycles is shown in Figure 13.

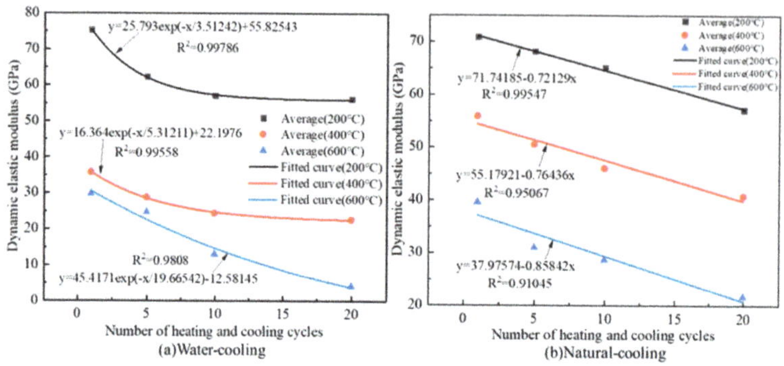

Figure 13. Dynamic elastic modulus changes law of granite specimens after different heating and cooling cycle treatments.

From Figure 13a, the dynamic modulus of elastic of granite specimens and the number of heating and cooling cycles under water cooling conditions satisfy the exponential function and have a negative correlation. At 400 °C, the average dynamic modulus of granite specimens decreased from 35.7 Gpa to 22.7 Gpa, decreasing by 36.4%; at 600 °C, the average dynamic modulus of granite specimens decreased from 29.8 Gpa to 4.2 Gpa, decreasing by 85.9%. Therefore, it can be seen that the higher the temperature, the greater the change in the dynamic modulus of elasticity of the specimens, which has an obvious temperature effect. From Figure 13b, the dynamic elastic modulus of the granite specimens and the number of heating and cooling cycles under the natural cooling condition satisfy the primary function relationship. Meanwhile, with the increase in heating temperature, the more significant the effect of heating and cooling cycles on the dynamic elastic modulus of granite specimens; the absolute value of the slope of the fitted curve gradually increases from 0.72192 to 0.85842.

4.4. Granite Strain Rate Variation Pattern

The variation relationship between the strain rate of granite specimens and the number of heating and cooling cycles is shown in Figure 14.

As can be seen from Figure 14, the strain rates of the granite specimens all increased with the increase in the number of heating and cooling cycles. Meanwhile, the change in strain rate has a strong temperature effect. With the increase in heating temperature, the effect of the number of heating-cooling cycles on the strain rate of granite was more significant; as can be seen from Figure 14a, the slope of the fitted curve increased from 0.6419 to 1.8579 with the increase in heating temperature from 200 °C to 600 °C, an increase of 189.6%. From Figure 14a,b, it can be seen that the effect of the number of heating and cooling cycles on the strain rate of granite is more significant under the water cooling method, and the slopes of the fitted curves for water cooling are 0.6419, 0.8975 and 1.8579 for the heating temperatures of 200 °C, 400 °C and 600 °C, respectively; the slopes of the fitted curves for natural cooling are 0.2510, 0.5347 and 1.5678, corresponding to 2.56, 1.68 and 1.19 times of the slope of the water-cooled fitted curve than that of the natural-cooled fitted curve, respectively. The trend of strain rate variation is consistent with the results obtained by Zhang et al. [12].

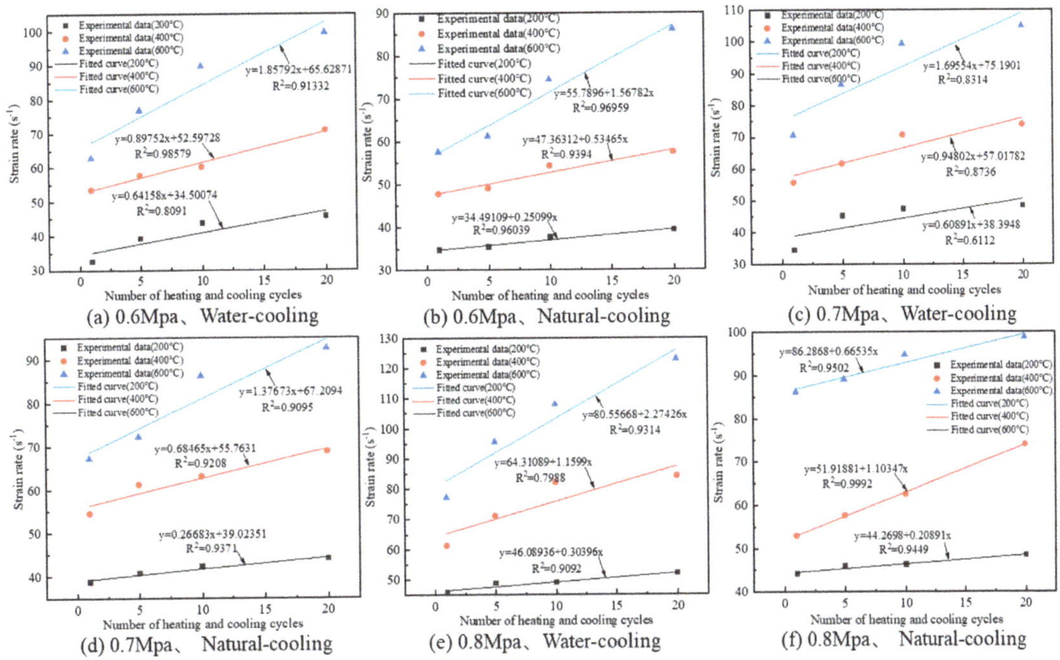

Figure 14. Strain rate variation law of granite specimens after different heating and cooling cycle treatments.

4.5. Granite Dynamic Crushing Characteristics

The crushing of granite specimens under different conditions is shown in Figure 15. With the increase in heating temperature, the increase in heating and cooling cycles and the increase in impact air pressure, the degree of granite specimen crushing gradually increased and the crushed particles gradually decreased.

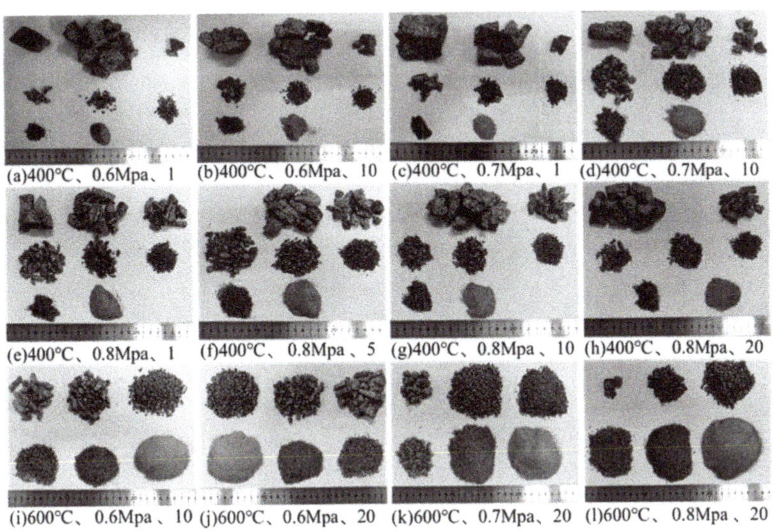

Figure 15. Crushing of granite specimens under different conditions.

5. Conclusions

In this study, heating and cooling cycle tests were conducted on a large number of fine-grained granite specimens at temperatures of 200~600 °C between one to twenty times using a box-type high-temperature muffle furnace, and the evolution of the basic physical parameters of fine-grained granite under the effect of heating and cooling cycle was summarized; dynamic impact tests were conducted on granite specimens after the effect of the heating and cooling cycle using the SHPB test system, and their dynamic compressive mechanical properties were analyzed.

(1) The mass, density and wave velocity of granite specimens show negative correlation with the heating temperature and the number of heating and cooling cycles, while the volume and porosity show positive correlation; the high temperature will cause a large temperature stress inside the granite specimen, so that its physical properties produce serious deterioration.

(2) The changes in the physical parameters have obvious temperature effects and are more sensitive to the water cooling method. A threshold value of 600 °C is determined for the changes in the physical parameters of granite, and the sensitivity of the physical parameters to the number of heating and cooling cycles is small before 600 °C; the sensitivity of the physical parameters to the number of heating and cooling cycles significantly increases when reaching 600 °C.

(3) With the increase in heating temperature and the increase in heating and cooling cycles, the surface color of granite specimens gradually changes from dark cyan to beige, mainly due to the loss of potassium and dehydration of water-rich dark minerals such as mica under high temperature conditions. According to SEM, it is known that the microscopic cracks of the specimens gradually sprouted and expanded until penetration, and finally formed macroscopic damage.

(4) The dynamic compression strength of granite specimens has an obvious temperature effect and a strain rate effect. Meanwhile, the number of heating and cooling cycles had some effects on the dynamic compressive strength, elastic modulus and strain rate of granite. The dynamic compressive strength and elastic modulus decreased with the increase in heating-cooling cycles, and the maximum decrease rate was 89.1% and 85.9%, and the strain rate linearly increased with the increase in heating-cooling cycles; the maximum strain rate reached 123 s^{-1}.

(5) With the increase in heating temperature, the increase in heating-cooling cycles and the increase in impact air pressure, the degree of granite specimen crushing gradually increased.

(6) The study of dynamic mechanical properties of granite after heating and cooling cycles is limited to the case of uniaxial dynamic loading, and the mechanical properties under the combined dynamic and static conditions must be investigated in the future, while numerical simulations can be used to further investigate the microstructure and mechanical properties of granite after heating and cooling cycles.

Author Contributions: Conceptualization, H.Z and. X.L.; methodology, H.Z; software, H.Z.; validation, H.Z., X.L. and X.H.; formal analysis, H.Z and X.H.; investigation, H.Z. and Y.W.; resources, L.H. and X.H.; data curation, Y.W.; writing—original draft preparation, H.Z.; writing—review and editing, H.Z.; visualization, H.Z. and X.H.; supervision, L.H. and X.H.; project administration, L.H. and Y.W.; funding acquisition, X.L. All authors have read and agreed to the published version of the manuscript.

Funding: This work was supported by the National Natural Science Foundation of China (51904335, 52174098, 51927808), and the Fundamental Research Funds for Central Universities of the Central South University (No. 2021zzts0276).

Institutional Review Board Statement: Not applicable.

Informed Consent Statement: Not applicable.

Data Availability Statement: The data presented in this study are available on request from the corresponding author.

Conflicts of Interest: The authors declare no conflict of interest.

Nomenclature

SEM	Scanning electron microscope;
D	Diameter of granite specimen, mm;
SHPB	Split Hopkinson pressure bar;
L1, L2	Horizontal distance from the upper and lower boundaries of the granite specimen to the central axis;
$\delta_m, \delta_v, \delta_\rho, \delta_p, \delta_{wv}$	Granite mass, volume, density, porosity, p-wave velocity variation rate.
N	Number of heating and cooling cycles
$\sigma_d, E_d, \dot{\varepsilon}_d$	Dynamic compression strength, dynamic elastic modulus, strain rate

References

1. Shao, S.; Wasantha, P.L.P.; Ranjith, P.G.; Chen, B.K. Effect of cooling rate on the mechanical behavior of heated Strathbogie granite with different grain sizes. *Int. J. Rock Mech. Min. Sci.* **2014**, *70*, 381–387. [CrossRef]
2. Chen, S.; Yang, C.; Wang, G. Evolution of thermal damage and permeability of Beishan granite. *Appl. Therm. Eng.* **2017**, *110*, 1533–1542. [CrossRef]
3. Li, Q.; Li, X.; Yin, T. Effect of microwave heating on fracture behavior of granite: An experimental investigation. *Eng. Fract. Mech.* **2021**, *250*, 107758. [CrossRef]
4. Wang, Y.; Luo, L.; Chen, J.; Liu, Z.; Geng, Y. Dynamic mechanical behavior and macroscopic and microscopic characteristics of granites subject to heating treatment. *Arab. J. Geosci.* **2023**, *16*, 112. [CrossRef]
5. Zhen, L.; Hong, W.; Jiang, Y.; Xin, S.; Yuan, G.; Qian, Y. Experimental investigation on damage characteristics and fracture behaviors of granite after high temperature exposure under different strain rates. *Theor. Appl. Fract. Mech.* **2020**, *110*, 102823.
6. Yin, T.; Yang, Z.; Yin, J. Effect of open fire on dynamic compression mechanical behavior of granite under different strain rates. *Arab. J. Geosci.* **2021**, *14*, 2144. [CrossRef]
7. Yin, T.; Chen, Y.; Li, X.; Li, Q. Effect of high temperature and strain rate on the elastic modulus of rocks: A review. *Int. J. Earth Sci.* **2021**, *110*, 2639–2660. [CrossRef]
8. Li, Y.; Zhai, Y.; Wang, C.; Meng, F.; Lu, M. Mechanical properties of Beishan granite under complex dynamic loads after thermal treatment. *Eng. Geol.* **2020**, *267*, 105481. [CrossRef]
9. Qiang, S.; Jh, D. Effects of heating on some physical properties of granite, Shandong, China. *J. Appl. Geophys.* **2021**, *193*, 104410.
10. Xu, J.J.; Zhang, Y.H.; Rutqvist, J.; Hu, M.S.; Wang, Z.Z.; Tang, X.H. Thermally induced microcracks in granite and their effect on the macroscale mechanical behavior. *J. Geophys. Res. Solid Earth* **2023**, *128*, e2022JB024920. [CrossRef]
11. Zhou, X.P.; Li, G.Q.; Ma, H.C. Real-time experiment investigations on the coupled thermomechanical and cracking behaviors in granite containing three pre-existing fissures. *Eng. Fract. Mech.* **2019**, *224*, 106797. [CrossRef]
12. Zhang, X.; Wei, M.; Lei, Z.; Chen, Y. A Multi-Scale Study on the Property Degradation of High-Temperature Treated Beishan Granite. *Minerals* **2022**, *13*, 27. [CrossRef]
13. Hao, J.; Qiao, L.; Li, Q. Study on cross-scale pores fractal characteristics of granite after high temperature and rock failure precursor under uniaxial compression. *Powder Technol.* **2022**, *401*, 117330. [CrossRef]
14. Ding, Q.-L.; Wang, P.; Cheng, Z. Influence of temperature and confining pressure on the mechanical properties of granite. *Powder Technol.* **2021**, *394*, 10–19. [CrossRef]
15. Wang, F.; Konietzky, H. Thermal damage evolution of granite under slow and high-speed heating conditions. *Comput. Geotech.* **2020**, *123*, 103590. [CrossRef]
16. Wang, F.; Frühwirt, T.; Konietzky, H.; Zhu, Q. Thermo-mechanical behaviour of granite during high-speed heating. *Eng. Geol.* **2019**, *260*, 105258. [CrossRef]
17. Xi, Y.; Wang, H.; Li, J.; Dong, W.; Li, H.; Guo, B. Experimental comparison of mechanical properties and fractal characteristics of geothermal reservoir rocks after different cooling treatments. *Energy Rep.* **2022**, *8*, 5158–5176. [CrossRef]
18. Zhao, F.; Sun, Q.; Ge, Z.; Dang, C. Laboratory investigation on engineering mechanics properties of granite after various heating/cooling treatments. *Environ. Sci. Pollut. Res.* **2022**, *30*, 12532–12544. [CrossRef]
19. Tang, Z.C.; Peng, M.H.; Xiao, S. Basic friction angle of granite fracture after heating and rapid cooling treatments. *Eng. Geol.* **2022**, *302*, 106626. [CrossRef]
20. Shao, Z.; Wang, Y.; Tang, X. The influences of heating and uniaxial loading on granite subjected to liquid nitrogen cooling. *Eng. Geol.* **2020**, *271*, 105614. [CrossRef]
21. Chen, Z.; Sha, S.; Xu, L.; Quan, J.; Rong, G.; Jiang, M. Damage Evaluation and Statistic Constitutive Model of High-Temperature Granites Subjected to Liquid Nitrogen Cold Shock. *Rock Mech. Rock Eng.* **2022**, *55*, 2299–2321. [CrossRef]

22. Rong, G.; Sha, S.; Li, B.; Chen, Z.; Zhang, Z. Experimental Investigation on Physical and Mechanical Properties of Granite Subjected to Cyclic Heating and Liquid Nitrogen Cooling. *Rock Mech. Rock Eng.* **2021**, *54*, 2383–2403. [CrossRef]
23. Yin, T.; Li, Q.; Li, X. Experimental investigation on mode I fracture characteristics of granite after cyclic heating and cooling treatments. *Eng. Fract. Mech.* **2019**, *222*, 106740. [CrossRef]
24. Li, Q.; Li, X.; Yin, T. Factors affecting pore structure of granite under cyclic heating and cooling: A nuclear magnetic resonance investigation. *Geothermics* **2021**, *96*, 102198. [CrossRef]
25. Chen, B.; Zhang, S.; Shen, B.; Li, Y.; Wang, C. Mechanical properties and failure characteristics of granite treated with a combined water–air cooling cycle. *Case Stud. Therm. Eng.* **2022**, *34*, 101964. [CrossRef]
26. Zhu, Z.; Ranjith, P.G.; Tian, H.; Jiang, G.; Dou, B.; Mei, G. Relationships between P-wave velocity and mechanical properties of granite after exposure to different cyclic heating and water cooling treatments. *Renew. Energy* **2021**, *168*, 375–392. [CrossRef]
27. Wu, X.; Huang, Z.; Cheng, Z.; Zhang, S.; Song, H.; Zhao, X. Effects of cyclic heating and LN2-cooling on the physical and mechanical properties of granite. *Appl. Therm. Eng.* **2019**, *156*, 99–110. [CrossRef]
28. Wang, F.; Frühwirt, T.; Konietzky, H. Influence of repeated heating on physical-mechanical properties and damage evolution of granite. *Int. J. Rock Mech. Min. Sci.* **2020**, *136*, 104514. [CrossRef]
29. Yin, Q.; Wu, J.; Zhu, C.; Wang, Q.; Zhang, Q.; Jing, H.; Xie, J. The role of multiple heating and water cooling cycles on physical and mechanical responses of granite rocks. *Géoméch. Geophys. Geo-Energy Geo-Resour.* **2021**, *7*, 69. [CrossRef]
30. Zhang, Y.; Zhao, G.-F. A multiphysics method for long-term deformation analysis of reservoir rock considering thermal damage in deep geothermal engineering. *Renew. Energy* **2023**, *204*, 432–448. [CrossRef]
31. Zhu, Z.; Yang, S.; Ranjith, P.G.; Tian, W.; Tian, H.; Zheng, J.; Jiang, G.; Dou, B. A comprehensive review on mechanical responses of granite in enhanced geothermal systems (EGSs). *J. Clean. Prod.* **2022**, *383*, 135378. [CrossRef]
32. Sun, Q.; Hu, J. The effect of rapid cooling on the thermal diffusivity of granite. *J. Appl. Geophys.* **2019**, *168*, 71–78. [CrossRef]
33. Ge, Z.; Sun, Q. Acoustic emission (AE) characteristics of granite after heating and cooling cycles. *Eng. Fract. Mech.* **2018**, *200*, 418–429. [CrossRef]
34. Jiang, Z.; Yin, Q.; Wu, J.; Zhu, Q.; Li, M.; Su, H.; Jing, H. The effect of cyclic heating and cooling on mechanical and deformation responses of granites under preset angle shearing. *Environ. Earth Sci.* **2023**, *82*, 29. [CrossRef]
35. Li, X.; Li, B.; Li, X.; Yin, T.; Wang, Y.; Dang, W. Thermal shock effects on the mechanical behavior of granite exposed to dynamic loading. *Arch. Civ. Mech. Eng.* **2020**, *20*, 66. [CrossRef]
36. *T/CSRME 001-2019*; Chinese Society of Rock Mechanics and Engineering. Test Procedure for Dynamic Properties of Rocks. China Standards Press: Beijing, China, 2019.
37. Yin, T.; Ma, J.; Wu, Y.; Zhuang, D.; Yang, Z. Effect of high temperature on the brittleness index of granite: An experimental investigation. *Bull. Eng. Geol. Environ.* **2022**, *81*, 476. [CrossRef]
38. Xie, H. *Introduction for Fractals-Rock Mechanics*; Science Press: Beijing, China, 1996. (In Chinese)
39. Wu, Y.; Huang, L.; Li, X.; Guo, Y.; Liu, H.; Wang, J. Effects of Strain Rate and Temperature on Physical Mechanical Properties and Energy Dissipation Features of Granite. *Mathematics* **2022**, *10*, 1521. [CrossRef]
40. Tarasov, B.G.; Stacey, T.R. Features of the energy balance and fragmentation mechanisms at spontaneous failure of class I and class II rocks. *Rock Mech. Rock Eng.* **2017**, *50*, 2563–2584. [CrossRef]

minerals

MDPI

Article

Triaxial Test Study on Energy Evolution of Marble after Thermal Cycle

Qi Wu [1,2,*], Bowen Li [1,2] and Xuehai Jiang [3]

1 State Key Laboratory of Water Resources and Hydropower Engineering Science, Wuhan University, Wuhan 430072, China
2 Key Laboratory of Rock Mechanics in Hydraulic Structural Engineering, Ministry of Education, Wuhan University, Wuhan 430072, China
3 Survey Wenzhou Branch, Zhejiang Geological and Mineral Exploration Institute, Hangzhou 310012, China
* Correspondence: wuqi0623@whu.edu.cn

Abstract: With the increasing requirements for the exploitation of underground resources, the subject of the physical and mechanical properties of rocks under high temperature and pressure needs to be studied urgently. In order to analyze the mechanical and energy characteristics of rocks under different thermal damages and confining pressures (c), a triaxial compression test is performed on 35 marble samples. The effects of thermal damage and high pressure are simulated with different thermal cycles and confining pressures. The results show that as the number of thermal cycles increases, the peak strain of marble gradually rises, but the peak stress and the elastic modulus (E) decrease by a degree, reaching 11.19‰, 39.53 MPa, 4.79 GPa, while there is no confining pressure applied at eight thermal cycles. At this point, the failure mode gradually changes from brittle fracture to plastic failure. When confining pressure rises, peak stress, peak strain, and elastic modulus all show an upward trend, reaching a maximum of 189.45 MPa, 13.39‰, 35.41 GPa, while the sample is undamaged at 30 MPa confining pressure. Moreover, peak stress increases linearly with confining pressure increase. The increased rate of the peak value of the total absorbed energy, elastic strain energy, and dissipated energy all show a convex trend. The dissipated energy gradually increases with the axial strain (ε_1) during the rock loading process. The elastic strain energy has an energy storage limit, but the rock fails when the value exceeds the limit. The limit increases first and then decreases with the number of thermal cycles. These results can provide important engineering references for mining underground resources.

Keywords: marble; thermal cycles; confining pressure; mechanical property; energy evolution

1. Introduction

With the rapid development of the world economy and the continuous advancement of urbanization, the demand for energy and underground space is increasing quickly. Underground mineral resources are steadily exploited, and this process must be closely related to geotechnical engineering, such as the mining of coal and oil in the deep crust; the treatment of radioactive nuclear waste; the development and utilization of geothermal energy, etc. [1], which are inseparable from geotechnical engineering. Moreover, geotechnical engineering is constantly developing deeper into the Earth's crust, and the deep part of the crust is under the physical conditions of high temperatures and high pressures, so it is very necessary to conduct experimental research on rocks under high temperatures and high pressures. Moreover, the failure of rock is an instability phenomenon driven ultimately by energy, and the failure of rock is the result of the rapid release of elastic deformation energy accumulated in rock when it reaches the elastic energy limit. Therefore, if the law of energy evolution in the process of rock deformation and failure can be analyzed in detail, plus the theory based on energy change can be established, the law of rock failure can be truly reflected, and the related engineering practice can be better served.

Many scholars have conducted systematic analysis and research on rock mechanical properties under high temperatures or pressures. After high-temperature treatment, the mechanical properties of rock appear to be weakened. The mechanical behavior of different rock types after heat treatment is quite different [2]. Many existing studies have verified this view. The peak stresses of both mudstone and sandstone increase first and then decrease with temperature; before 400 °C, the peak stress is not significantly affected by temperature [3–5], but at 600 °C, the peak stress drops rapidly as the temperature rises [6], and the peak stress under a uniaxial compression test and temperature even show a linear trend [7]. When the heat treatment temperature increases, the limestone samples gradually show obvious ductile failure characteristics [8]. The strengths of marble and sandstone depend directly on the confining pressure and are opposite to temperature [9].

In a triaxial compression test, when the temperature of coarse sandstone and the confining pressure are below a certain threshold, rock strength is positively correlated with temperature and confining pressure, and when the value is exceeded, the trend turns negative [10]. As the temperature increases, the peak stress of sandstone gradually decreases. However, as the temperature continues to rise, the strength of rocks under different confining pressures tends to be the same [11]. For slate, a high confining pressure will weaken the thermal effect of the rock's mechanical properties, while a higher temperature will enhance the effect of confining pressure [12].

When the temperature gradually increases from 20 °C to 600 °C, the mechanical properties of marble decrease significantly, and the critical temperature of marble is 600 °C [13]. Another scholar's research indicates that the uniaxial compressive strength of marble intermittently increases in the temperature range of 300 °C to 500 °C, and then decreases rapidly from temperatures above 400 °C [14]. Thermal cycling will change the physical properties of rocks and weaken their mechanical properties [15].

When the temperature rises from room temperature to 1000 °C, the failure mode of granite changes from single-crack failure to ideal shear failure and double-shear failure [16]. When the confining pressure is 15–30 MPa, the failure of granite samples is tensile-shear failure, but as the temperature increases, tensile failure will dominate [17]. When mudstone, after being subjected to high temperature, is compressed, its failure type is the same as that at room temperature, and brittle cracks occupy a leading position [18]. The effect of thermal stress caused by temperature increases leads to severe expansion and deformation of granite structures [19]. Under low temperature and confining pressure, multiple vertical cracks will generate along the axial stress (σ_1) direction during the failure of granite; with the increase in temperature and confining pressure, the direction of crack propagation gradually changes from a vertical to a tilt direction. After confining pressure increases, marble has a tendency to change from brittleness to ductility, but heating increases the ductility to a certain extent.

Some scholars believe that the failure evolution process and ultimate failure of rocks are driven by energy conversion [20]. Energy evolution reflects the deformation and failure process of rocks, which can further reveal the underlying mechanism of rock failure [21,22]. Both the elastic strain energy and dissipation energy in evolution rise with the increase in stress [23]. Moreover, there is a strong linear relationship between the elastic strain energy, dissipation energy, and total input energy [24]. As the thickness of the coal seam increases, the elastic energy, dissipation energy, and total input energy all decrease first and then rise [25]. When the coal-to-rock ratio is higher, the difference between energy accumulation and dissipation is larger, and the energy dissipation of the crack is smaller [26]. As the depth increases, the elastic energy of coal accumulates faster, and the dissipation energy increases faster too.

However, these studies have only analyzed rock failure from a mechanical point of view, and scholars have rarely been involved in the energy analysis of the failure process of marble. In this study, an analysis of rock deformation and failure is carried out from the perspective of energy evolution. This method can break through the original limitations of stress-strain analysis only and brings a new perspective and analysis method to understand rock's mechanical behavior. Moreover, marble is often used in underground engineering. Few scholars have analyzed the energy and mechanical evolution process of marble under thermal damage conditions at the same time. This paper can also provide some help for mining and underground engineering safety.

In this paper, a triaxial compression test of marble after thermal cycling is carried out, and stress-strain curves and energy evolution curves are drawn based on the experimental data. Based on the concept of energy change, the reason for the mechanical properties of marble, after different thermal cycles and under different confining pressures, is analyzed. The study of the evolution of mechanical properties of rocks under thermal cycling conditions has a certain reference value in the construction of underground engineering oriented to a situation with high temperatures.

2. Materials and Experimental Methods

2.1. Rock Sample Preparation

Marble was taken as the research object to conduct mechanical tests under different thermal cycles and confining pressures, and the rocks came from Henan province in China. According to ISRM [27], the marble was processed into a cylindrical sample with a diameter of 50 mm and a height of 100 mm, as shown in Figure 1. The fine-grained marble was milky white and had good homogeneity and massive structure, among which the contents of dolomite, calcite, and albite were 96.04%, 3.58%, and 0.38% under each test condition, and three samples were set to take the average value. The physical and mechanical parameters of marble are shown in Table 1, including the mean value and range of each parameter.

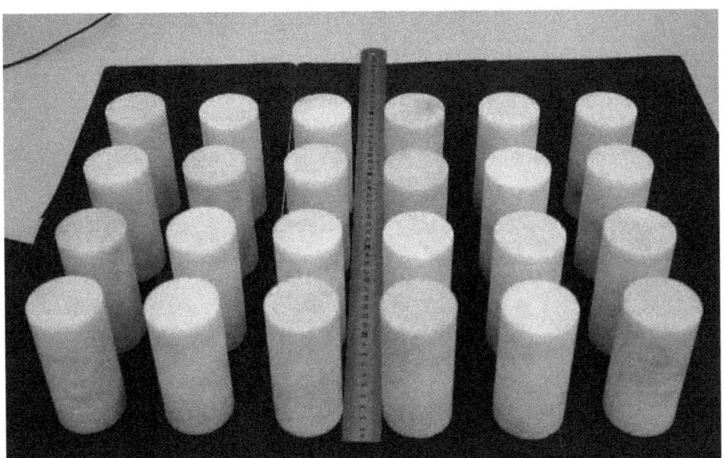

Figure 1. Photograph of the prepared marble specimens.

Table 1. Physical and mechanical parameters of marble samples.

Marble Parameter	Density (kg/m^3)	Uniaxial Compressive Strength (MPa)	E (GPa)	P-Wave Velocity (V_p) (km/s)	Poisson's Ratio
Mean value	2714.31	109.04	26.18	4.74	0.18
Value range	2472.17~2958.55	92.47~120.88	23.46~29.01	4.43~5.02	0.13~0.24

2.2. Thermal Cycle Treatment

The equipment shown in Figure 2 was used for this test. An SX2-10-12 high-temperature box-type resistance furnace (made in Changchun Rising Sun Testing Instruments Co. Ltd, Changchun, China), as shown in Figure 2a, was used as sample heating equipment. The furnace has a rated power and temperature of 10 kW and 1200 °C. All the samples were placed in the furnace and then heated to 400 °C at a rate of 10 °C/min. The temperature of the furnace was kept for 4 h to ensure all the samples were heated completely and uniformly; then, the furnace was opened, and the samples were left to cool naturally to room temperature. The steps above are called a thermal cycle (Figure 3). A cycle went through about 15 h. The number of thermal cycles set in this test was 0, 1, 2, 4, and 8. Three samples were set for each test.

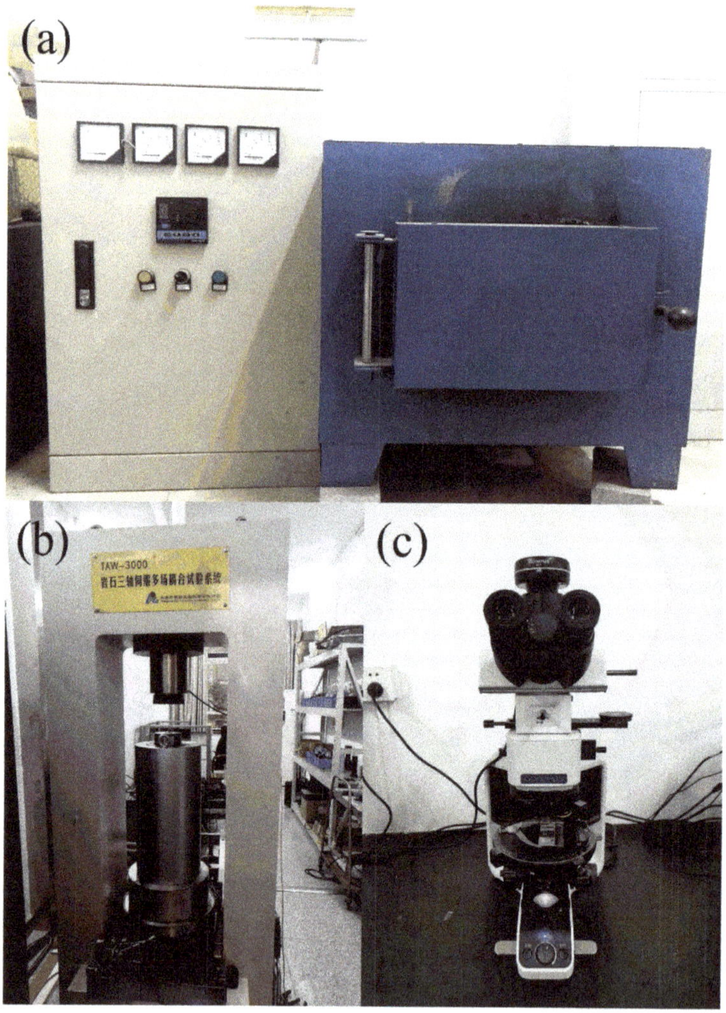

Figure 2. Test equipment: (**a**) SX2-10-12 high-temperature box-type resistance furnace; (**b**) TAW-3000 rock triaxial servo multi-field coupling test system; (**c**) photograph of Olympus BX53M polarizing microscope.

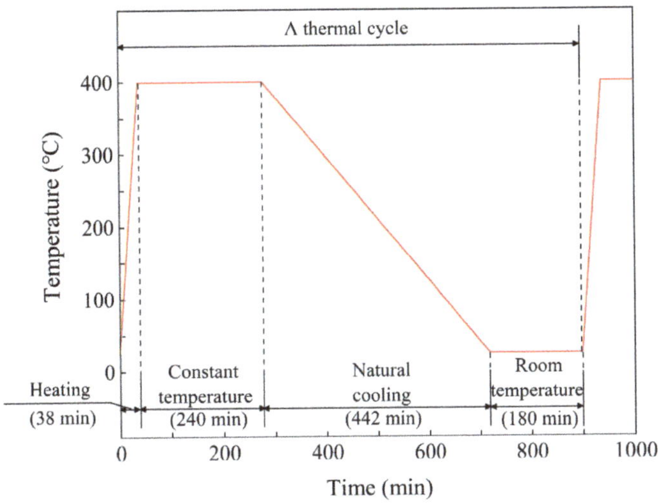

Figure 3. Temperature change curves during thermal cycle process.

2.3. *Triaxial Compression Test*

A triaxial compression test was carried out on a TAW-3000 rock servo multi-field coupling test system, as shown in Figure 2b. The maximum axial test force and confining pressure of the system were 3000 kN and 100 MPa.

In this test, 7 confining pressure levels of 0, 5, 10, 15, 20, 25, and 30 MPa were set.

2.4. *Microscopic Observation*

The rock slices were identified using a polarized light microscope. The microscope shown in Figure 2c can observe transparent minerals under transmitted or reflected light conditions to identify their cleavage, crystal shape, and optical properties, such as pleochroism, refractive index, reflectivity, extinction angle, and so on.

In order to study the influence of thermal cycle on the crack growth law by the equipment shown in Figure 2c, samples experiencing a different number of thermal cycles were made into sections with a thickness of 0.03 mm from the same position.

3. Test Results and Analysis

The data in the following figure and table in Section 3 use the mean value of each group of tests.

3.1. *Physical-Mechanical Properties*

3.1.1. P-Wave Velocity

P-wave velocity is extremely sensitive to microcracks in rock thermal damage, which can be used to quantitatively evaluate the thermal damage degree. The velocities of fine-grained marble after different thermal cycles are shown in Table 2 and Figure 4.

Table 2. Variation in V_p of marble with the number of thermal cycles.

P-Wave Velocity (km/s)	0	1 Cycle	2 Cycles	4 Cycles	8 Cycles
The small value	4.43	1.51	1.44	1.22	1.21
The middle value	4.77	1.81	1.51	1.39	1.24
The larger value	5.02	1.91	1.58	1.5	1.39
Mean value	4.74	1.77	1.51	1.37	1.28

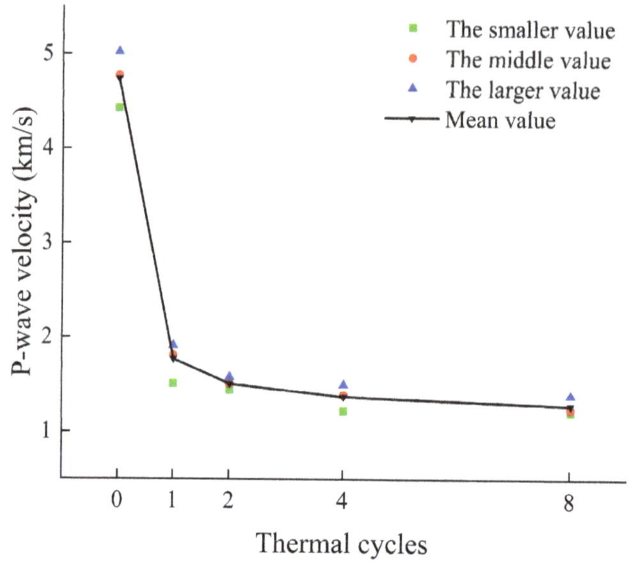

Figure 4. Variation in V_P of marble with the number of thermal cycles.

The results show that under the same set of test conditions, the amplitude of wave velocity change was small; moreover, after the effect of the thermal cycle, V_P decreased sharply, then slowly with the increase in thermal cycles. The velocity was 1.77 km/s, which was about 37% of that of the marble without thermal damage. Subsequently, with the increase in the number of thermal cycles, V_P declined slowly, and the velocity of marble after eight thermal cycles decreased to 27% of the undamaged rate.

3.1.2. Stress-Strain Curve

Figure 5 shows the stress-strain curves of all tests. It can be seen from the figure that at a constant number of thermal cycles, the triaxial compression stress-strain curves of marble almost all went through four stages: crack closure, linear elastic, nonlinear deformation, and failure [28].

Crack closure stage: The stress rises slowly, while the strain increases rapidly. At this time, the stress-strain curve tends to be concave. However, this trend becomes inconspicuous when confining pressure increases, which can hardly be observed when the confining pressure is very high. Linear elastic stage: after the initial crack closure stage, the stress-strain curve approximates a straight line, showing the elastic characteristics of marble samples. Nonlinear deformation stage: at this stage, the elastic modulus of the rock sample decreases with the increase in strain, which results in the rock yield phenomenon. Failure stage: the internal cracks of the rock sample are connected to form macroscopic cracks, and the ability to bear loads is lost.

It can be seen intuitively from the figure that the peak stress and peak strain of the rock sample increased significantly with the confining pressure. When the number of thermal cycles rose to four, the ductility characteristics of the rock sample began to show more obviously. Moreover, when the number of thermal cycles was low, the curve under high confining pressure also reflects ductility characteristics.

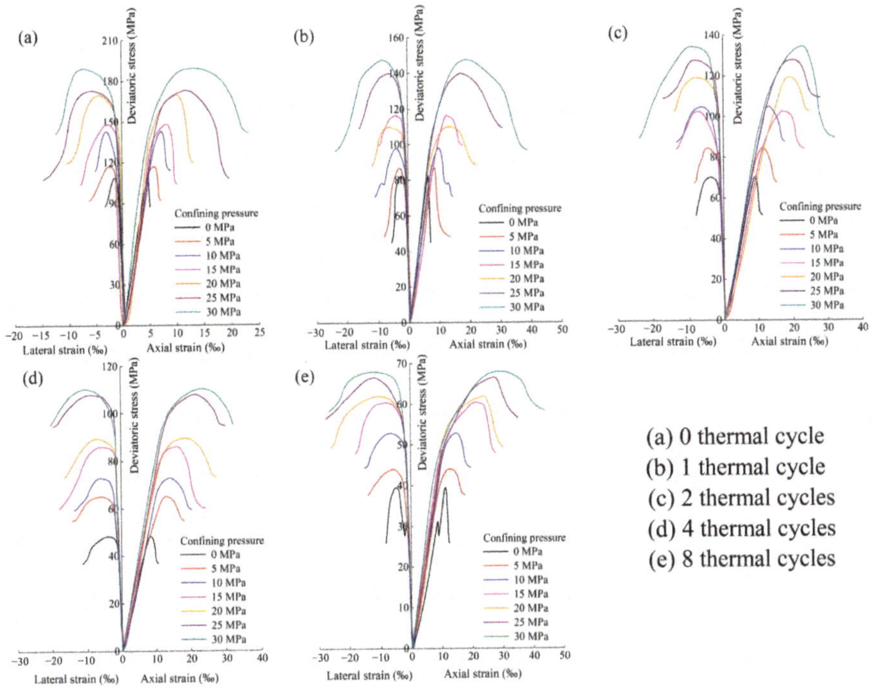

Figure 5. Complete stress-strain curves of marble specimens in triaxial compression under different thermal cycles.

3.1.3. Strength Characteristics

Figure 6 shows the relationship between the peak stress and confining pressure under the same number of thermal cycles. The figure presents a good positive linear relationship, and the phenomenon is consistent with the Coulomb criterion. In summary, peak stress decreases when the number of cycles increases and increases when confining pressure rises. When it comes to the same number of thermal cycles, the peak stress under 30 MPa confining pressure is about 1.74–2.29 times that of UCS.

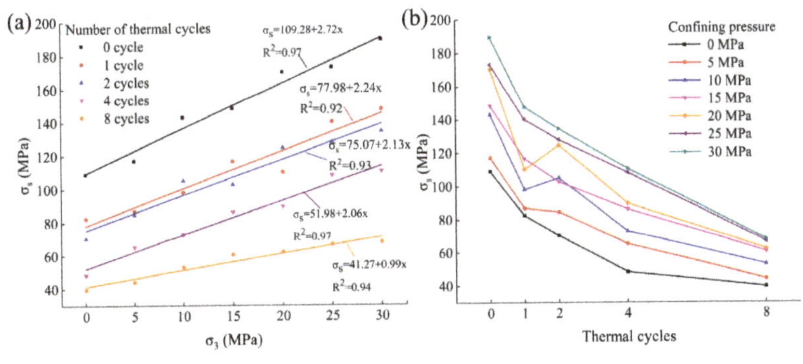

Figure 6. Relationship between the triaxial compressive strength of marble and confining pressure and number of thermal cycles. (**a**) σ_s (Peak stress)-σ_3 under different thermal cycles; (**b**) σ_s-thermal cycles under different confining pressure.

3.1.4. Analysis of Deformation Characteristics

The peak strain under different conditions is shown in Figure 7. Under the same number of thermal cycles, the peak strain rises with the increase in confining pressure. With the increase in the number of thermal cycles, the peak strain of fine-grained marble increases gradually, and the growth rate shows a decreasing trend. The interval of the maximum peak strain growth rate is from zero to one thermal cycle. When the confining pressure is the same, the peak strain of the samples after an increased number of thermal cycles is always greater than a lower number of thermal cycles, except for individual discrete data points. Thermal damage and confining pressure have a great influence on peak strain. For example, the peak strain was 4.72‰ under a uniaxial compression test without thermal damage. However, under 30 MPa confining pressure after eight thermal cycles, the strain reached 30.03‰.

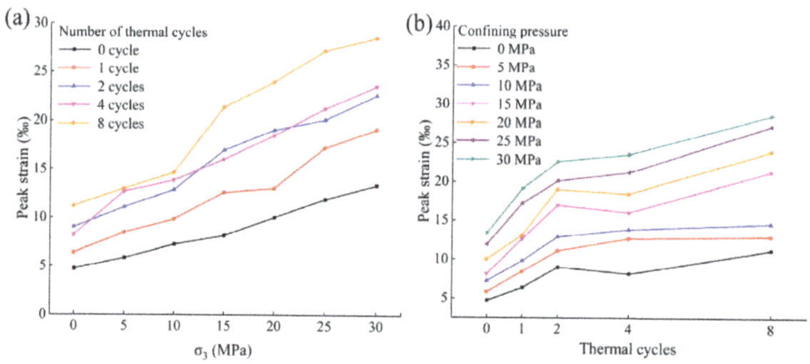

Figure 7. Relationship between peak strain of marble and the confining pressure and the number of thermal cycles. (**a**) Peak strain-σ_3 under different thermal cycles; (**b**) Peak strain-thermal cycles under different confining pressure.

In this paper, the strength evolution model of thermocycling marble was established by statistical analysis of the data based on indoor tests according to Section 3.1.2. The effect of thermal cycling causes the deterioration of the macroscopic mechanical characteristics of the rocks, resulting in a decrease in mechanical property indexes such as strength.

The strength evolution model under different thermal cycling times was constructed considering the influence of the confining pressure. In a three-dimensional cartesian coordinate system, the *x*-axis is the number of thermal cycles, the *y*-axis is the confining pressure, and the *z*-axis is the triaxial compressive strength. By fitting the surface to the test data (Figure 8), the triaxial compressive strength evolution model of marble with different numbers of thermal cycles was obtained as follows with $R^2 = 0.95815$:

$$\sigma_c = 102.20278 - 18.41821n + 3.01708c + 1.32175n^2 - 0.01358c^2 - 0.19291cn \quad (1)$$

where σ_c is the triaxial compressive strength (MPa); n is the number of thermal cycles; and c is the confining pressure (MPa).

The elastic modulus of rock is an important index to reflect its properties. The elastic modulus of the samples, after experiencing a different number of thermal cycles and under different confining pressures, is shown in Figure 9. The change rule of the elastic modulus is similar to peak stress. At zero confining pressure, the elastic modulus under eight thermal cycles was 18.30% of the undamaged one. The elastic modulus under 30 MPa confining pressure was 1.35 times that of the uniaxial compression test while there was no heat damage. According to the law of growth rate, it can be roughly divided into two sections; namely, the one with low confining pressure (0–15 MPa) increasing rapidly and the other with medium-high confining pressure (15–30 MPa) increasing slowly, which reflects the strengthening effect of the confining pressure on the elastic modulus of marble. After the first thermal cycle, the elastic modulus of marble decreased significantly, then the rate of decline slowed down.

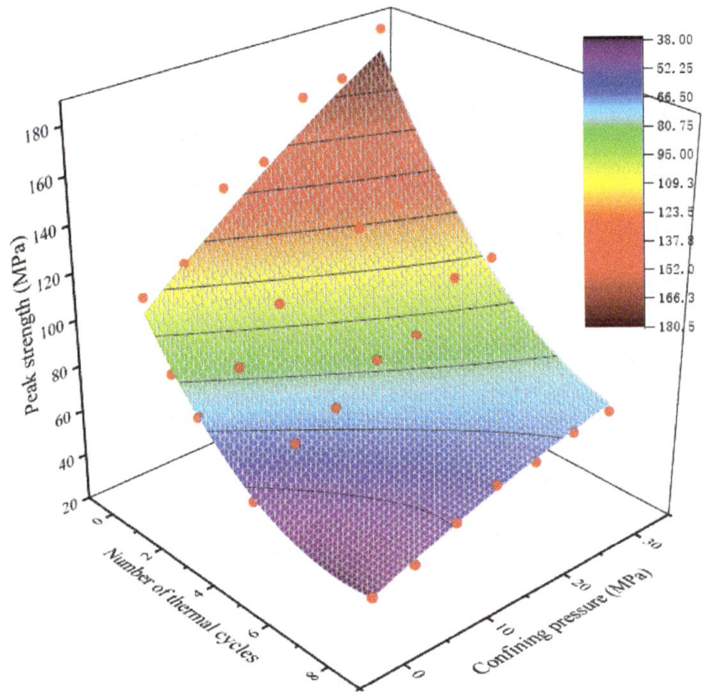

Figure 8. Triaxial compressive strength evolution of marble under thermal cycles.

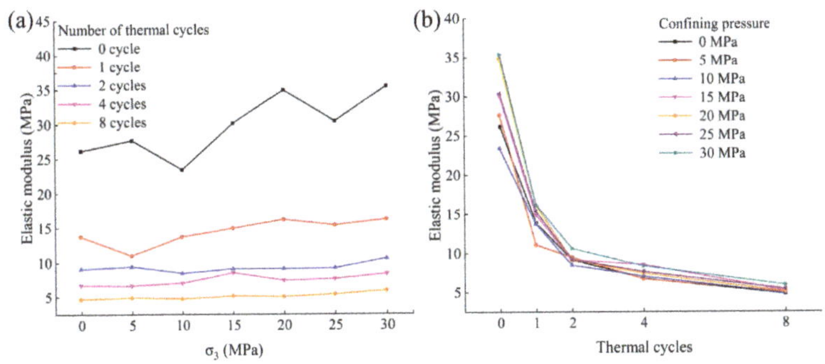

Figure 9. Relationship between elastic modulus of marble and the confining pressure and the number of thermal cycles. (**a**) Elastic modulus-σ_3 under different thermal cycles; (**b**) Elastic modulus-thermal cycles under different confining pressure.

Considering the influence of confining pressure, the elastic modulus evolution model under different thermal cycles was constructed. In the three-dimensional cartesian coordinate system, the x-axis is the number of thermal cycles, the y-axis is the confining pressure, and the z-axis is the elastic modulus. Through surface fitting to the test data (Figure 10), the triaxial compressive strength evolution model of marble with different thermal cycles was obtained as follows with $R^2 = 0.98039$:

$$E = \frac{26.38718 + 0.50408n - 0.47751c + 0.04709c^2 - (9.30699 \times 10^{-4})c^3}{1 + 1.16128n - 0.11915n^2 + 0.0066n^3 - 0.00652c + (5.2046 \times 10^{-5})c^2} \tag{2}$$

where E is the elastic modulus (GPa); n is the number of thermal cycles; and c is the confining pressure (MPa)

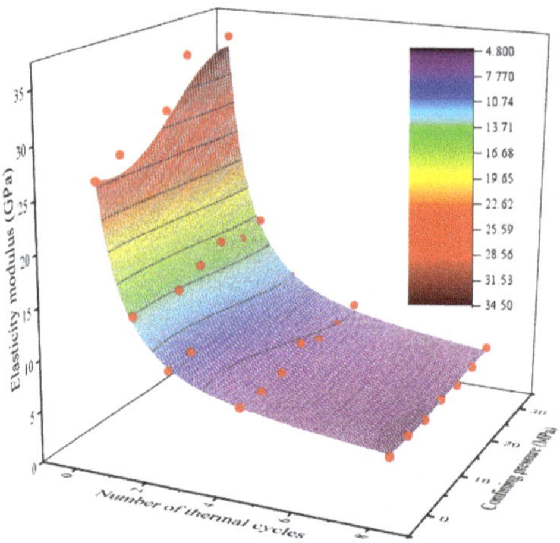

Figure 10. Elastic modulus evolution of marble under triaxial compression under thermal cycles.

3.1.5. Failure Mode

Figure 11 shows the failure mode of the marble that underwent thermal cycles under different confining pressures, and the red line highlights the surface crack of the rock sample.

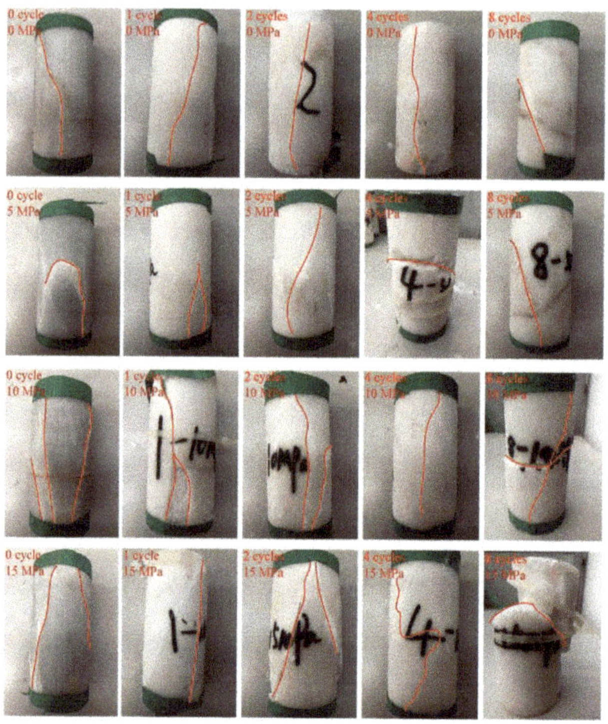

Figure 11. *Cont.*

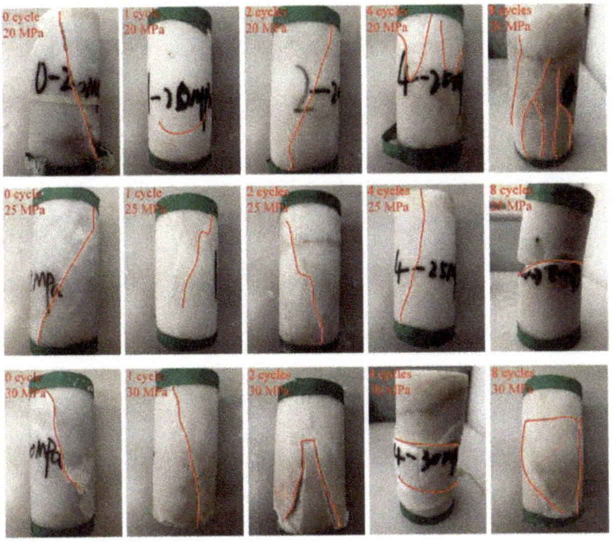

Figure 11. Fracture modes of marble under different confining pressures and different thermal cycles.

The samples showed brittle fracture failure before two thermal cycles, and the surface fractures of the rock samples were not obvious. After two thermal cycles, cracks appeared on the surface of the samples more clearly, indicating that the rock experienced plastic failure.

When confining pressure was applied, the samples that underwent one thermal cycle treatment showed brittle fracture failure. As for the rock samples treated with two or four thermal cycles, they exhibited brittle fracture failure or brittle shear failure under a confining pressure of less than 30 MPa and ductile failure at 30 MPa.

3.2. Energy Evolution Characteristics

Rock loading is a process of energy input, accumulation, dissipation, release, and transformation. Assuming that no heat exchange occurs, then part of the work performed by external forces accumulates in the elastic deformation energy in the rock, and the other part is dissipated in the plastic deformation energy and damage energy.

When the elastic deformation accumulated inside reaches the energy storage limit of the rock, the rock will be damaged and releases energy to the outside. The process from rock deformation to unstable failure is manifested as strain energy generated by the work of external forces, energy conversion, and the balance of the rock interior and external output. It is significant to understand the damage evolution of rock and the change rules of the total input energy, elastic strain energy, and dissipated energy in the deformation and failure process of the rock; their densities are named U^t, U^e, and U^d. At the same time, describing the mechanical properties and damage characteristics of rock from the perspective of energy is necessary.

The rock element deforms under external forces, assuming that there is no heat exchange between the loading process and the outside world [20,21]. According to the first law of thermodynamics:

$$U^t = U^d + U^e \tag{3}$$

For the triaxial compression loading test, the above formula can be written as:

$$U^e = \frac{1}{2E^t}\left[\sigma_1^2 + 2\sigma_3^2 - 2v^t\left(2\sigma_1\sigma_3 + \sigma_3^2\right)\right] \tag{4}$$

The triaxial loading test shows that after reaching the determined hydrostatic stress, σ_1 will increase and continuously generate compression deformation and positive work on the rock sample. However, the expansion deformation mainly occurs in the lateral direction, and the work performed by σ_3 to the rock sample is negative. U^t converted into rock sample during the test is:

$$U^t = \int_0^{\varepsilon_1^t} \sigma_1 d\varepsilon_1 + 2\int_0^{\varepsilon_3^t} \sigma_3 d\varepsilon_3 + \frac{3(1-2\nu)}{2E_t}(\sigma_3)^2 \tag{5}$$

It is worth noting that the energy density calculated here is input, accumulated or dissipated in the axial and lateral directions, and is calculated from the hydrostatic stress state. In fact, from the initial state to the hydrostatic stress state, there are also energy evolution processes in the rock samples. Considering that the energies do not directly lead to the failure of rock samples and facilitate analysis, the aspect of energy evolution is not analyzed here. So, U^t can be calculated as follows:

$$U^t = \int_0^{\varepsilon_1^t} \sigma_1 d\varepsilon_1 + 2\int_0^{\varepsilon_3^t} \sigma_3 d\varepsilon_3 \tag{6}$$

3.2.1. Energy Evolution Characteristics under Thermal Cycles

Because the amount of data is too large (Table 3), a typical example of the uniaxial compression test was used to analyze the energy evolution law under a different number of thermal cycles. U^t, U^e, and U^d under different thermal cycles are calculated according to the stress-strain curve and the formula above. Plotting these data into Figure 12, the evolution law of energy is obtained with the increase in axial strain.

Table 3. Energy characteristic values of marble triaxial test under different numbers of thermal cycles.

Thermal Cycles	Confining Pressure	U_e^{peak}	U_d^{peak}	U^{peak}	Peak Stress	Elastic Modulus
0	0	227.06	159.10	305.37	109.04	26.18
0	5	260.53	313.83	475.93	117.15	27.68
0	10	475.64	438.80	744.59	142.99	23.42
0	15	415.47	676.96	886.87	148.31	30.12
0	20	482.10	1139.16	1384.18	170.15	34.89
0	25	603.68	2043.72	2291.34	173.10	30.40
0	30	618.05	3176.89	3526.32	189.45	35.41
1	0	245.63	236.28	310.02	82.40	13.82
1	5	367.43	542.84	662.72	86.80	11.10
1	10	397.451	631.47	844.06	98.07	13.75
1	15	537.93	939.09	1331.35	116.50	14.96
1	20	488.95	1548.50	1861.71	110.01	16.20
1	25	825.94	2837.10	3340.77	140.07	15.37
1	30	871.87	3843.56	4232.42	147.63	16.21
2	0	269.46	291.96	437.92	70.30	9.17
2	5	408.25	560.19	825.06	84.54	9.51
2	10	745.93	556.46	1079.80	105.05	8.52
2	15	719.75	1045.94	1540.10	102.74	9.16
2	20	1045.47	1045.05	1805.11	124.51	9.17
2	25	1218.65	1508.91	2376.53	127.91	9.23
2	30	1188.68	2001.38	2548.22	134.44	10.61
4	0	171.62	201.75	302.20	48.31	6.80
4	5	350.82	497.73	748.86	65.22	6.74
4	10	462.17	637.31	945.08	72.83	7.08
4	15	560.91	981.79	1264.03	86.08	8.58
4	20	738.28	1235.58	1715.32	89.52	7.48
4	25	1057.05	1376.60	2153.06	108.02	7.66
4	30	1055.06	1727.63	2466.24	110.46	8.39
8	0	162.12	178.08	248.30	39.53	4.79
8	5	228.89	304.52	472.17	44.07	5.01
8	10	390.54	324.12	594.91	52.89	4.83
8	15	515.27	804.83	1120.32	60.50	5.20
8	20	629.66	585.78	961.61	62.03	5.09
8	25	712.67	792.85	1253.19	66.63	5.42
8	30	696.74	1666.30	2117.04	68.04	5.97

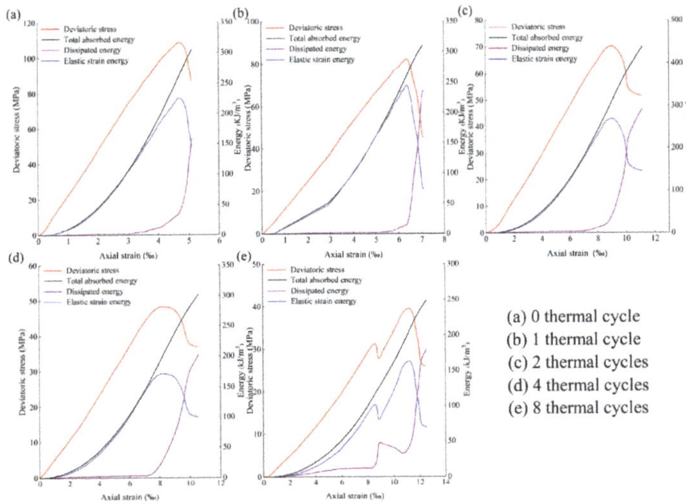

Figure 12. Energy evolution law under different numbers of thermal cycles.

For the pre-peak stage, U^t and U^d increased gradually with the rise in the axial strain at different times during the thermal cycle tests. The TAW-3000 rock test system transmits the mechanical energy to the rock samples continuously. On the one hand, the rock sample accumulates it; on the other hand, it dissipates and causes a change in the sample's structure. The increase in elastic strain energy was much greater than dissipated, indicating that the energy behavior of the rock in this stage was mainly reflected in energy accumulation. Furthermore, the growth rate of U^e rose slowly in the crack closure stage of loading, which is due to the compression of the original void volume in the rock sample, the small primary stiffness, and the low energy conversion efficiency. When it was about to reach the failure stage, the growth rate of U^d increased suddenly, while the growth rate of U^e decreased, which means that the rock structure changed greatly, and the cracks expanded and converged significantly.

For the post-peak stage, U^e decreased sharply and U^d increased greatly. There was a peak value in both energy curves. The value of U^e corresponds to peak stress of axial stress-strain curve, and U^d increases gradually with axial strain. Therefore, the peak point of U^d corresponds to the maximum axial strain, and U^e in this stage will be converted into energy dissipation and release.

In order to compare the influence of thermal cycles on the energy evolution behavior of the rock more intuitively, the relationship between the change in U^e and U^d with the axial strain under different confining pressures is plotted in Figures 13 and 14.

U^e changes under the influence of U^t and U^d. When the number of thermal cycles was low, there were few cracks in the heat-treated rock sample, so it absorbed little energy in the crack closure stage. However, due to the increase in U^t and the slight growth of damage energy, the energy storage limit presented an increasing trend when the number of thermal cycles was relatively low. When the number was higher, the thermal damage of the rock samples was aggravated, and U^d increased obviously in the crack closure stage and linear elastic stage; moreover, the lateral strain (ε_3) increased, so U^t and the peak of U^e showed a downward trend.

Similar to U^e, under different thermal cycles, the curves of U^d almost overlap before reaching their peak value. As the strain grew slowly, U^d increased sharply when the rock approached failure, which indicates the occurrence of macroscopic fracture, and then, it maintained the trend and increased gradually with the strain.

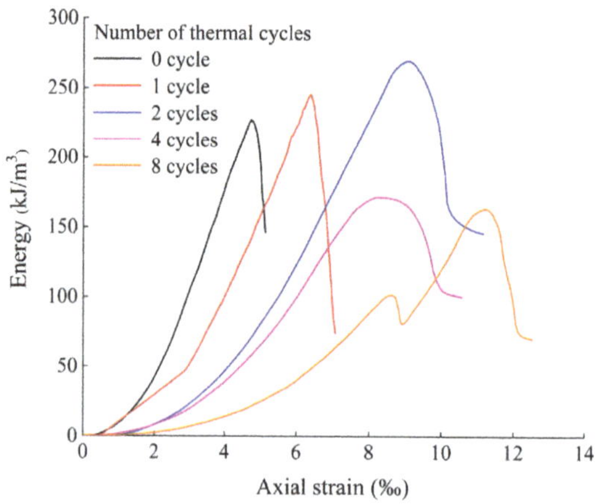

Figure 13. Variation in the elastic strain energy of marble with axial strain during different numbers of thermal cycles under 0 confining pressure.

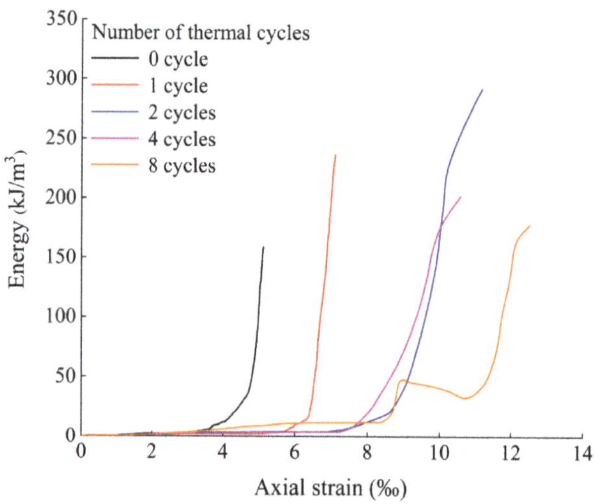

Figure 14. Variation in the dissipated energy of marble with axial strain during different numbers of thermal cycles under 0 confining pressure.

According to Table 2, for marble, its energy evolution parameters (such as peak elastic energy and peak dissipation energy) and physical and mechanical parameters (such as peak strength and elastic modulus) do not follow a strict co-directional relationship. The energy evolution parameters have a good positive correlation with the confining pressure. In particular, the marble samples under different confining pressures all reached the peak elastic energy of the test group when the number of thermal cycles was two.

3.2.2. Energy Evolution Characteristics under Different Confining Pressures

According to the data collected during this test, the evolution law of zero thermal cycles is taken as a typical example to analyze the energy evolution law under different confining pressures, as shown in Figure 15.

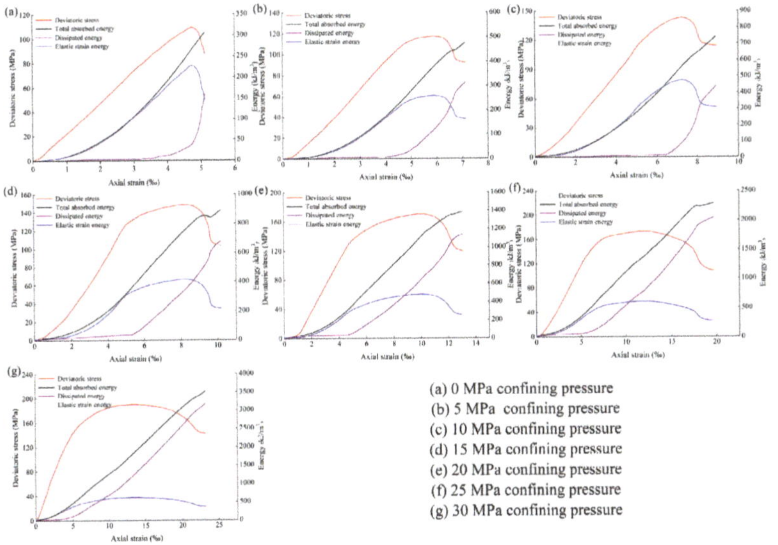

Figure 15. Energy evolution law under different confining pressures.

Crack closure stage: The growth rate of U^e is relatively slow, and U^d almost presents a horizontal trend. This is because the primary crack in the sample is compressed, and the energy conversion efficiency is low.

Linear elastic stage: With the increase in load, U^t and U^e gradually rise. At this stage, damage and plastic deformation inside the rock are small, and U^d is small, so the curve shows a gentle trend. Most of the energy absorbed by the marble from the servo machine is stored as U^e, and the curve's trend is almost consistent with the stress-strain curve.

Nonlinear deformation stage: At this time, the crack inside the rock becomes larger, U^d increases, and U^e decreases, but the energy inside the rock is still dominated by the stored elastic strain energy. When approaching failure, the rock structure changes greatly and fractures expand significantly, the growth rate of U^d increases obviously, but U^e decreases. U^e reaches the maximum value when axial stress reaches its peak; then, the rock is destroyed. So, U^e is converted into U^d rapidly, and the rock damage process is mainly energy release and dissipation.

Failure stage: The rock still has the capacity to bear a load, and the crack continues to grow with axial stress under the influence of confining pressure. The energy absorbed by the rock sample at this time is mainly used for the energy dissipation of crack propagation. U^e stored in the rock at this stage after rock failure is maintained at a low level, and U^d continues to increase with the extension of the crack.

The relationship between energy density and axial strain is plotted in Figures 16 and 17, respectively, in order to directly compare the effects of different confining pressures on rock energy evolution behavior.

We can infer from the figures that before rock failure, U^e under different confining pressures almost coincides with the changing curve of strain. This shows that the growth law of U^e before the peak does not change with confining pressure and that the limit value of U^e is closely related to the pressure. The peak value is defined as the rock energy storage limit, and the value after the release of elastic strain energy after the peak is the residual elastic energy. With an increase in confining pressure, the energy storage limit of the rock increases greatly. Before 25 MPa, it increases significantly with confining pressure, and when it reaches 25 MPa, it tends to be stable with elastic strain energy.

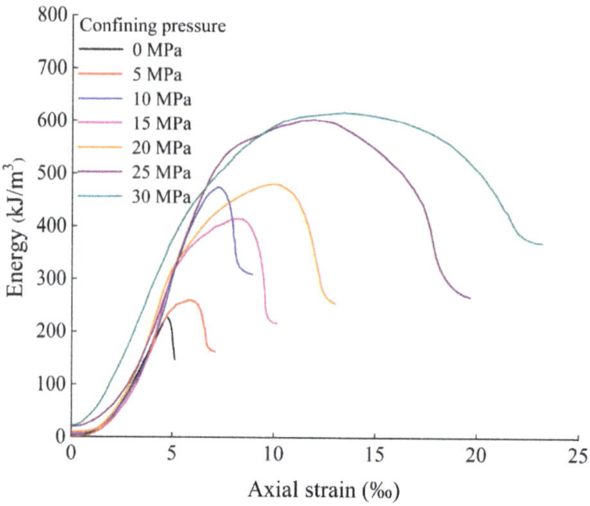

Figure 16. Variation in the elastic strain energy of marble with axial strain under different confining pressures at 0 thermal cycles.

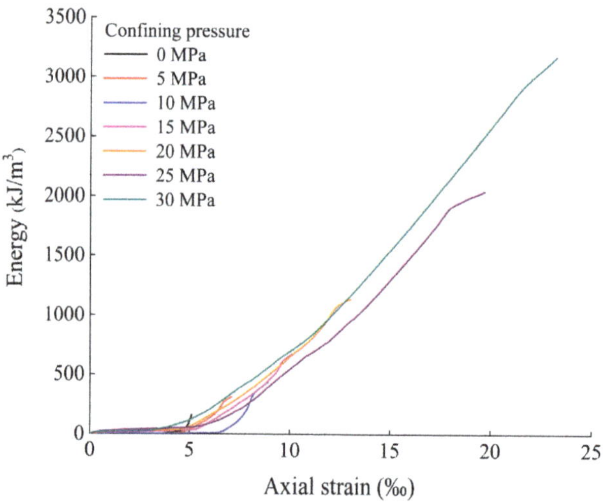

Figure 17. Variation law of the dissipated energy of marble with axial strain under different confining pressures at 0 thermal cycles.

The trend before the peak is very similar under various confining pressures. As the strain grows slowly to the point of approaching failure, U^d increases rapidly; this is because the higher the confining pressure, the smaller the fracture surface forms after rock failure. So, when the confining pressure is high, most of the dissipated energy is used for plastic deformation.

3.2.3. Peak Elastic Energy Evolution of Marble under Triaxial Compression and Thermal Cycling Conditions

The peak elastic energy evolution model under different thermal cycles was constructed considering the influence of confining pressure [29]. In the three-dimensional cartesian coordinate system, the x-axis is the number of thermal cycles, the y-axis is confining pressure, and the z-axis is the peak elastic energy. Through surface fitting to the test data (Figure 18), the model of the peak elastic energy evolution of marble under triaxial compression with different thermal cycles was obtained as follows with $R^2 = 0.93805$:

$$U^e = \frac{217.39255 - 10.70847n + 5.68309c - 0.54128c^2 + 0.01242c^3}{1 - 0.22541n + 0.05745n^2 - 0.00411n^3 - 0.04425c + (8.4146e-4)c^2} \quad (7)$$

where U^e is the peak elastic energy (kJ/m^3); n is the number of thermal cycles; and c is confining pressure (MPa).

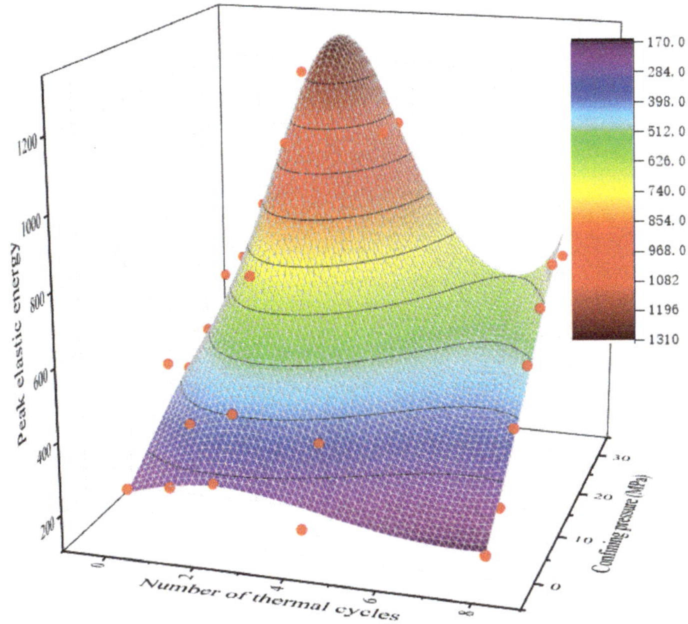

Figure 18. Peak elastic energy evolution of marble under triaxial compression and thermal cycling conditions.

3.3. Microstructure

Three types of cracks are induced by temperature: grain boundary crack, intragranular crack, and transgranular crack [30]. Since the distinction between intragranular and transgranular cracks is not obvious in recent research, this paper unifies them as intragranular cracks. The microstructure of rock samples undergoing different thermal cycles is shown in Figure 19. The results indicate that the microscopic morphology and crack development changed significantly after different numbers of thermal cycles, and the color of mineral particles gradually darkened at high temperatures. For samples without heat treatment, the structure was compact, the mineral boundary contact was good, and no obvious microcracks could be observed.

After one thermal cycle, a large number of cracks occurred at the boundary of mineral particles, some of the particles were separated, and the opening of the mineral boundary increased greatly. This is mainly due to the uneven thermal expansion coefficient of different mineral particles, which generates thermal stress at the boundary after heating. Additionally, the stress exceeded the yield strength of the mineral so that microcracks appeared, and the cracks developed randomly and directionless. As the number of cycles continuously rose, the number and opening of cracks continued to develop. The development of grain boundary cracks was the main trend, and cracks began to appear inside the mineral. After four thermal cycles, the crack opening increased significantly, among which the grain boundary cracks reached a maximum of 14 μm. This is because the small cracks at the particle boundary converged to form a crack zone under the influence of high temperature. At the same time, a smooth failure district appeared at the grain boundary, which gradually expanded with the increase in the number of thermal cycles. The location and direction of the damaged district are related to the shape and size of the mineral particles and are also influenced by the dissociation that occurs at the mineral boundary. After eight thermal cycles, cracks continued to increase, and mineral boundary openings continued to expand. The maximum opening of grain boundary cracks reached 29 μm, and the maximum opening of intragranular cracks increased to 9 μm.

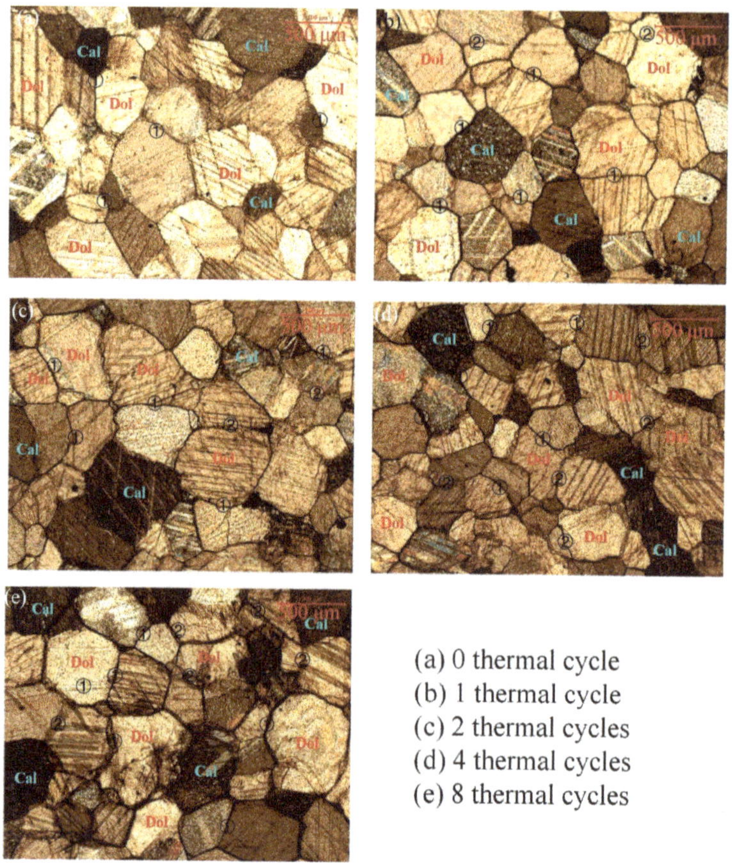

Figure 19. Microstructure of rock specimens after different cycles of thermal cycle treatment (① grain boundary microcracks, ② intra-granular microcracks).

The marble used in the experiment was recrystallized by metamorphism, and the grains were densely arranged. Thermal-damaged cracks caused by uneven expansion easily form. The continuous increase in cracks was mainly due to the further development of the original cracks after heat treatment and the formation of new small cracks, which then extended, converged, and transfixed into large cracks.

3.4. Damage Mechanism Analysis

Considering the form of a crack, the difference in the failure mode is mainly determined by the interaction between tensile and shear cracks. When the damage degree is low, the influence of tensile cracks on the rock sample's fracture is dominant, and the tensile crack formed is parallel to the loading direction, so the type of failure is axial cleavage. With the increase in the thermal damage degree or confining pressure, shear cracks gradually take a dominant position and act together with tensile cracks to form a main shear plane when the rock sample is broken.

Considering the internal structure of the rock, when the degree of thermal damage increases, the mineral composition and crystal structure of the rock sample will change, and the cohesion (C) between the crystals will decrease. Then, a dislocation glide occurs, which will degrade the physical and mechanical properties. Because rock is a heterogeneous and anisotropic material, the strengths of the materials that make up the minerals are different. During the loading process, the different bearing capacities of minerals on the same surface result in different strengths of the materials in the section. When the axial stress increases, the material with low load-bearing strength yields first, causing the loading stress to shift and the actual bearing stress of the unyielding material to increase. Therefore, shear cracks will appear along the yield surface. As the cracks propagate, merge, and transfix, the shear failure mode is formed finally.

Considering the energy, the damage to rock is mainly caused by the release of elastic strain energy, and this can be attributed to loading or high temperature. The greater the release of elastic strain energy, the greater the damage caused. Macroscopically, the rock sample has more fracture surfaces or cracks. Therefore, the fracture surface loading after thermal cycling or confining pressure is higher than those without. However, after a certain degree of thermal damage, the elastic energy storage limit of the rock will rise slightly, and, ultimately, the energy will not be fully released.

4. Conclusions

In order to study the influence of thermal cycles and confining pressures on the mechanical properties and energy characteristics of marble, this study conducted tests on marble after different numbers of thermal cycles of uniaxial compression and triaxial compression. According to the test results, the following conclusions can be drawn:

1. After the same number of thermal cycles, the peak strength, peak strain, and elastic modulus of the marble sample are positively correlated with the confining pressure; under the same confining pressure, the peak strain of marble samples increases with the increase in thermal cycles, while the peak strength and elastic modulus both show a downward trend.
2. Different minerals in the marble sample have different degrees of expansion after being heated, which causes the rock structure to crack. In addition, the microcracks inside the rock develop significantly, so the ductility of the rock sample increases and the mechanical properties deteriorate.
3. When the number of thermal cycles is constant, the samples have different energy evolution trends at each stage under different confining pressures. However, under the influence of different thermal cycles at a set confining pressure, the energy characteristics behaves similarly.
4. The curves of the peaks of the total absorbed energy, dissipated energy, and elastic strain energy all show a convex trend, and the peaks of these three characteristic energy values under different confining pressures all appear at one or two thermal cycles.
5. The energy evolution and the stress-strain curve in the rock loading process show a good corresponding relationship. In the crack closure stage, the energy conversion rate of the original crack compaction process is low; in the linear elastic deformation stage, the total absorbed energy and elastic strain energy increase significantly, the energy dissipated by internal damage and plastic deformation is small, and the energy converted by external force is mainly elastic strain energy storage; in the nonlinear deformation stage, the dissipated energy significantly increases; and in the failure stage, the accumulated elastic strain energy reaches the rock energy storage limit, and then, the macroscopic failure of rock occurs. Moreover, the elastic strain energy is quickly transformed into dissipated energy for rock failure.

6. The dissipated energy in the marble loading process gradually increases with the accumulation of axial strain. In the crack closure and linear elastic stages, dissipated energy increases slowly at a low level as the axial strain increases; in the nonlinear deformation stage, the internal cracks expand and transfix, and the energy dissipation rate increases rapidly as the axial strain increases. The stress limit of the rock in the failure stage is reduced, but the energy dissipation remains at a high level because of continuous crack propagation.
7. The energy evolution parameters (such as the peak elastic energy and peak dissipation energy) have a positive correlation with the confining pressure. In particular, the marble samples under different confining pressures all reach the peak elastic energy of the test group when the number of thermal cycles is two.

The conclusions drawn in this study explain the mechanical properties and energy evolution characteristics of rocks under different confining pressures and after a different number of thermal cycles, which are helpful in providing references for the design and construction of underground rock projects involving high temperatures and high pressures. However, because the number of thermal cycles and confining pressure settings are very few, a corresponding numerical model cannot be established. In the future, more tests will be required at various temperatures and different confining pressures to establish an accurate constitutive model for quantitative calculation and deformation prediction in actual underground engineering.

Author Contributions: Process design, writing, experiment operation, Q.W.; data processing, chart processing, B.L.; test material, X.J. All authors have read and agreed to the published version of the manuscript.

Funding: This research received no external funding.

Data Availability Statement: Not applicable.

Conflicts of Interest: The authors declare no conflict of interest.

References

1. Majer, E.L.; Baria, R.; Stark, M.; Oates, S.; Bommer, J.; Smith, B.; Asanuma, H. Induced seismicity associated with enhanced geothermal systems. *Geothermics* **2007**, *36*, 185–222. [CrossRef]
2. Peng, J.; Yang, S. Comparison of mechanical behavior and acoustic emission characteristics of three thermally-damaged rocks. *Energies* **2018**, *11*, 2350. [CrossRef]
3. Su, H.; Jing, H.; Du, M.; Wang, C. Experimental investigation on tensile strength and its loading rate effect of sandstone after high temperature treatment. *Arab. J. Geosci.* **2016**, *9*, 616. [CrossRef]
4. An, H.; Zeng, T.; Zhang, Z.; Liu, L. Experimental study of the rock mechanism under coupled high temperatures and dynamic loads. *Adv. Civ. Eng.* **2020**, *2020*, 8866621. [CrossRef]
5. Zhang, L.; Mao, X.; Liu, R.; Guo, X.; Ma, D. The mechanical properties of mudstone at high temperatures: An experimental study. *Rock Mech. Rock Eng.* **2014**, *47*, 1479–1484. [CrossRef]
6. Liu, S.; Xu, J. Analysis on damage mechanical characteristics of marble exposed to high temperature. *Int. J. Damage Mech.* **2015**, *24*, 1180–1193. [CrossRef]
7. Rao, Q.; Wang, Z.; Xie, H.; Xie, Q. Experimental study of mechanical properties of sandstone at high temperature. *J. Cent. South Univ. Technol.* **2007**, *14*, 478–483. [CrossRef]
8. Meng, Q.; Qian, W.; Liu, J.; Zhang, M.; Lu, M.; Wu, Y. Analysis of triaxial compression deformation and strength characteristics of limestone after high temperature. *Arab. J. Geosci.* **2020**, *13*, 153. [CrossRef]
9. Zhang, P.; Mishra, B.; Heasley, K.A. Experimental investigation on the influence of high pressure and high temperature on the mechanical properties of deep reservoir rocks. *Rock Mech. Rock Eng.* **2015**, *48*, 2197–2211. [CrossRef]
10. Wei, S.; Yang, Y.; Su, C.; Cardosh, S.R.; Wang, H. Experimental study of the effect of high temperature on the mechanical properties of coarse sandstone. *Appl. Sci.* **2019**, *9*, 2424. [CrossRef]
11. Deng, Y.; Deng, H. Experimental study on failure criterion of deep tight sandstone under coupling effects of temperature and pressure. *Arab. J. Geosci.* **2019**, *12*, 575. [CrossRef]
12. Meng, L.; Li, T.; Cai, G. Temperature effects on the mechanical properties of slates in triaxial compression test. *J. Mt. Sci-Eng.* **2017**, *14*, 2581–2588. [CrossRef]
13. Su, H.; Jing, H.; Yin, Q.; Yu, L.; Wang, Y.; Wu, X. Strength and deformation behaviors of veined marble specimens after vacuum heat treatment under conventional triaxial compression. *Acta Mech. Sinica-Prc.* **2017**, *33*, 886–898. [CrossRef]

14. Idris, M.A. *Effects of Elevated Temperature on Physical and Mechanical Properties of Carbonate Rocks in South-Southern Nigeria*; Dnipro University of Technology: Dnipropetrovsk Oblast, Ukraine, 2018.
15. Rong, G.; Peng, J.; Cai, M.; Yao, M.; Zhou, C.; Sha, S. Experimental investigation of thermal cycling effect on physical and mechanical properties of bedrocks in geothermal fields. *Appl. Therm. Eng.* **2018**, *141*, 174–185. [CrossRef]
16. Xu, X.L.; Zhang, Z.Z. Acoustic Emission and Damage Characteristics of Granite Subjected to High Temperature. *Adv. Mater. Sci. Eng.* **2018**, *2018*, 8149870. [CrossRef]
17. Chen, G.; Li, T.; Li, G.; Qin, C.A.; He, Y. Influence of temperature on the brittle failure of granite in deep tunnels determined from triaxial unloading tests. *Eur. J. Environ. Civ. Eng.* **2018**, *22*, s269–s285. [CrossRef]
18. Luo, J.A.; Wang, L. High-Temperature Mechanical Properties of Mudstone in the Process of Underground Coal Gasification. *Rock Mech. Rock Eng.* **2011**, *44*, 749–754. [CrossRef]
19. Meng, X.; Liu, W.; Meng, T. Experimental Investigation of Thermal Cracking and Permeability Evolution of Granite with Varying Initial Damage under High Temperature and Triaxial Compression. *Adv. Mater. Sci. Eng.* **2018**, *2018*, 8149870. [CrossRef]
20. Chen, Z.; He, C.; Ma, G.; Xu, G.; Ma, C. Energy Damage Evolution Mechanism of Rock and Its Application to Brittleness Evaluation. *Rock Mech. Rock Eng.* **2019**, *52*, 1265–1274. [CrossRef]
21. Luo, Y.; Wang, G.; Li, X.; Liu, T.; Mandal, A.K.; Xu, M.; Xu, K. Analysis of energy dissipation and crack evolution law of sandstone under impact load. *Int. J. Rock Mech. Min.* **2020**, *132*, 104359. [CrossRef]
22. He, M.M.; Pang, F.; Wang, H.T.; Zhu, J.W.; Chen, Y.S. Energy Dissipation-Based Method for Strength Determination of Rock under Uniaxial Compression. *Shock Vib.* **2020**, *2020*, 8865958. [CrossRef]
23. Li, M.; Zhang, J.; Zhou, N.; Huang, Y. Effect of Particle Size on the Energy Evolution of Crushed Waste Rock in Coal Mines. *Rock Mech. Rock Eng.* **2017**, *50*, 1347–1354. [CrossRef]
24. Gong, F.; Luo, S.; Yan, J. Energy Storage and Dissipation Evolution Process and Characteristics of Marble in Three Tension-Type Failure Tests. *Rock Mech. Rock Eng.* **2018**, *51*, 3613–3624. [CrossRef]
25. Ma, Q.; Tan, Y.; Liu, X.; Gu, Q.; Li, X. Effect of coal thicknesses on energy evolution characteristics of roof rock-coal-floor rock sandwich composite structure and its damage constitutive model. *Compos. Part B Eng.* **2020**, *198*, 108086. [CrossRef]
26. Zhang, H.; Lu, C.; Liu, B.; Liu, Y.; Zhang, N.; Wang, H.Y. Numerical investigation on crack development and energy evolution of stressed coal-rock combination. *Int. J. Rock Mech. Min.* **2020**, *133*, 104417. [CrossRef]
27. Hudson, J.A.; Cornet, F.H.; Christiansson, R. ISRM Suggested Methods for rock stress estimation—Part 1: Strategy for rock stress estimation. *Int. J. Rock Mech. Min.* **2003**, *40*, 991–998. [CrossRef]
28. Liu, H. Hydrogen assisted intergranular cracking in steels. *Eng. Fract. Mech.* **2011**, *78*, 2563–2571. [CrossRef]
29. Zhao, K.; Yu, X.; Zhou, Y.; Wang, Q.; Wang, J.; Hao, J. Energy evolution of brittle granite under different loading rates. *Int. J. Rock Mech. Min. Sci.* **2020**, *132*, 104392. [CrossRef]
30. Xie, S.; Han, Z.; Shu, R.; Chen, Y.; Feng, F. A new method to determine the crack closure stress based on stress difference. *Theor. Appl. Fract. Mech.* **2022**, *119*, 103337. [CrossRef]

applied sciences

MDPI

Article

Damping and Stiffness Responses of Silica Rock under Constant Amplitude and Variable Rate Cyclic Loading

Yunfeng Wu [1,2], Yu Wang [1,2,*], Changhong Li [1,2], Baokun Zhou [3], Zicheng Tian [1,2], Changkun Sun [1,2] and Youdong Zhu [1,2]

1 Department of Civil Engineering, School of Civil & Resource Engineering, University of Science & Technology Beijing, Beijing 100083, China; laowu412825@163.com (C.S.)
2 Key Laboratory of Ministry of Education for High-Efficient Mining and Safety of Metal Mines, University of Science & Technology Beijing, Beijing 10083, China
3 School of Energy and Mining Engineering, China University of Mining and Technology-Beijing, Beijing 100083, China; 18841202983@163.com
* Correspondence: wyzhou@ustb.edu.cn

Abstract: In this paper, the shear modulus and damping ratio of silica rock under cyclic loading were experimentally analyzed using two loading modes, constant amplitude and increasing amplitude, combined with three increasing loading rates. Observations have indicated a decrease in the shear modulus of specimens as the number of cycles increased during the loading and unloading phases and an overall increase with larger amplitude intervals. The change in loading rate significantly affects the damping ratio of the specimens, leading to a stepwise decrease within the same cyclic group, while the damping ratio of a single specimen exhibits a 'concave' distribution throughout the cyclic interval. Based on the axial strain and dissipation energy, this paper develops two damage models that can effectively predict the damage accumulation process in rocks under cyclic loading. These findings have significant implications for a deeper understanding of the mechanical behavior of rocks under dynamic loading and offer theoretical guidance and technical support for rock engineering.

Keywords: cyclic loading; loading rate; shear modulus; damping ratio; damage model

1. Introduction

Earthquakes are among the most destructive geological hazards, with the resulting ground motion characteristics being closely related to the dynamic parameters of the rock mass [1], including shear modulus [2] (modulus of elasticity) and damping ratio [3], which are essential for exploring the intrinsic connections between earthquakes and rock behavior. Shear modulus measures a rock mass's resistance to deformation under seismic forces, and this relationship is proportional; that is, a larger shear modulus indicates higher material stiffness [4–6]. Theoretical analyses indicate that seismic waves propagate with reduced energy attenuation in materials of higher stiffness, consequently increasing the amplitude of vertical ground fluctuations [7–9]. The damping ratio quantifies a material's energy dissipation during vibration; a higher damping ratio indicates effective energy dissipation, which suppresses the amplitude of vibrations caused by earthquakes and reduces their destructive impact on rock structures [10–12]. Studying the dynamic properties of rocks under various cyclic modes holds both theoretical value and practical significance for enhancing the safety of underground engineering constructions.

In mechanical tests, the shear modulus and damping ratio of rocks are influenced by factors such as material composition, stress amplitude, and loading frequency [13–15]. Scholars have conducted extensive theoretical and experimental research on the fatigue deformation [16], strength, and damage of intact rocks. In studies on the fatigue of soft rocks, such as sandstone, shale, and salt rock, water content and stress amplitude significantly influence the dynamic shear modulus and damping characteristics. Under conditions of low strain amplitude, the effect of water content on the dynamic shear modulus is subtle, but it significantly affects the damping ratio and damping coefficient [17]. However, as stress amplitude increases, the impact of water content on the dynamic shear modulus becomes more pronounced. This indicates a complex interaction between water content and stress amplitude that collectively determines the dynamic response characteristics of soft rocks [18,19]. Under graded cyclic loading, the damping parameter of soft rocks increases with rising stress amplitude, and this trend varies under different stress paths. Specifically, the damping parameter values measured under stress paths with incrementally increasing stress levels are higher than those under paths with incrementally increasing stress amplitudes. This may be attributed to the nonlinear response of the internal structure of the rock and the associated damage accumulation mechanism [20,21]. Under constant amplitude cyclic loading, the damping ratio and damping coefficient of soft rock change as the number of cycles increases. These changes adhere to a specific evolutionary law, particularly when the stress level surpasses the fatigue damage threshold of the rock. The evolution of the damping parameter exhibits a three-stage semi-U curve, encapsulating the entire process from the rock's initial elastic response through the accumulation of fatigue damage to its ultimate destruction. This phenomenon encapsulates the entire process from the rock's initial elastic response to the accumulation of fatigue damage, culminating in its destruction [22–24].

For hard rocks, such as marble, granite, and schist, their mechanical fatigue properties are influenced by cyclic loading conditions. Hard rocks typically exhibit higher shear moduli and lower damping ratios, attributed to their internal structural compactness and mineral composition. Under cyclic loading, the fatigue damage and damping properties of hard rocks demonstrate an evolutionary pattern distinct from that of soft rocks. The fatigue damage process in hard rocks is often more abrupt, with stress–strain curves displaying a more pronounced linear elastic phase before the onset of damage. Moreover, the damping behavior of hard rocks can be influenced by internal defects, like cracks and voids, potentially enhancing their damping capacity [25–27]. In summary, research on soft and hard rocks primarily focuses on the effects of rock type, stress level, stress path, and number of cycles, frequently overlooking the link between loading rate and the shear modulus and damping ratio of the test material.

This study focuses on the dynamic properties of hard rock (silica) under variable-rate cyclic loading, analyzing the effects of loading rate and stress amplitude on the evolution patterns of the dynamic shear modulus and damping ratio of silica.

2. Materials and Methods

2.1. Material Characterization

The test material originated from a lead–zinc mine in Yunnan, China. Drill a 110 mm diameter borehole into the roof of the rock drilling chamber at the top of the mining site to obtain core samples. Then, cut, grind, and polish the core samples to prepare standard specimens. Given that the mineral composition and structure are critical in influencing the rock's mechanical properties, X-ray powder diffraction (XRD) and polariscope tests were conducted on the material prior to testing to identify the microscopic composition of the specimen. As shown in Figure 1a, the primary structural components of the material include actinolite, garnet, diopside, and quartz particles that are cemented together. Figure 1b displays the XRD analysis results of the crushed specimen; the principal components are stilbite, actinolite, and calcite, comprising 26.1%, 42.6%, and 16.9% respectively.

The mechanical behavior of siliceous mudstone is significantly influenced by the stiffness of its primary mineral grains, particularly tourmaline and epidote, which are stiffer and comprise 60% of the rock. The presence of these hard minerals effectively impedes microcrack propagation. Under monotonic loading, the intergranular tension within the rock can lead to the formation of vertical cracks. Under cyclic loading, misalignment displacements among mineral grains of differing stiffnesses lead to stress concentrations, particularly at the edges of the mineral grains. As the load cycles, damage accumulates in these stress concentration areas, leading to the coalescence of cracks and the formation of macroscopic shear fractures along the rock's diagonals, ultimately leading to rock failure.

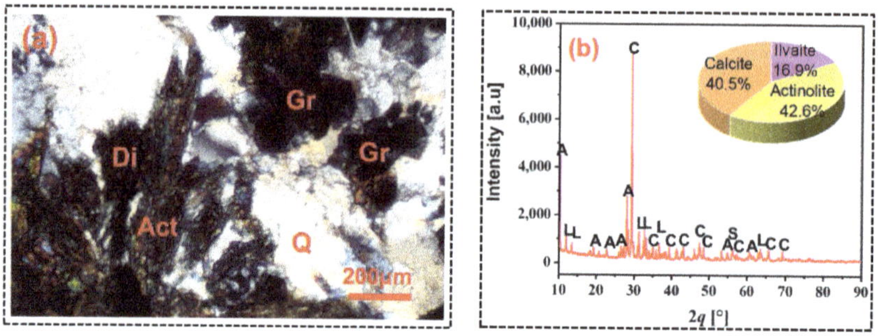

Figure 1. Analysis of the microstructure of skarn rock. (**a**) Polarization microscope micrograph of the sample crushed block processed into a cast thin film, Gr (Garnet), Di (Diopside), Act (Actinolite), Q (Quartz); (**b**) analysis results of the powder XRD test after sample failure.

2.2. Sample Preparation

The tester has strict requirements for the morphology of the test material. In order to ensure the scientific rigor and validity of the rock mechanics test, it is recommended to carry out the pre-preparation of the test specimens according to ISRM requirements so as to improve the quality of the test data and provide a reliable basis for the study of the mechanical properties of the rock and its engineering applications.

Based on the purpose and requirements of the test, samples free of cracks, voids, or other visible defects were selected from the parent core, and the shape and size of the specimens were required to meet the specifications of the experimental equipment to avoid boundary effects (the top and bottom of the specimens were maintained to be less than 0.02 mm non-parallel). Figure 2a illustrates the selection of test specimens, with three specimens chosen as representatives due to their density and wave speed being near the batch mean; another four specimens were randomly selected for the fatigue test. Figure 2b–d present the measurements of wave speed, mass, and size of the specimens, respectively, which are foundational for selecting the test specimens. The measuring equipment is an NM-4A non-metallic ultrasonic testing analyzer. The physical parameters of the specimens are detailed in Table 1.

Table 1. Physical parameters of diorite specimens (experimental part).

No.	Mass [g]	High [mm]	Diameter [mm]	Density [g/m^3]	Axial Longitudinal Wave Velocity [m/s]	Radial Longitudinal Wave Velocity [m/s]
X10	675.1	100.29	49.82	3.45	4811	3906
X15	675.6	99.63	50.09	3.44	4630	3676
X49	676.0	100.15	50.19	3.41	4690	3788
X43	718.9	100.30	50.23	3.62	4314	3472
X48	635.9	100.15	50.19	3.41	4690	3788
X53	645.6	99.78	50.03	3.29	4606	3472
X54	658.8	100.57	49.58	3.39	4438	3205

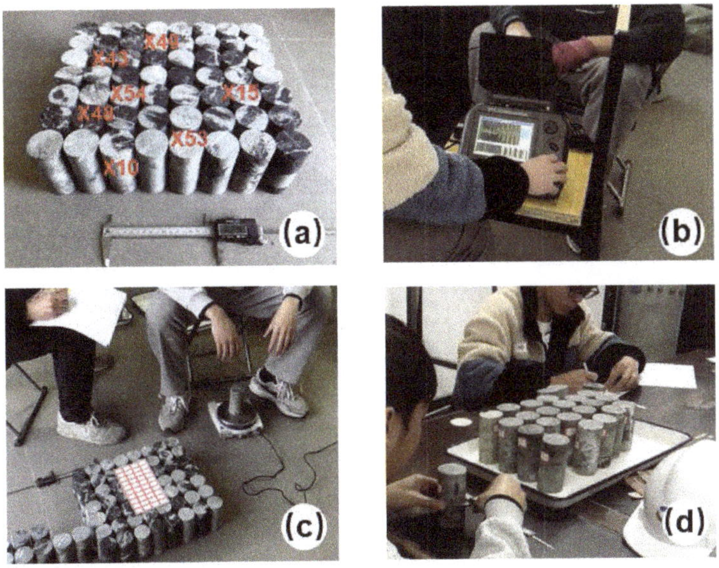

Figure 2. Specimen pre-treatment: (**a**) selection of specimens; (**b**) wave velocity measurement; (**c**) mass measurement; (**d**) dimensional measurement.

2.3. Experimental Setup

The test equipment is the YAW-2000 servo rigid mechanical testing machine(Chaoyang Testing Machine Co., Ltd, Changchun, China), which has a stiffness of more than 10GN/m and a maximum load capacity of 2000 kN, making it suitable for uniaxial and fatigue testing of hard rock materials, such as granite and marble. The test path and control are commanded by the German DOLI electronic system, which allows for free switching between multimodal controls, such as strain, displacement, and load. The axial and radial deformations are measured by the model YYSJ50 extensometer(Beijing Zhongxi Huada Science and Technology Co, Beijing, China). The test equipment is depicted in Figure 3.

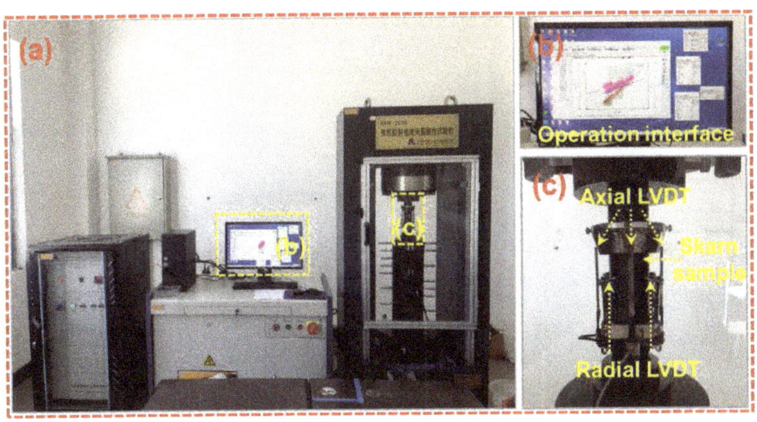

Figure 3. Experimental setup: (**a**) overview of the test loading system; (**b**) control interface of the operating system; (**c**) schematic of the installation of the system for measuring the deformation of rock samples.

2.4. Testing Scheme

2.4.1. Uniaxial Compressive Strength Test

According to the pre-test preparation, three specimens, X10, X45, and X49, were selected for uniaxial compressive strength testing, with the results serving as a benchmark for the design of the cyclic loading test path. The test was controlled by stress mode, with the contact force between the testing machine and the specimen surface maintained at 2 kN. The specimen was loaded at a rate of 0.2 kN/s until its failure. The peak strengths of the three specimens were 180.86 MPa, 155.95 MPa, and 167.06 MPa, respectively, yielding an average peak strength of 168 MPa. The stress–strain curves are depicted in Figure 4.

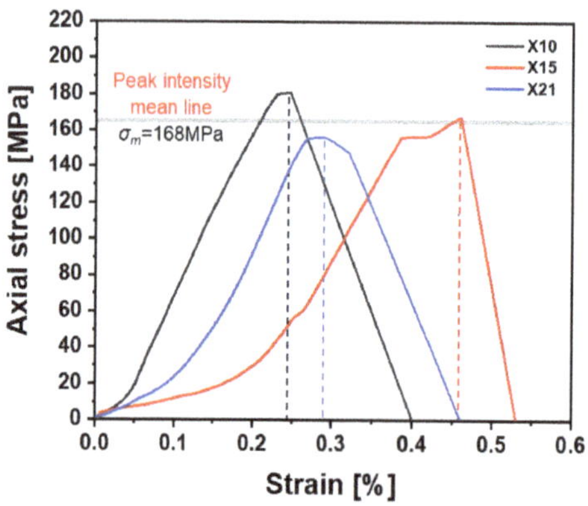

Figure 4. Uniaxial stress–strain curve.

2.4.2. Cyclic Loading Test

This part of the test was divided into a constant amplitude cyclic loading test and an increased amplitude cyclic loading test, with 15 cycles in each cyclic group. It featured rates of 2 kN/s (f = 0.006 Hz) for the first five cycles, 4 kN/s (f = 0.012 Hz) for the middle five cycles, and 6 kN/s (f = 0.018 Hz) for the last five cycles. The amplitude intervals for the three specimens of the constant amplitude loading test were as follows: 50%σ_m–100%σ_m for mode 1, 60%σ_m–110%σ_m for mode 2, and 70%σ_m–120%σ_m for mode 3; each was loaded monotonically at a rate of 0.2 kN/s until specimen damage occurred after surviving more than six cycle sets within the designated cyclic interval. The increasing value cyclic loading test was mode 4, with the lower amplitude starting at a constant 10%σ_m and increasing by 15%σ_m, continuing until specimen destruction. The peak strengths of the four specimens were 206.8 MPa, 258.8 MPa, 260.3 MPa, and 163.5 MPa, respectively. The loading paths for the four modes are depicted in Figure 5.

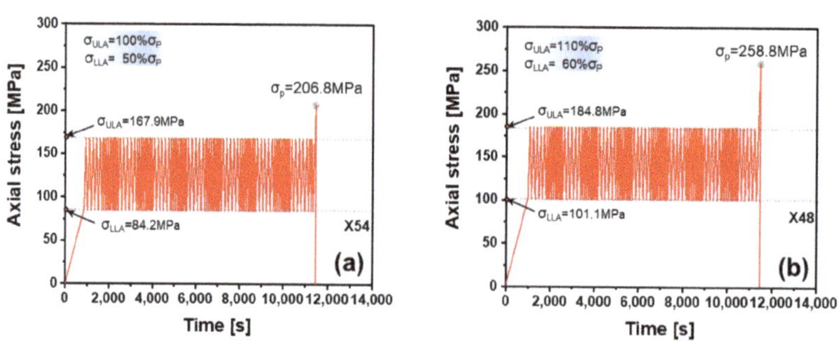

Figure 5. *Cont.*

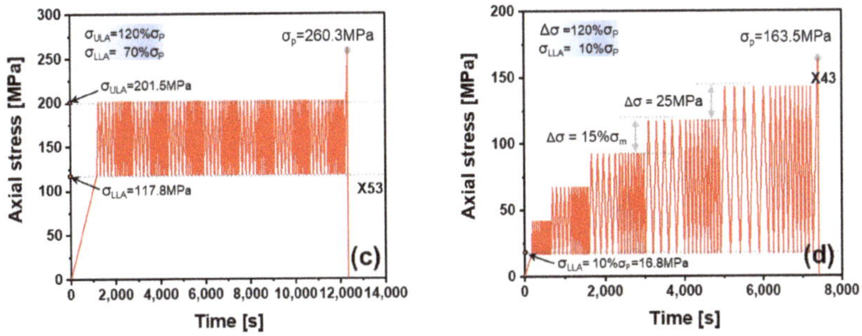

Figure 5. Cyclic loading path diagram: (**a**) mode 1; (**b**) mode 2; (**c**) mode 3; (**d**) mode 4.

3. Testing Results

3.1. Determination of the Shear Modulus and Damping Ratio

As a multiphase material, the interaction between mineral particles, such as stilbite, actinolite, and quartz, as well as the dynamic reorganization of the microstructure, significantly influences the shear modulus. Under cyclic loading conditions, the relative positions and contact states between particles alter, leading to the transfer and redistribution of stress concentration areas. Such changes may lead to variations in the trend of the shear modulus. Specifically, the shifting of stress concentration, changes in the contact state, microstructural reorganization, damage accumulation, and changes in the inter-particle forces significantly affect the shear modulus of rocks.

In cyclic loading and unloading tests, the loading and unloading curves do not coincide and usually form a hysteresis loop, indicating irreversible mechanical actions that lead to energy dissipation [28,29] during a single cyclic loading cycle. The shear modulus and damping ratio are usually determined from the stress–strain curves of rock specimens, reflecting the nonlinearity and hysteresis of the material [30]. In this test, the modulus of elasticity E is defined as the ratio of the stress difference (the peak stress minus the minimum stress during the loading stage) to the strain difference (the strain corresponding to the peak stress minus the strain corresponding to the minimum stress during the loading stage), calculated over a single cycle. The damping ratio relates to the hysteresis loop formed by the stress–strain curve, and the magnitude of this ratio intuitively reflects the material's energy dissipation under loading effects due to damping. The dynamic shear modulus G and damping ratio λ are calculated as follows:

$$E = \frac{(\sigma_{\max} - \sigma_{\min})}{(\varepsilon_{\max} - \varepsilon_{\min})} \tag{1}$$

where σ_{max} is the maximum stress in a single cycle, ε_{max} is the strain corresponding to the maximum stress, σ_{min} is the minimum stress during unloading in this cycle, and ε_{min} is the strain corresponding to the minimum stress.

$$G = \frac{E}{2(1+\nu)} \tag{2}$$

where ν is Poisson's ratio.

$$\lambda = \frac{A_R}{4\pi A_S} \tag{3}$$

where A_R is the area of ABCDA, and A_S is the area of AOE. The horizontal coordinate in Figure 6 is ε_d, and the vertical coordinate is the stress σ_d.

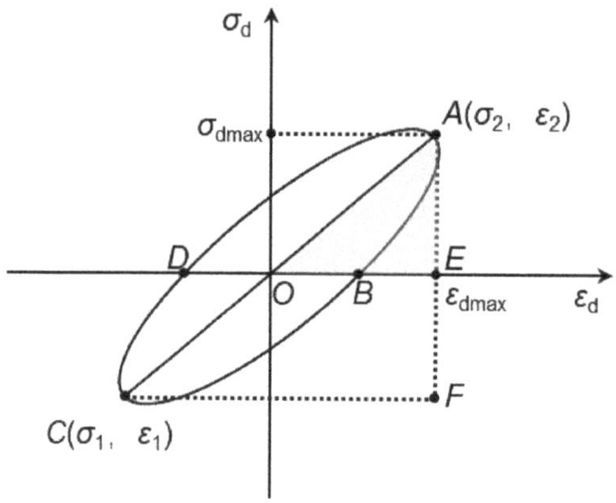

Figure 6. Hysteresis loop of axial stress $\sigma_d - \varepsilon_d$.

3.2. Shear Modulus Analysis

The shear modulus of the rock shows a corresponding trend with the accumulation of cyclic loading and unloading times, attributed primarily to the gradual expansion and coalescence of microcracks and fissures within the rock during the loading process. This results in internal damage that progressively diminishes the rock's strength and stiffness, ultimately leading to its destruction [31–33]. The changing trend of the shear modulus of sillimanite in cyclic loading and unloading tests primarily exhibits characteristics of initial decrease, subsequent stabilization, and eventual attenuation, influenced by numerous factors such as the rock's microstructure, loading frequency, and stress level. Understanding this change pattern deeply is crucial for predicting the engineering behavior of the rock.

Figure 7 displays the shear modulus of the four specimens versus the number of cycles. Analysis of the four graphs reveals that both the loading and unloading shear moduli increased equally as the amplitude interval was incremented. During the loading phase, the average shear modulus was 28.22 GPa for X43, 34.78 GPa for X54, 40.91 GPa for X48, and 45.04 GPa for X53. The growth rates relative to the X43 specimen were 23.3%, 45%, and 59.5%, respectively. In the unloading stage, the average shear modulus was 27.58 GPa for X43, 34.01 GPa for X54, 40.07 GPa for X48, and 44.04 GPa for X53. The growth rates relative to the X43 specimen were 23.3%, 45.3%, and 59.7%, respectively. In the growth rate test, the loading and unloading shear moduli at the same cycle level decrease as the number of cycles increases, and the transverse value test indicates a general decline in both shear moduli throughout the stage.

Gr is defined as the ratio of the loaded cutline modulus to the unloaded cutline modulus, as illustrated in Figure 7a. Initially, Gr rapidly increases with the number of cycles, and within the same cycle group, the ratio increases as the loading rate increases. Between two adjacent cycle groups, Gr suddenly drops when the loading rate changes from 6 kN/s to 2 kN/s, highlighting that the loading rate significantly impacts the cutline modulus. This phenomenon is consistent across the three specimens undergoing transverse value loading. In the increased load test, Gr's trend at the same cycle level mirrors the transverse amplitude load's performance, initially growing and then stabilizing after a certain value is reached. At different cycle levels, Gr initially shows a marked increase, which gradually diminishes as the cycle level increases.

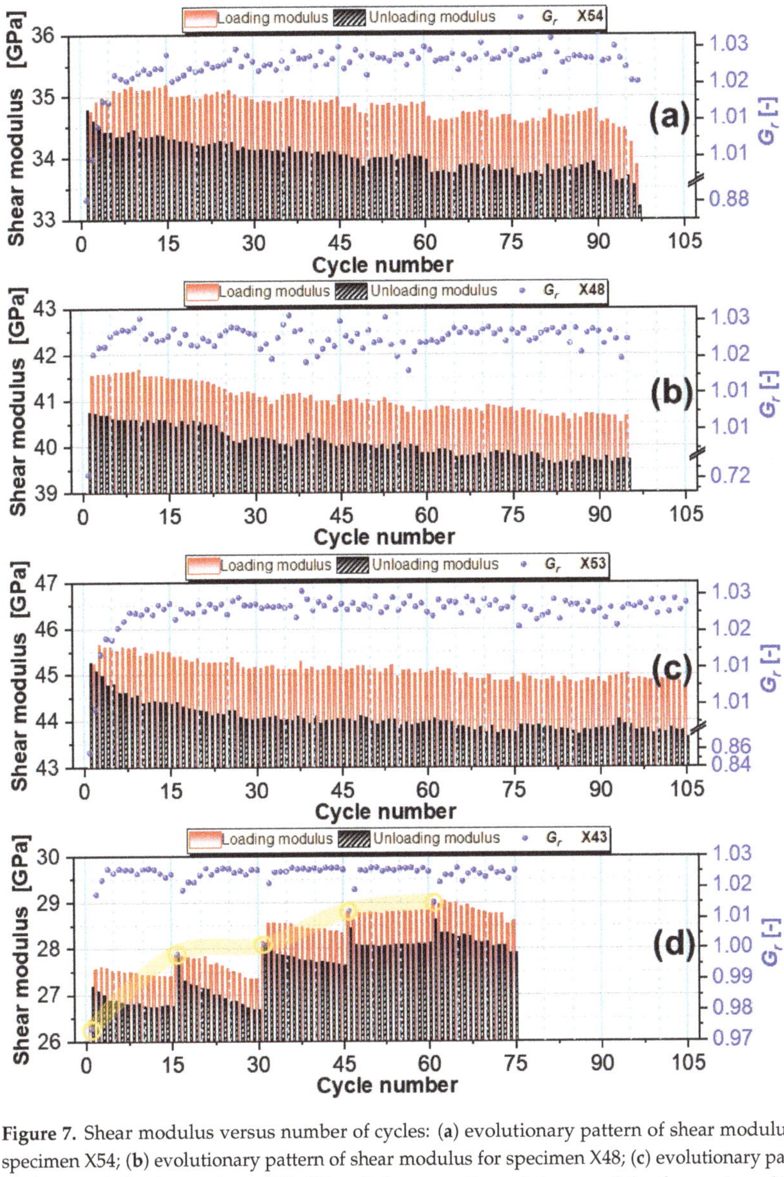

Figure 7. Shear modulus versus number of cycles: (**a**) evolutionary pattern of shear modulus for specimen X54; (**b**) evolutionary pattern of shear modulus for specimen X48; (**c**) evolutionary pattern of shear modulus for specimen X53; (**d**) evolutionary pattern of shear modulus for specimen X43.

3.3. Damping Ratio Analysis

In the process of cyclic loading and unloading, the sillimanite specimen has already shown obvious local damage, resulting in irrecoverable macroscopic cracks. This may promote the expansion of these cracks during further cyclic loading and unloading, leading to a decreased damping ratio as crack expansion diminishes the rock's overall energy absorption capacity [34,35]. As the number of cyclic loadings increases, the damping ratio of silica rocks gradually decreases, attributed to the progressive expansion and coalescence of microscopic cracks and fissures within the rocks. This process results in a gradual reduction in the rocks' stiffness and strength, subsequently leading to a decrease in the damping ratio. After reaching a certain number of cyclic loadings, the damping ratio of skarn tends to stabilize. At this stage, the cracks and microstructure within the rock become relatively stable, and the development and coalescence rate of the cracks slow, resulting in slower changes to the damping ratio. Prolonged cyclic loading may lead to continual expansion [36] and coalescence of cracks within the rock, drastically reducing its strength and stiffness, which typically leads to a sharp decline in the damping ratio as the rock nears a state of damage, and this decline may coincide with the rock's destruction. Under cyclic loading, the damping ratio of the rock will follow a 'concave' pattern, decreasing and then gradually increasing, which reflects the rock's progression from compaction and elastic deformation to crack expansion and destruction.

Figure 8 illustrates the relationship between the damping ratio and the number of cycles for the four specimens, and the comparison indicates that the damping ratio of silica exhibits a trend to first decrease and then increase with the increase in the number of cycles throughout the cycling stage. As illustrated in Figure 8a, the damping ratio initially decreases in the first five-cycle stage, and with the increase in the loading rate to 4 kN/s, the damping ratio experiences a sharp decrease and then stabilizes, and the damping ratio experiences a discontinuous decrease when the rate reaches 6 kN/s. Observations of the damping ratio change in the subsequent unused cycle group indicate that the rate-dependent behavior is consistent with that of the first cycle group, and this behavior is also observed in the other three specimens. The red bar graph in Figure 8 represents the mean damping ratio at each rate, whether at constant or increasing amplitude. The mean damping ratio at the three rates in each cycle group shows a stepwise decrease, suggesting that the loading rate significantly influences the damping ratio; specifically, higher rates result in lower damping ratios and correspondingly lower energy dissipation. In summary, the trends in the damping ratios of silica rocks under cyclic loading correlate with internal damage and crack development, as well as the loading rate. These trends are crucial for understanding and predicting the long-term stability of rocks under dynamic loading.

Table 2 displays the fitting equation for the mean damping ratio of a single sample at a consistent rate. The fitting curve represents a logarithmic function, and the correlation coefficients listed in the table suggest a high degree of fit between them.

Table 2. Fitting of the mean damping ratio of a single sample at the same rate.

No.	Loading Rate	Fitting Equation	R^2
X43	2 kN/s	$y = -0.063\ln(x) + 0.1167$	0.772
	4 kN/s	$y = -0.05\ln(x) + 0.0895$	0.718
	6 kN/s	$y = -0.044\ln(x) + 0.0789$	0.677
X48	2 kN/s	$y = 0.0071\ln(x) + 0.0181$	0.833
	4 kN/s	$y = 0.0096\ln(x) + 0.0084$	0.968
	6 kN/s	$y = 0.0104\ln(x) + 0.006$	0.961
X53	2 kN/s	$y = -0.018\ln(x) + 0.0549$	0.520
	4 kN/s	$y = 0.0021\ln(x) + 0.0151$	0.752
	6 kN/s	$y = 0.003\ln(x) + 0.0095$	0.983
X54	2 kN/s	$y = -0.02\ln(x) + 0.0553$	0.768
	4 kN/s	$y = 0.0071\ln(x) + 0.0181$	0.992
	6 kN/s	$y = -0.004\ln(x) + 0.017$	0.987

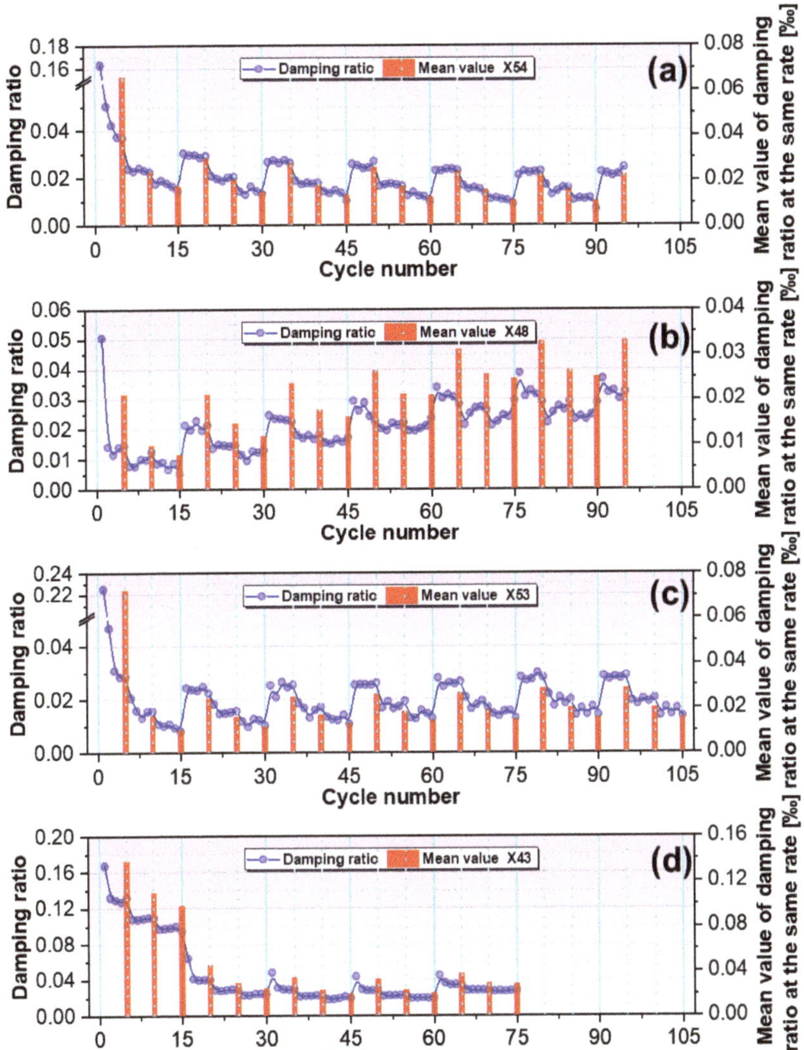

Figure 8. Damping ratio versus number of cycles: (**a**) evolutionary pattern of the damping ratio for specimen X54; (**b**) evolutionary pattern of the damping ratio for specimen X48; (**c**) evolutionary pattern of the damping ratio for specimen X53; (**d**) evolutionary pattern of the damping ratio for specimen X43.

4. Discussion

Modeling the damage evolution of rocks [37] under cyclic loading and unloading conditions is complex, covering the entire course of deformation, damage accumulation, and eventual destruction. The damage variable, a core concept in rock mechanics, quantifies the degradation of the rock from integrity to damage. Under cyclic loading, microstructural changes, including microcrack initiation, extension, and penetration, are manifested as macroscopic damage through cumulative damage effects. Damage evolution models aim to capture these behaviors, defining the damage variable D from 0 (no damage) to 1 (complete damage) [38]. The model is constructed based on experimental observations and validated through theoretical analyses and numerical simulations to ensure accuracy and reliability in rock mechanics applications.

4.1. Damage Characterization Analysis Based on Axial Strain

At the beginning of cyclic loading, the strain hardening phenomenon in rock is evident as a decrease in the axial strain increment, even at a constant stress level, suggesting that the rock possesses damage resistance. However, as the loading cycles increase, the cumulative effect of damage begins to dominate the rock's behavior, leading to the strain softening phase. In this stage, the axial strain increment increases at the same stress level, indicating an accelerated damage accumulation within the rock and a degradation of its macroscopic mechanical properties. This transition from hardening to softening [39] is a critical feature of the damage evolution process in rocks under cyclic loading, enhancing the understanding and prediction of rock fatigue life. In this study, rock damage is quantified in terms of axial strain using the following equation:

$$D = \frac{\varepsilon_p}{\varepsilon} \frac{\varepsilon - \varepsilon_0}{\varepsilon_p - \varepsilon_0} \tag{4}$$

where D is the damage variable, ε represents the strain corresponding to the minimum unloaded stress in a single cycle, ε_0 is the axial initial strain in the first cycle, and ε_p is the strain at the axial minimum unloaded stress in the last cycle.

Based on the calculation results in Equation (4), Figure 9 illustrates the trend of damage evolution in skarn specimens under fatigue loading. The S-shaped curve in the figure demonstrates the correlation between the damage value and the relative number of cycles, defined as the ratio n/N of the number of cycles n to the total number of cycles N. In the early stages of cyclic loading, compaction of the rock's internal pores leads to a rapid increase in the damage value. As the number of cycles accumulates, the expansion of cracks and further damage accumulation become the predominant processes within the rock. When damage accumulates to a certain extent, the growth of the damage value significantly accelerates in the late stages of cycling, typically indicating the rock's critical state prior to undergoing significant damage. As shown in Figure 9a, within a single cyclic group, the damage effects on the rock vary with different loading rates. The data points marked by rectangular boxes show a trend of increasing damage with the number of cycles, with higher damage values observed under low-rate loading conditions, suggesting the impact of the loading rate on the damage accumulation process. Comparing adjacent cycle groups, a loading rate of 6 kN/s versus 2 kN/s shows that the latter induces a more significant damage increment, highlighting the role of the loading rate in rock damage behavior. Additionally, the effect of an increasing loading rate on rock damage in the cyclic group shown in Figure 9d aligns with the previously described trends. As the stress amplitude increases, the damage increment exhibits a decreasing trend, possibly due to the rock's greater deformation capacity and damage tolerance at higher stress levels, indicating that stress amplitude is another crucial factor influencing rock damage evolution.

In this study, we develop a mathematical model based on the stress–strain relationship of rocks to describe the damage evolution of rocks under cyclic loading. The model is derived through parameter fitting and can quantify and digitally characterize the damage state of rocks. The expression for the proposed mathematical model of rock damage evolution is presented as follows:

$$D = 1 - \left(1 - \left(\frac{n}{N}\right)^a\right)^b \tag{5}$$

where n is the number of cycles, N is fatigue life, and a and b are material-related parameters.

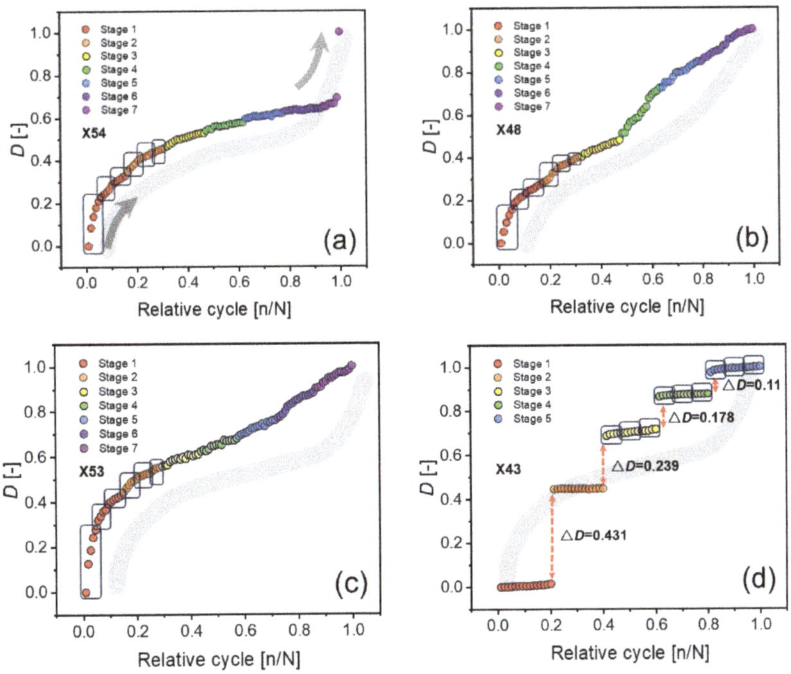

Figure 9. Trend of rock damage evolution under cyclic loading: (**a**) specimen X54; (**b**) specimen X48; (**c**) specimen X53; (**d**) specimen X43.

Figure 10 details the relationship between maximum damage and the relative number of cycles for four specimens across different cycle sets, as well as the evolutionary trend of damage accumulation versus the relative number of cycles. By fitting an S-function to these data points, the dynamics of the damage variable as the number of cycles increased were captured, with the parameters of the fitted results listed in the boxed squares in the upper left corner of the Figure 10a–d. The fitted correlation coefficients for all cyclic groups exceed 0.97, indicating that the strain-based proposed fatigue damage model is highly reliable and accurate for describing the damage evolution of rocks under cyclic loading.

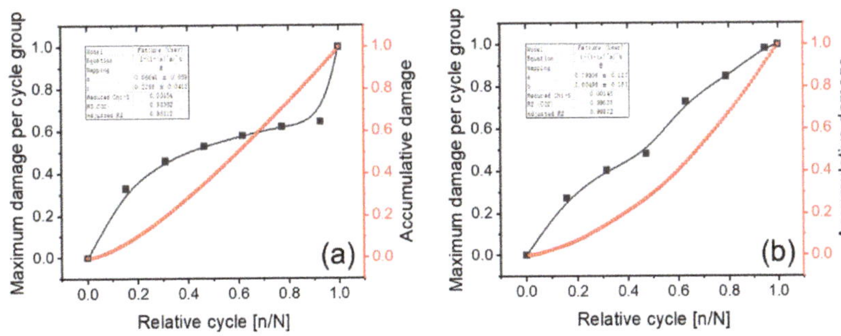

Figure 10. *Cont.*

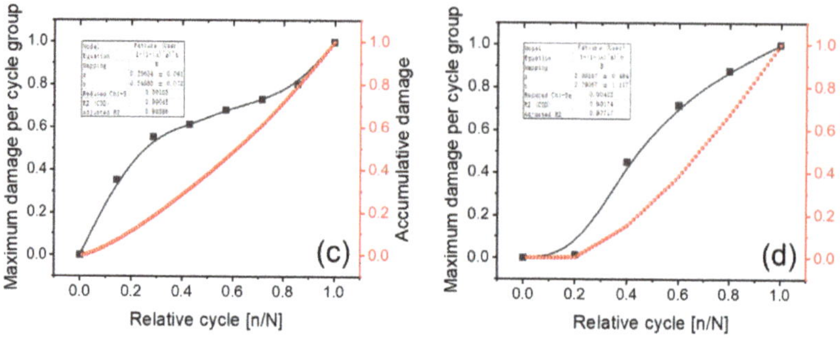

Figure 10. Cumulative damage evolution model for rocks: (**a**) specimen X54; (**b**) specimen X48; (**c**) specimen X53; (**d**) specimen X43.

4.2. Damage Characterization Analysis Based on Dissipative Energy

The damage process in rocks is usually accompanied by energy dissipation, released in the form of heat, acoustic energy, etc. [40], indicating the degradation of the internal structure of the rock. Under cyclic loading conditions, the damage variable of a rock accumulates with increasing loading cycles, and its growth is closely related to the initiation, extension, and penetration of microcracks within the rock. The accumulation of dissipated energy, a key indicator [41,42] for evaluating the evolution of rock damage, reflects the intensification of damage within the rock. By monitoring changes in dissipated energy, the damage state of rocks can be quantitatively analyzed, providing a scientific basis for predicting fatigue life and assessing the engineering stability of rock materials.

To quantify the degree of damage in rock samples, many early researchers adopted the concept of dissipated energy density. Based on the cumulative change characteristics of dissipated energy density, the damage variable D is defined as the ratio of dissipated energy density to the cumulative dissipated energy density, expressed as follows:

$$\mathrm{D_e} = \frac{\sum_{\mathrm{i}=1}^{\mathrm{n}} \mathrm{E_D}}{\sum_{\mathrm{i}=1}^{\mathrm{N}} \mathrm{E_D}} \tag{6}$$

where D_e denotes the dissipated energy damage variable, n denotes the nth cycle in cyclic loading and unloading, and E_D is the dissipated energy in a single cycle.

Using the method defined in Equation (6), the dissipated energy damage variables of the four rock specimens were calculated in this study, and the results are displayed in Figure 11. The trends of the dissipated energy damage variables under different amplitude intervals were consistent, showing an increasing trend with the increase in the relative number of cycles. When comparing the same cycle groups, lower loading rates resulted in higher levels of dissipated energy, which indicated an increase in the degree of rock damage. Additionally, for a given relative number of cycles, the increase in dissipated energy damage variables is more pronounced at higher amplitude intervals, as shown in Figure 11d, indicating the contribution of higher stress amplitude intervals to the accumulation of rock damage.

By integrating the relationship between dissipated energy and rock damage, a model can be developed for predicting rock damage, which is used to calculate the cumulative damage of rocks under multi-stage cyclic loading and unloading. The model can quantify the cumulative process of rock damage and predict the evolution of rock damage over the number of cyclic loading cycles. The damage accumulation model is expressed mathematically as follows:

$$\mathrm{D} = \mathrm{a}\left(\frac{\mathrm{n}}{\mathrm{N}}\right)^2 + \mathrm{b}\left(\frac{\mathrm{n}}{\mathrm{N}}\right) + \mathrm{D_0} \tag{7}$$

where n is the number of cycles, N is fatigue life, a and b are the fitting coefficients, and D_0 is the initial damage.

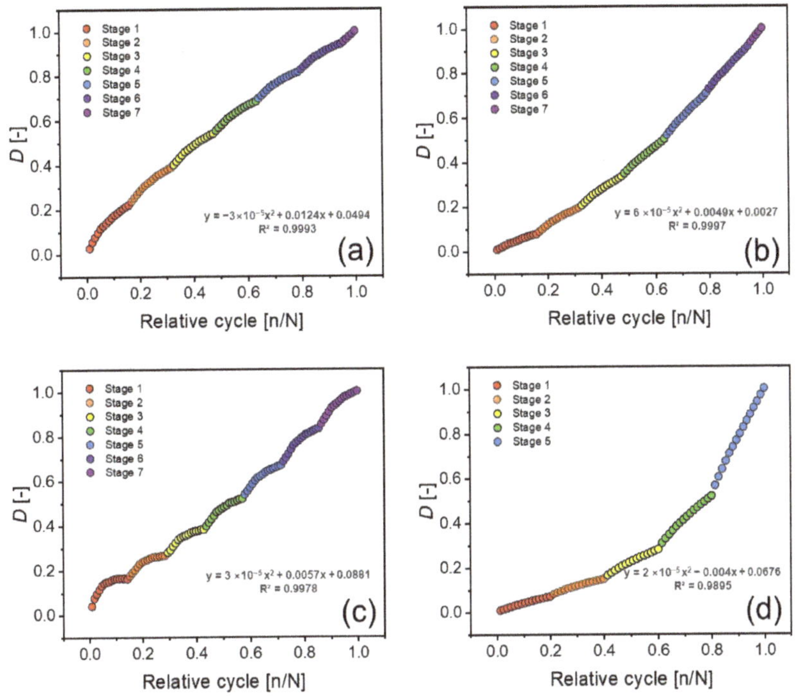

Figure 11. Evolution of dissipative energy damage variables with relative number of cycles: (**a**) specimen X54; (**b**) specimen X48; (**c**) specimen X53; (**d**) specimen X43.

In this study, Equation (7) was utilized to fit and quantify the dissipated energy damage variable, enabling the quantification of the cumulative damage of rocks under multi-stage cyclic loading and unloading. As shown in Figure 11, the fitting results of this equation exhibit high consistency with the experimental data for specimens of different amplitude intervals, and the coefficients of determination of all the fitted curves exceed 0.98, indicating a very high goodness of fit. This result validates the applicability of Equation (7) in describing the variation of dissipated energy damage variables and provides a reliable mathematical tool for the numerical analysis of rock damage evolution. Using this model, the damage accumulation process of rocks under cyclic loading can be more accurately predicted.

Figure 12 demonstrates the damage pattern of the silica specimens after testing, where the cracks are mainly distributed along the diagonal in a shear manner, representing a typical phenomenon of shear damage characteristics in rock compression tests. As the cyclic loading amplitude increased, a tendency for the width of primary cracks to become narrower in the four specimens was noted. This narrowing may be attributed to the closure or rearrangement of microcracks at higher stress levels. Furthermore, the number of secondary cracks decreased with increasing cyclic loading amplitude, likely due to the merging of microcracks or the increase in rock fragmentation at higher stress levels, resulting in less observable secondary cracks.

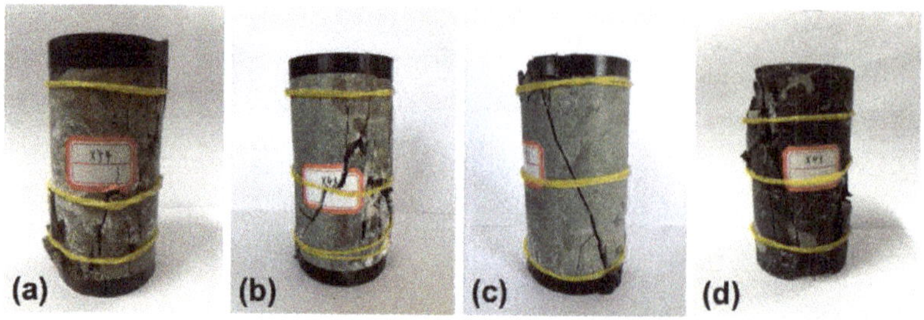

Figure 12. Specimen damage form diagram: (**a**) specimen X54; (**b**) specimen X48; (**c**) specimen X53; (**d**) specimen X43.

5. Conclusions

In this study, three different loading rates were designed, and cyclic loading tests were conducted under two paths of increasing and constant amplitude. The evolution of the dynamic shear modulus and damping ratio of silica rock under different strain amplitudes and loading rates was thoroughly investigated, and two damage models were subsequently established. The findings of this study demonstrate a complex relationship between the loading rate, damping ratio, and the mechanical behavior of rocks, which is crucial for understanding the response of rocks under various loading conditions. In particular, gaining a deeper understanding of the relationship between the loading rate and damping ratio is crucial for assessing the response of rocks under dynamic loads, such as earthquakes, and enhancing the safety of earthquake engineering. Future studies should continue to explore how multiple factors, including the loading rate, amplitude, and frequency, comprehensively influence the mechanical behavior of rocks to develop a more accurate rock mechanics model. This model can serve as a scientific basis for engineering design in earthquake engineering and related fields. The conclusions of thist study are as follows:

1. Under constant amplitude loading conditions, the shear modulus tended to increase during both loading and unloading as the amplitude interval increased. Conversely, the shear modulus in the transverse amplitude test exhibited a decreasing trend throughout the test phase. Additionally, the shear modulus ratios (Gr) of two neighboring cyclic groups exhibited a sudden decrease, suggesting that higher loading rates could result in larger Gr values.
2. The damping ratio of the diorite initially decreases and then increases with the number of cycles during the cycling process, a trend observed under both constant amplitude and increasing value loading paths. The mean values of the damping ratios for each cyclic group at the three different rates exhibited a stepwise decrease in response to the loading rate.
3. The strain-based damage model revealed that the rock damage was larger under low-rate loading conditions, suggesting that the loading rate had a significant effect on the damage accumulation process. The model fitting captures that the damage variables dynamically change with an increasing number of cycles following an S-shaped function, offering a quantitative description of the damage evolution process.
4. Drawing on the correlation between the calculated dissipated energy and the relative cyclic number, a new damage model was proposed in this study. The model accurately predicts the dissipated energy density damage accumulation process of silica rock under cyclic loading, offering an effective tool for assessing the damage of rocks under cyclic loading.

Author Contributions: Conceptualization, Y.W. (Yunfeng Wu); methodology, C.L.; software, B.Z.; formal analysis, C.S.; resources, Z.T.; data curation, Y.Z.; writing—original draft preparation, Y.W. (Yunfeng Wu); writing—review and editing, Y.W. (Yu Wang); funding acquisition, Y.W. (Yu Wang). All authors have read and agreed to the published version of the manuscript.

Funding: This article is funded by the National Natural Science Foundation of China (52174069), the Beijing Natural Science Foundation (8202033), and the Henan Key Laboratory for Green and Efficient Mining & Comprehensive Utilization of Mineral Resources (Henan Polytechnic University) (KCF2203).

Institutional Review Board Statement: Not applicable.

Informed Consent Statement: Not applicable.

Data Availability Statement: The original contributions presented in the study are included in the article, further inquiries can be directed to the corresponding author.

Conflicts of Interest: The authors declare no conflicts of interest.

References

1. Luo, J.; Pei, X.; Evans, S.G.; Huang, R. Mechanics of the earthquake-induced Hongshiyan landslide in the 2014 Mw 6.2 Ludian earthquake, Yunnan, China. *Eng. Geol.* **2019**, *251*, 197–213. [CrossRef]
2. He, M.M.; Zhang, Z.Q.; Li, N. Experimental investigation and empirical model to determine the damping and shear stiffness properties of soft rock under multistage cyclic loading. *Soil Dyn. Earthq. Eng.* **2021**, *147*, 106818. [CrossRef]
3. Huang, J.; Chen, J.; Ke, W.; Zhong, Y.; Lu, Y.; Yi, S. Damping ratio evolution of saturated Ningbo clays under cyclic confining pressure. *Soil Dyn. Earthq. Eng.* **2021**, *143*, 106581. [CrossRef]
4. Pan, K.; Yuan, Z.H.; Zhao, C.F.; Tong, J.H.; Yang, Z.X. Undrained shear and stiffness degradation of intact marine clay under monotonic and cyclic loading. *Eng. Geol.* **2022**, *297*, 106502. [CrossRef]
5. Likitlersuang, S.; Teachavorasinskun, S.; Surarak, C.; Oh, E.; Balasubramaniam, A. Small strain stiffness and stiffness degradation curve of Bangkok Clays. *Soils Found.* **2013**, *53*, 498–509. [CrossRef]
6. Yan, J.; Kong, L.; Wang, J. Evolution law of small strain shear modulus of expansive soil: From a damage perspective. *Eng. Geol.* **2023**, *315*, 107017. [CrossRef]
7. Li, H.; Gheibi, A.; Hedayat, A.; Yin, J.; Deng, J. Experimental validation of a velocity discontinuity model for prediction of the seismic and static shear stiffness of rock joints. *Int. J. Rock Mech. Min. Sci.* **2023**, *165*, 105360. [CrossRef]
8. Wu, X.; Zheng, H.; Jiang, Y. Study on the evolution law of rock joint shear stiffness during shearing process through loading-unloading tests. *Tunn. Undergr. Space Technol.* **2022**, *12*, 1045847. [CrossRef]
9. Kasyap, S.S.; Senetakis, K. Small-scale size effects on the shearing behavior and sliding stiffness reduction-displacement curves of rock-plastic gouge simulant under low normal stresses. *J. Struct. Geol.* **2022**, *157*, 104553. [CrossRef]
10. Corkum, A.G.; Lorig, L.J.; Varun, V. Rock mass permeability evolution during triaxial shearing based on three-dimensional distinct element model (*3DEC*) simulation. *Comput. Geotech.* **2024**, *168*, 106150. [CrossRef]
11. Gu, Q.; Zhang, Q.; Ye, S.; Dai, W.; Li, T. Shear fracture behavior and damage constitutive model of rock joints considering the effect of pre-peak cyclic loading. *Theor. Appl. Fract. Mech.* **2024**, *130*, 104289. [CrossRef]
12. Wang, Y.; Yi, X.; Long, D.; Xia, Y.; Li, P.; Cai, M. Energy-Based Fatigue–Creep Damage and Failure Pattern of Rock–Backfill Composite Structure Material in Mine Stopes Under High Static Stress: Disturbed Amplitude Effect. *Rock Mech. Rock Eng.* **2024**, *57*, 3021–3042. [CrossRef]
13. Senetakis, K.; Payan, M.; Li, H.; Zamanian, M. Nonlinear stiffness and damping characteristics of gravelly crushed rock: Developing generic curves and attempting multi-scale insights. *Transp. Geotech.* **2021**, *31*, 100668. [CrossRef]
14. Jindal, S.; Rahmanli, U.; Aleem, M.; Cui, L.; Jalbi, S.; Bhattacharya, S. Identification of soil damping for stiff piles based on experiments: Methodology and results. *Ocean Eng.* **2024**, *301*, 117124. [CrossRef]
15. Shear mechanical properties and energy evolution of rock-like samples containing multiple combinations of non-persistent joints. *J. Rock Mech. Geotech. Eng.* **2023**, *15*, 1651–1670. [CrossRef]
16. Skrzypkowski, K.; Zagórski, K.; Zagórska, A. Determination of the Extent of the Rock Destruction Zones around a Gasification Channel on the Basis of Strength Tests of Sandstone and Claystone Samples Heated at High Temperatures up to 1200 °C and Exposed to Water. *Energies* **2021**, *14*, 6464. [CrossRef]
17. Wang, S.; Zhu, Y.; Ma, W.; Wang, Z.; Li, G. Effects of rock block content and confining pressure on dynamic characteristics of soil-rock mixtures. *Eng. Geol.* **2021**, *280*, 105963. [CrossRef]
18. Ram, A.K.; Mohanty, S. Laboratory investigation on damping characteristics of homogeneous and stratified soil-ash system. *J. Rock Mech. Geotech. Eng.* **2023**, *15*, 2757–2777. [CrossRef]
19. Qin, Y.; Xu, X.; Yan, C.; Wen, L.; Wang, Z.; Luo, Z. Dynamic damping ratio of mudded intercalations with small and medium strain during cyclic dynamic loading. *Eng. Geol.* **2021**, *280*, 105952. [CrossRef]
20. Araei, A.A.; Razeghi, H.R.; Ghalandarzadeh, A.; Tabatabaei, S.H. Effects of loading rate and initial stress state on stress-strain behavior of rock fill materials under monotonic and cyclic loading conditions. *Sci. Iran.* **2012**, *19*, 1220–1235. [CrossRef]

21. Zhang, D.; Li, Q.; Liu, E.; Liu, X.; Zhang, G.; Song, B. Dynamic properties of frozen silty soils with different coarse-grained contents subjected to cyclic triaxial loading. *Cold Reg. Sci. Technol.* **2019**, *157*, 64–85. [CrossRef]
22. Peellage, W.H.; Fatahi, B.; Rasekh, H. Stiffness and damping characteristics of jointed rocks under cyclic triaxial loading subjected to prolonged cyclic loading. *Int. J. Fatigue* **2024**, *181*, 108121. [CrossRef]
23. Wang, Y.; Yi, Y.F.; Li, C.H.; Han, J.Q. Anisotropic fracture and energy characteristics of a Tibet marble exposed to multi-level constant-amplitude (MLCA) cyclic loads: A lab-scale testing. *Eng. Fract. Mech.* **2021**, *244*, 107550. [CrossRef]
24. Zhu, C.; Karakus, M.; He, M.; Meng, Q.; Shang, J.; Wang, Y.; Yin, Q. Volumetric deformation and damage evolution of Tibet interbedded skarn under multistage constant-amplitude-cyclic loading. *Int. J. Rock Mech. Min. Sci.* **2022**, *152*, 105066. [CrossRef]
25. Pires, V.; Rosa, L.G.; Dionisio, A. Implications of exposure to high temperatures for stone cladding requirements of three Portuguese granites regarding the use of dowel-hole anchoring systems. *Constr. Build. Mater.* **2014**, *64*, 440–450. [CrossRef]
26. You, W.; Dai, F.; Liu, Y. Experimental and numerical investigation on the mechanical responses and cracking mechanism of 3D confined single-flawed rocks under dynamic loading. *J. Rock Mech. Geotech. Eng.* **2022**, *14*, 477–493. [CrossRef]
27. Song, Y.; Li, K.; Tian, W. Mechanical response of sandstone exposed to monotonic and multilevel fatigue loading: Insights from deformation, energy and acoustic emission characteristics. *Theor. Appl. Fract. Mech.* **2023**, *128*, 104181. [CrossRef]
28. Liu, J.; Lyu, X.; Liu, Y.; Zhang, P. Energy evolution and macro-micro failure mechanisms of frozen weakly cemented sandstone under uniaxial cyclic loading and unloading. *Cold Reg. Sci. Technol.* **2023**, *214*, 103947. [CrossRef]
29. Meng, Q.; Liu, J.; Pu, H.; Huang, B.; Zhang, Z.; Wu, J. Effects of cyclic loading and unloading rates on the energy evolution of rocks with different lithology. *Geomech. Energy Environ.* **2023**, *34*, 100455. [CrossRef]
30. Yoshinaka, R.; Tran, T.V.; Osada, M. Non-linear, stress- and strain-dependent behavior of soft rocks under cyclic triaxial conditions. *Int. J. Rock Mech. Min. Sci.* **1998**, *35*, 941–955. [CrossRef]
31. Chen, Y.; Sheng, B.; Xie, S.; Cao, R.; Wang, Y.; Zhao, Y.; Lin, H. Crack propagation and scale effect of random fractured rock under compression-shear loading. *J. Mater. Res. Technol.* **2023**, *23*, 5164–5180. [CrossRef]
32. Cai, W.; Li, Y.; Gao, K.; Wang, K. Crack propagation mechanism in rock-like specimens containing intermittent flaws under shear loading. *Theor. Appl. Fract. Mech.* **2022**, *117*, 103187. [CrossRef]
33. Zhao, Z.; Liu, H.; Ma, Q.; Shang, J. Micro-macro damage, deterioration and cracking of heterogeneous composite rock masses with non-penetrating cracks under uniaxial compression. *Theor. Appl. Fract. Mech.* **2023**, *125*, 103919. [CrossRef]
34. Sharma, R.; Das, K.K.; Pei, J.; Jang, J.G. Influence of sillimanite sand and corundum sand on the mechanical property, durability and pore structure of cement mortar. *J. Build. Eng.* **2023**, *80*, 108091. [CrossRef]
35. Yue, C.; Liang, K.; Xu, C.; Du, X. Experimental study on cyclic deformation properties of saturated Fujian sand at different loading frequencies. *Undergr. Space* **2023**, *13*, 150–165. [CrossRef]
36. Wang, Y.; Yi, X.; Long, D.; Mao, T. On the fracture and energy characteristics of rock–backfill composite structure specimens exposed to fatigue–creep interaction loading. *Fatigue Fract. Eng. Mater. Struct.* **2023**, *47*, 153–169. [CrossRef]
37. Li, T.; Chen, C.; Peng, F.; Ma, C.; Li, M.; Wang, Y. Creep damage constitutive model of rock based on the mechanisms of crack-initiated damage and extended damage. *Undergr. Space* **2024**. [CrossRef]
38. Wang, Y.; Yi, X.; Long, D.; Li, P.; Cai, M. Macro-meso fracture and instability of rock-backfill composite structure (RBCS) specimens subjected to fatigue-creep interaction loading: Cement-tailing ratio effect. *Constr. Build. Mater.* **2024**, *420*, 135615. [CrossRef]
39. Xiong, X.B.; Wang, X.J. A cohesive strengthening/damage-plasticity model and its application for DEM numerical investigation in hardening/softening of rock and soil. *Comput. Geotech.* **2023**, *159*, 105500. [CrossRef]
40. Li, B.; He, Y.; Li, L.; Zhang, J.; Shi, Z.; Zhang, Y. Damage evolution of rock containing prefabricated cracks based on infrared radiation and energy dissipation. *Theor. Appl. Fract. Mech.* **2023**, *125*, 103853. [CrossRef]
41. Chen, S.; Zhang, H.; Wang, L.; Yuan, C.; Meng, X.; Yang, G.; Shen, Y.; Lu, Y. Experimental study on the impact disturbance damage of weakly cemented rock based on fractal characteristics and energy dissipation regulation. *Theor. Appl. Fract. Mech.* **2022**, *122*, 103665. [CrossRef]
42. Wang, Y.; Chen, Z.; Yi, X.; Long, D. Macro-meso damage cracking and energy dissipation of rock-backfill composites: Effect of cyclic disturbance frequency. *Fatigue Fract. Eng. Mater. Struct.* **2024**. [CrossRef]

materials

MDPI

Article

Piecewise Linear Strength Models for Analyzing Multiple Failure Mechanisms in Rocks Materials

Shiqi Li, Yuan Li *, Dongjue Fan, Liang Zhao and Litian Zhang

School of Civil and Resource Engineering, University of Science and Technology Beijing, Beijing 100083, China; m202221602@xs.ustb.edu.cn (S.L.); 18104835851@163.com (D.F.); m15833200311@163.com (L.Z.); zzyyllt@163.com (L.Z.)

* Correspondence: liyuan@ustb.edu.cn

Abstract: Rock materials failures are accompanied by the co-existence of various failure mechanisms, including rock fracturing, shearing, and compaction yield. These mechanisms manifest macroscopically as multiple failure modes and nonlinear strength characteristics related to stress levels. Considering the limitations of current rock mechanics strength theories, which are primarily derived from single failure mechanisms, this study evaluates the applicability of alternative strength theories. Based on the extensional-strain criterion and the PMC (Paul-Mohr-Coulomb) model, a piecewise linear strength model was proposed that is suitable for analyzing multiple failure mechanisms in rocks, revealing the intrinsic mechanisms of multi-mechanism rock material failure. A multiple failure mechanism strength model in the form of inequalities was proposed, using the generalized shear stress, mean stress, and stress Lode angle as parameters. Strength tests conducted on sandstone and granite rock material samples under different stress conditions revealed distinct piecewise linear strength characteristics for both rock types, validating the rationality and applicability of the multiple failure mechanism model. The findings construct a multi-mechanism failure model for rocks, providing enhanced predictive capabilities and aiding in the prevention of rock structural failures.

Keywords: transformation of failure mode; rock material strength criterion; extensional-strain criterion; Paul-Mohr-Coulomb model

1. Introduction

Rock failure processes can exhibit significant nonlinear behavior and characteristic failure mechanism transitions. During these processes, various failure mechanisms such as rock extension, shearing, and compaction yields can co-exist. Furthermore, rocks exhibit different failure modes under different confining pressures [1]. For example, under tensile stress or low-pressure conditions, rocks may exhibit tensile failure. As the pressure gradually increases, the failure mode transitions from tensile to shear failure. Similarly, under conditions of high temperatures and water exposure, rock materials such as sandstone and claystone exhibit significant changes in strength and failure mechanisms, as explored by Skrzypkowski et al. [2]. At higher stress levels in deeper regions, the failure mode shifts from conventional shear failure to pore collapse failure [3]. Thus, rocks transition between multiple failure modes, including tensile failure, shear failure, and compaction-shear yield.

Paul (1968) considered the influence of the intermediate principal stress on the Mohr-Coulomb model and proposed an expression involving three principal stresses [4,5]. Three parameter constants were determined for the stress states of uniaxial compression, uniform triaxial (isotropic) extension, and uniaxial extension. Meyer and Labuz further developed this into the PMC strength model [6]. They discussed the transformation characteristics of failure mechanisms in rock materials based on the PMC model and examined the shape of dodecagonal PMC failure surfaces. Two failure planes were fit to describe the nonlinear failure surface, which led to the development of the dodecahedral PMC model. This model allows for the observation of changes in the shape of the failure surface with increasing mean stress, characterizing the transition of rock material failure from tensile failure to shear failure as the stress conditions increase.

For the material parameters, Makhnenko et al. proposed a plane fitting method that minimizes the sum of the squared orthogonal distances between data points and the failure surface [7], whereas Zeng et al. suggested using the least squares method to fit parameters [8]. Asem et al. introduced a simplified PMC model that assumes the same friction angle under both tensile and compressive conditions and reduces the dodecagonal PMC model by two parameters; in other words, the resulting model has only four parameters [9].

Under tensile stress or low-pressure conditions, the rock failure mode is a tensile failure. Rocks undergo splitting failure under uniaxial stress or low confining pressure conditions with no tensile stress on the failure plane. Stacey (1981) proposed the extensional-strain criterion to explain the brittle failure mechanisms of deep quartzites in South Africa [10]. The results indicate that for materials exhibiting linear deformation behavior, the onset and depth of failure can be correlated with extensional strain.

This study characterizes rock failure under tensile and tensile-shear conditions using the extensional-strain criterion and describes compressive-shear failure based on the PMC model. A unified strength model under the influence of multiple failure mechanisms was then derived. By analyzing field and laboratory test data, a theoretical model suitable for multiple failure mechanisms in rocks was proposed, which revealed intrinsic low-strength rock failure mechanisms.

2. Related Work

2.1. PMC Model

In 1968, Paul considered the influence of the intermediate principal stress and proposed the PMC model [4,5]. In 2013, Meyer and Labuz developed this concept into the PMC strength model, and the expression for principal stresses is as follows [6]:

$$A\sigma_1 + B\sigma_2 + C\sigma_3 = 1 \tag{1}$$

where A, B, and C are material constants, and σ_1, σ_2, and σ_3 represent the three principal stresses. When $B = 0$, Equation (1) becomes the MC (Mohr-Coulomb) strength criterion.

$$A = \frac{1 - sin\varphi_c}{2sin\varphi_c} \cdot \frac{1}{V_0} \tag{2}$$

$$B = \frac{sin\varphi_c - sin\varphi_e}{2sin\varphi_c sin\varphi_e} \cdot \frac{1}{V_0} \tag{3}$$

$$C = -\frac{1 + sin\varphi_e}{2sin\varphi_e} \cdot \frac{1}{V_0} \tag{4}$$

where φ_c and φ_e are the internal friction angles during compression and extension, respectively, and V_0 is the theoretical triaxial tensile strength.

By substituting the three parameters A, B, and C into Equation (1), they can be expressed in terms of the two internal friction angles and the vertex as follows:

$$\left[\frac{1 - sin\varphi_c}{2sin\varphi_c}\right] \cdot \frac{\sigma_1}{V_0} + \left[\frac{sin\varphi_c - sin\varphi_e}{2sin\varphi_c sin\varphi_e}\right] \cdot \frac{\sigma_2}{V_0} - \left[\frac{1 + sin\varphi_e}{2sin\varphi_e}\right] \cdot \frac{\sigma_3}{V_0} = 1 \tag{5}$$

The PMC failure model constructs a failure surface in the principal stress space with a common vertex V_0. The PMC criterion accounts for the influence of the intermediate principal stress by using the different friction angles of rock materials under compression and tension. The multiaxial linear failure criterion includes three principal stresses. These three material parameters can be obtained using conventional triaxial tests on rocks. The PMC model failure envelope is shown in Figure 1.

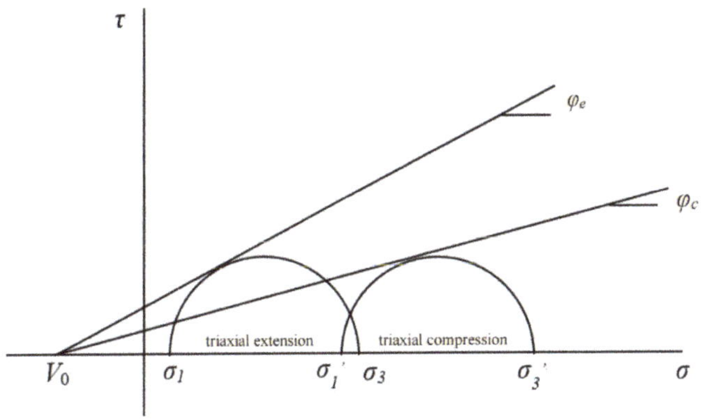

Figure 1. PMC model failure envelope. The internal friction angles during compression and extension are φ_c and φ_e, respectively. V_0 is the theoretical triaxial tensile strength. The three principal stresses are σ_1, σ_2, and σ_3.

To capture the transition characteristics of failure mechanisms in geotechnical materials, the Labuz team developed a dodecahedral PMC model based on the PMC framework [6]. The PMC failure criterion constructs a dodecagonal failure surface in a piecewise linear manner by fitting two planes (Figure 2), which can be approximately regarded as a nonlinear failure surface. This model characterizes the mechanism by which rock materials transition from tensile to shear failure under increasing stress conditions. When the constitutive model is applied to numerical simulations, the flow rule needs to be established [7].

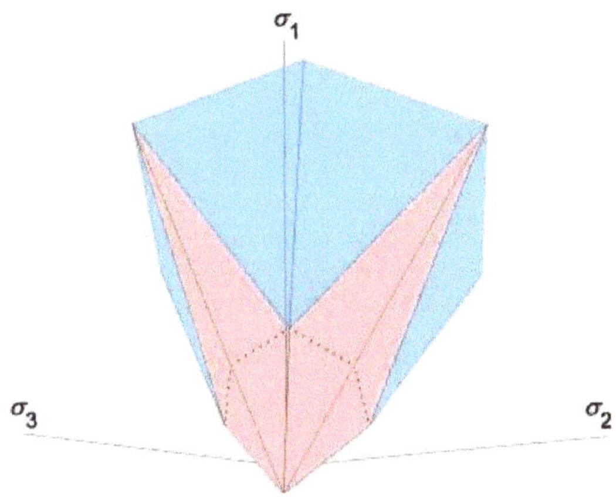

Figure 2. Dodecagonal PMC model. The dodecagonal failure surface, constructed in a piecewise linear manner by fitting two failure planes. The three principal stresses are σ_1, σ_2, and σ_3.

2.2. Extensional-Strain Criterion and Brittle Failure

Rocks exhibit different failure modes under different confining pressures. Under tensile stress or low-pressure conditions, rocks primarily undergo tensile failure, not shear failure. However, the MC and PMC models only include failure criteria under compressive and pure shear stress conditions and do not address failure criteria under tensile-shear conditions [11]. Therefore, using the MC criterion to calculate rock failure under tensile or low-stress levels often results in significant errors. To address this issue, Paul (1961) modified the single inclined line of the MC criterion to an inclined line and a vertical line tangent to the uniaxial tensile stress circle, forming a two-segment linear strength curve known as the tensile cutoff criterion. The modified failure criterion is shown in Figure 3 [12], where σ_t represents the measured tensile strength. However, while the dodecagonal PMC model can characterize the mechanism of rock material transitioning from tensile failure to shear failure as stress conditions increase, it cannot represent the failure mechanism of rocks under tensile-shear conditions. Therefore, it is necessary to study the tensile-shear failure criterion of rocks and establish a connection with the PMC strength model.

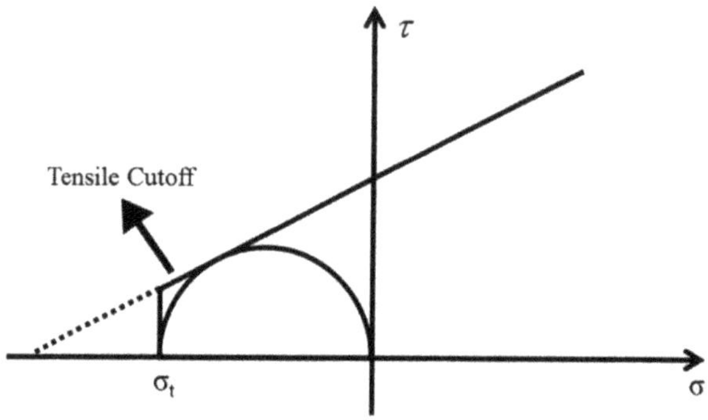

Figure 3. Schematic diagram of stretch truncation. The single inclined line of the MC criterion is modified to an inclined line and a vertical line tangent to the uniaxial tensile stress circle, forming a two-segment linear strength curve known as the tensile cutoff criterion. σ_t represents the measured tensile strength.

Numerous scholars have investigated the strength criteria of rocks under tensile-shear conditions, and the resulting models can be roughly classified into hyperbolic models, parabolic models, and models based on the Hoek-Brown criterion [13–19]. However, relatively few studies have been conducted on this topic, and consequently, experimental data are lacking. Therefore, the tensile-shear constitutive models and failure criteria currently lack adequate support. Stacey (1981) predicted the brittle failure mechanisms of deep massive quartzites in South Africa using the extensional-strain criterion [10]. The results indicate that for materials exhibiting linear deformation behavior, failure may be related to extensional strain. The extensional strain of rocks can be calculated using the generalized Hooke's law, as follows:

$$\begin{cases} \varepsilon_1 = \frac{1}{E}\left[\sigma_1 - \mu(\sigma_3 + \sigma_2)\right] \\ \varepsilon_2 = \frac{1}{E}\left[\sigma_2 - \mu(\sigma_1 + \sigma_3)\right] \\ \varepsilon_3 = \frac{1}{E}\left[\sigma_3 - \mu(\sigma_1 + \sigma_2)\right] \end{cases} \tag{6}$$

where μ is the Poisson's ratio; E is the elastic modulus; and ε_1, ε_2, and ε_3 are the first, second, and third principal strains, respectively. According to the extensional-strain criterion, when the minimum principal stress-strain exceeds the critical strain, the rock undergoes tensile failure. When $\varepsilon_3 < 0$, the specimen experiences tensile strain, and when $\varepsilon_3 > 0$, the specimen experiences compressive strain.

3. Methodology

3.1. Proposal of the PMC Model Considering Extensional Strain

Based on the dodecagonal PMC model, a failure surface was added that considers the extensional-strain criterion. A three-segment strength formula that accounts for both tensile-shear and compressive-shear failures is proposed. This model can be expressed using the three principal stresses as follows:

$$A'\sigma_I + B'\sigma_{II} + C'\sigma_{III} = 1 \quad (7)$$

where, A', B', and C' are material constants; σ_1, σ_2, and σ_3 are the maximum, intermediate, and minimum principal stresses, respectively. For the failure surface, based on the generalized Hooke's law, the failure conditions can be obtained as follows:

$$\varepsilon_1 \ or \ \varepsilon_2 \ or \ \varepsilon_3 = \varepsilon_c \quad (8)$$

where the critical strain $\varepsilon_c = -\sigma_t/E$ and is the tensile strength, μ is the Poisson's ratio, and E is the elastic modulus. When $\sigma_{III} - \mu(\sigma_I + \sigma_{II}) < 0$, the rock is in the tensile failure stage or transitioning from tensile to shear failure. According to the extensional-strain criterion, when $\varepsilon_1 = \varepsilon_c$, the rock undergoes tensile failure, and the following can be derived:

$$-\sigma_t = \sigma_{III} - \mu(\sigma_I + \sigma_{II}) \quad (9)$$

According to Equations (8) and (9):

$$A' = B' = -\frac{\mu}{\sigma_t} \quad (10)$$

$$C' = -\frac{1}{\sigma_t} \quad (11)$$

The three-segment linear form of the strength criterion can fundamentally satisfy the fitting of the transformation of rock failure mechanisms. Tensile or tensile-shear failure under tensile stress or low pressure is characterized by the extensional-strain criterion; as the pressure gradually increases, the rock failure mode transitions from tensile failure to shear failure; under higher stress levels, the rock failure mode transitions to conventional shear failure, as shown in Figure 4.

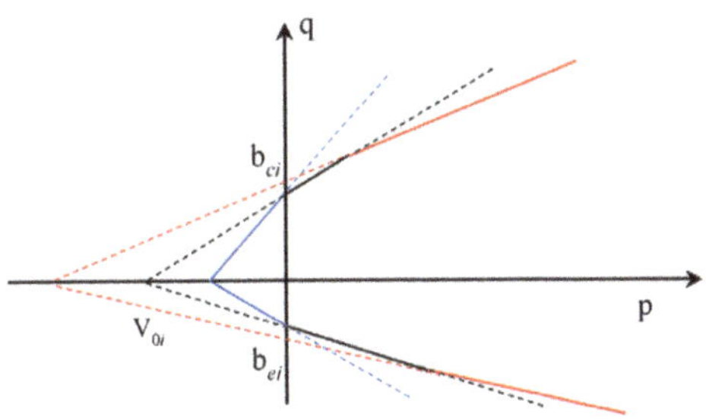

Figure 4. Extensional-strain criterion *p*-*q* plane schematic diagram. The linear forms represent the rock failure modes, from left to right: tensile failure, transition from tensile failure to shear failure, and conventional shear failure. The parameters in the figure, such as root and intercept, do not have specific meanings.

In the *p*-*q* plane, where the horizontal axis represents the mean stress p and the vertical axis represents the generalized shear stress q, the conventional PMC linear strength intersects the q axis at b_{ci} and b_{ei} and intersects the p axis at V_{0i}. In the *p*-*q* plane, Equation (7) can be written as follows:

$$q = \frac{b_{\theta i}}{V_{0i}} p + b_{\theta i} = b_{\theta i}\left(\frac{p}{V_{0i}} + 1\right) \quad (12)$$

where $b_{\theta i}$ is the intercept of the conventional PMC linear strength line in the *p*-*q* plane, and V_{0i} is its root.

By fitting the failure curve using a piecewise linear approach, a PMC strength formula that considers extensional strain can be obtained. In the *p*-*q* plane, the formula can be written as follows:

$$q_c{}^{(n)} = min\left[b_{cn}\left(\frac{p}{V_{0i}}+1\right)\right] \quad (13)$$

$$q_e{}^{(n)} = min\left[b_{en}\left(\frac{p}{V_{0i}}+1\right)\right] \quad (14)$$

where n represents the failure mechanism number and $q_c{}^{(n)}$ and $q_e{}^{(n)}$ are the critical shear stress values under different failure mechanisms. Similarly, b_{cn} and b_{en} represent the intercepts of the PMC linear strength lines on the *p*-*q* plane under different failure mechanisms, and V_{0i} represents their roots.

3.2. Unified Equation of the Multiple Failure Mechanism Model

Equation (14) is presented in the form of intercepts and other data, without specific expressive significance. For spatial loading, the influence of the Lode angle should be considered. Therefore, under the premise of spatial loading, deriving a model for multiple failure mechanisms is more effective. Based on the PMC criterion, a general model is proposed to obtain different solutions under various failure mechanisms.

In the PMC model, the strength from the meridional plane is extended to the stress space as follows:

$$f = \sigma_1\left(\frac{1-sin\varphi_c}{2sin\varphi_c}\right) + \sigma_2\left(\frac{sin\varphi_c - sin\varphi_e}{2sin\varphi_c sin\varphi_e}\right) - \sigma_3\left(\frac{1+sin\varphi_e}{2sin\varphi_e}\right) - V_0 = 0 \quad (15)$$

It should be noted that this model assumes static loading conditions and does not account for dynamic loads. The loading rate in the experiment was referenced according to the method recommended by the International Society for Rock Mechanics [10]. Where V_0 is the assumed strength under triaxial equal tensile conditions; φ_c is the internal friction angle under triaxial compression; and φ_e is the internal friction angle under triaxial extension. The triaxial compression and extension strength lines at the same V_0 point can be considered a set, where n denotes the failure mechanism number. The definition of the stress Lode angle is shown in Figure 5. The Lode angle ($0° \leq \theta \leq 60°$) is the rotation angle from the maximum principal stress axis (θ = 0°) to the minimum principal stress axis.

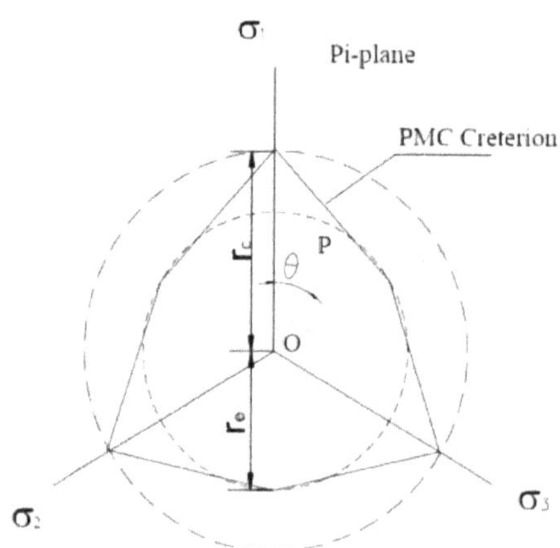

Figure 5. Relationship of the stress Lode angle and the π-plane. The maximum radius on the plane corresponds to the r_c value under triaxial compression, and the minimum radius corresponds to the r_e value under triaxial extension. θ represents the Lode angle.

The maximum radius on the π-plane corresponds to the r_c value under triaxial compression, and the minimum radius corresponds to the r_e value under triaxial extension. Let the stress state at any point on this π-plane be (p, q, θ), where p is the mean stress, q is the generalized shear stress, and θ is the Lode angle. The relationship with the principal stresses is as follows:

$$p = \frac{\sigma_1 + \sigma_2 + \sigma_3}{3} \tag{16}$$

$$q = \sqrt{\frac{(\sigma_1 - \sigma_2)^2 + (\sigma_2 - \sigma_3)^2 + (\sigma_3 - \sigma_1)^2}{2}} \tag{17}$$

As shown in Figure 5, with the centroid O of the π-plane as the center, the distance from any point on the plane to the center represents the magnitude of the deviatoric stress in the stress space. Taking the meridional plane as an example, its radius is determined as follows:

$$r_\theta = (p + V_0)tan\varphi_\theta \tag{18}$$

where r_θ and φ_θ are the radius and the corresponding internal friction angle on the π-plane when the Lode angle is θ, and the connection line between the stress point and the far point is the slope line. When the stress state is triaxial compression, $\varphi_\theta = \varphi_c$; when it is triaxial extension, $\varphi_\theta = \varphi_e$.

On the π-plane, through the stress point (p, q, θ), perpendicular lines are drawn to the three coordinate axes, intersecting the axes at the points $(\sigma_I', 0, 0)$, $(0, \sigma_{II}', 0)$, and $(0, 0, \sigma_{III}')$. Here, σ_I', σ_{II}', and σ_{III}' represent the principal stresses, but their magnitudes are not determined. In the quadrant shown in Figure 5, the order of the coordinates on the three axes is $\sigma_1 > \sigma_2 > \sigma_3$. Roman numerals were used to indicate the order of the principal stress values. Thus:

$$\sigma_I' = r_\theta cos\theta \tag{19}$$

$$\sigma_{II}' = r_\theta cos(60^\circ - \theta) \tag{20}$$

The relationship between the line segment on the π-plane and the principal stress space coordinates is shown in Figure 6. The length of the AH line segment is the height h of the equilateral triangle of the slope plane, and the intercepts of the slope plane along the three coordinate axes are all $3p$. The geometric relationship yields:

$$\frac{\sigma_1}{\sigma_I' + \frac{h}{3}} = \frac{3p}{h} \tag{21}$$

Since the π-plane is an equilateral plane, $h = \frac{3\sqrt{6}}{2}p$. Substituting this into the equation yields $\sigma_1 = \frac{\sqrt{6}}{3}\sigma_I' + p$, and substituting into Equation (19) results in the following:

$$\sigma_1 = \frac{\sqrt{6}}{3} r_\theta cos\theta + p \tag{22}$$

Similarly, the following can be obtained:

$$\sigma_2 = p - \frac{\sqrt{6}}{3}\sigma_{III}' = p - \frac{\sqrt{6}}{3} r_\theta cos(60^\circ - \theta) \tag{23}$$

In the equation, p represents the mean stress. Since the principal stresses on the π-plane are equal, the stress value on the σ_3 axis corresponds to the intermediate principal stress, which is as follows:

$$\sigma_3 = 3p - \sigma_1 - \sigma_2 = p - \frac{\sqrt{6}}{3} r_\theta \left[cos\theta - cos\left(60^\circ - \theta\right)\right] = p - \frac{\sqrt{6}}{3} r_\theta cos\left(60^\circ + \theta\right) \tag{24}$$

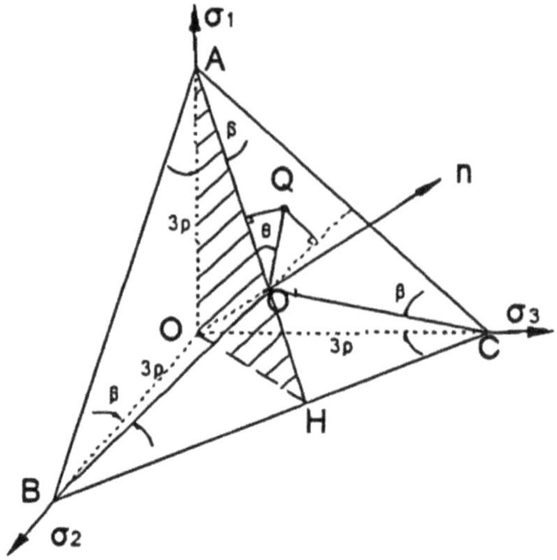

Figure 6. Relationship between the spatial and deviatoric stress radius on the π-plane. The length of the AH line segment is the height h of the equilateral triangle of the slope plane, and the intercepts of the slope plane along the three coordinate axes are all $3p$. θ represents the Lode angle.

Substituting Equations (22)–(24) into Equation (15) and rearranging yields results in the following:

$$\frac{\sqrt{6}}{3}r_\theta\left[cos\theta\cdot\left(\frac{1-sin\varphi_c}{2sin\varphi_c}\right)-cos\left(60^\circ+\theta\right)\cdot\left(\frac{sin\varphi_c-sin\varphi_e}{2sin\varphi_c sin\varphi_e}\right)+cos\left(60^\circ-\theta\right)\cdot\left(\frac{1+sin\varphi_e}{2sin\varphi_e}\right)\right]-(p+V_0)=0 \quad (25)$$

The relationship between the π-plane and the principal stress space shows that:

$$r_\theta=(p+V_0)tan\varphi_\theta=\sqrt{\frac{2}{3}}q \quad (26)$$

In the equation, q is the generalized shear stress. Substituting Equation (26) into Equation (25) yields:

$$\frac{2}{3}q\left[cos\theta\cdot\left(\frac{1-sin\varphi_c}{2sin\varphi_c}\right)-cos\left(60^\circ+\theta\right)\cdot\left(\frac{sin\varphi_c-sin\varphi_e}{2sin\varphi_c sin\varphi_e}\right)+cos\left(60^\circ-\theta\right)\cdot\left(\frac{1+sin\varphi_e}{2sin\varphi_e}\right)\right]-(p+V_0)=0 \quad (27)$$

Which can be rearranged as follows:

$$q=\frac{6sin\varphi_c sin\varphi_e(p+V_0)}{2\sqrt{3}sin\varphi_c sin\theta-2sin\varphi_c sin\varphi_e cos(60^\circ+\theta)+2\sqrt{3}sin\varphi_e sin(60^\circ-\theta)} \quad (28)$$

When θ = 0°, the stress state corresponds to conventional triaxial compression. From Equation (28), the following expression is obtained:

$$q=\frac{6sin\varphi_c}{3-sin\varphi_c}(p+V_0) \quad (29)$$

When θ = 60°, the stress state corresponds to triaxial extension. From Equation (28), the following is obtained:

$$q=\frac{6sin\varphi_e}{3+sin\varphi_e}(p+V_0) \quad (30)$$

The calculation results are consistent with those obtained from the PMC strength Equation (15). Therefore, based on Equation (28), the general formula for the strength criterion in stress space under the coupled influence of multiple failure mechanisms can be derived as:

$$q = min\left[\frac{6sin\varphi_c^i sin\varphi_e^i (p + V_0)}{2\sqrt{3}sin\varphi_c^i sin\theta - 2sin\varphi_c^i sin\varphi_e^i cos(60^\circ + \theta) + 2\sqrt{3}sin\varphi_e^i sin(60^\circ - \theta)}\right](i = 1, \ldots, n) \quad (31)$$

where q is the generalized shear stress; p is the mean stress; θ is the stress Lode angle (with the axis of the maximum principal stress as 0°); φ_c is the internal friction angle under triaxial compression; φ_e is the internal friction angle under triaxial extension; V_0 is the assumed strength under triaxial equal tensile conditions; and n is the failure mechanism group.

When $n = 1$, the formula corresponds to the conventional PMC model; when $n \geq 2$, the formula represents the multiple failure mechanism model. Specifically, when forming a yield model similar to a "cap", the shear strength envelope can effectively explain the pore collapse phenomenon in porous rocks. Based on the data reported by Folta et al., the data fitting curve can be obtained as shown in Figure 7 [20]:

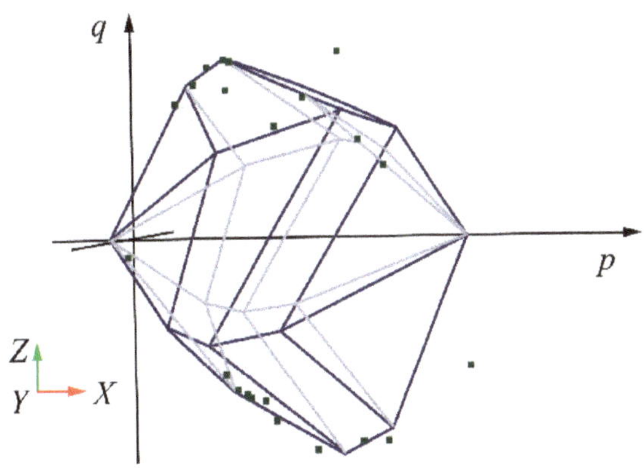

Figure 7. Model shape and data fitting in *p-q* space. The "cap"-shaped shear strength envelope during pore collapse in rocks.

3.3. *Parameter Fitting Method*

Zeng et al. proposed a new fitting method based on the transformation of stress invariant relationships in the *p-q* plane [8,21]. They derived a parameter acquisition method for a multiple failure mechanism model based on this approach.

In a π-plane, any given stress state can be represented in cylindrical coordinates (r_θ, θ, ρ). A schematic of the PMC failure surface in the π-plane when $p = 0$ is shown in Figure 8.

The intercept b_θ needs to be defined, and a relationship should be established between b_θ and r_θ. The failure envelope line $y = kx + r_c$ in the rectangular coordinate system can be expressed in terms of r_θ and θ as follows:

$$r_\theta cos\theta = k \cdot r_\theta sin\theta + r_c \quad (32)$$

where θ represents the Lode angle and r_θ is the radius of the π-plane corresponding to the Lode angle θ and the slope line's apical angle. r_c is the intercept in the axisymmetric compression line. In the π-plane, k is the slope of the line, and r_θ is as follows:

$$r_\theta = \frac{r_c}{cos\theta - ksin\theta} \quad (33)$$

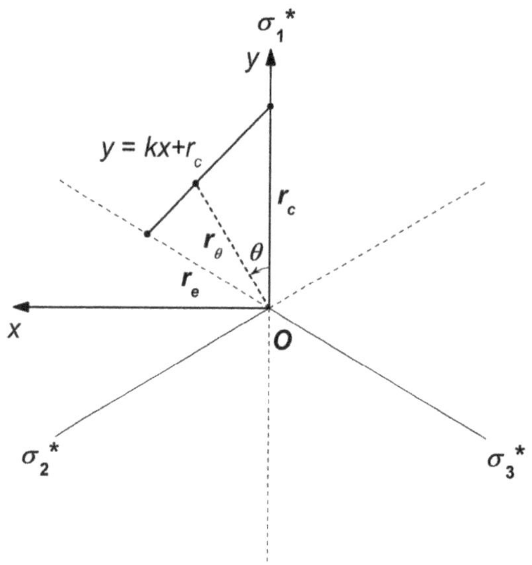

Figure 8. Schematic showing the PMC failure surface in the π plane with polar and orthogonal coordinate systems when $p = 0$. $y = kx + r_c$ represents the failure envelope line. θ represents the Lode angle, and r_θ is the radius of the π-plane corresponding to the Lode angle θ and the slope line's apical angle. r_c is the intercept in the axisymmetric compression line, and r_e represents the intercept in the multiaxial line. σ_1*, σ_2*, and σ_3* represent the stress components in the transformed or cylindrical coordinate system within the π-plane. The * symbol indicates a specific stress state in the transformed cylindrical coordinate system within the π-plane.

In any π-plane, the relationship between r_θ and q_θ can be obtained through the second stress invariant J_2 as follows:

$$r_\theta = \sqrt{\frac{2}{3}} q_\theta \tag{34}$$

where q_θ is the shear stress in the direction of r_θ. In the multiaxial line, $q_\theta = b_\theta$; in the axisymmetric compression line, $q_\theta = b_c$, where b_c is the shear stress intercept in the axisymmetric compression line. In the axisymmetric extension line, $q_\theta = b_\theta$. Therefore, r_θ can be written as:

$$\begin{cases} r_\theta = \sqrt{\frac{2}{3}} b_\theta \\ r_c = \sqrt{\frac{2}{3}} b_c \\ r_e = \sqrt{\frac{2}{3}} b_e \end{cases} \tag{35}$$

In the formula, b_e represents the shear stress intercept in the axisymmetric extension line, and r_e represents the intercept in the multiaxial line. Substituting Equation (35) into Equation (33) yields b_e as follows:

$$b_\theta = \frac{b_c}{cos\theta - ksin\theta} \tag{36}$$

In the p-q plane, Equation (36) represents the functional relationship of the intercept b_θ and the unknown variable k needs to be determined. In the π-plane, where the failure line passes through the axisymmetric extension point ($r_\theta = r_e$, $\theta = 60°$), Equation (32) can be written as:

$$r_e = \sqrt{3} \cdot k \cdot r_e + 2r_c \tag{37}$$

Considering Equation (35), the ratio $\alpha = \frac{r_c}{r_e}$ can be written as:

$$\alpha = \frac{r_c}{r_e} = \frac{\sqrt{\frac{2}{3}} \cdot b_c}{\sqrt{\frac{2}{3}} \cdot b_e} = \frac{b_c}{b_e} \tag{38}$$

Substituting Equation (38) into Equation (37) yields:

$$k = \frac{1 - 2\alpha}{\sqrt{3}} \tag{39}$$

Knowing the expressions for b_θ and k, the general Equation (39) for the PMC criterion in the p-q plane can be written as:

$$\frac{b_c}{V_0} p + kqsin\theta + b_c = qcos\theta \tag{40}$$

According to Equation (40), a linear equation system can be constructed using the data from axial compression, extension, and multiaxial tests, with θ under various conditions. In some studies, θ is equivalent to the Lode angle, but differences exist. The original definition of the Lode angle ranges from $-30°$ to $30°$, but in the coordinate system, θ ranges from $0°$ to $360°$. However, θ ranging from $0°$ to $60°$ is sufficient for isotropic expressions. Whereas $\theta = 0°$ corresponds to axisymmetric compression and $\theta = 60°$corresponds to axisymmetric extension. The θ value in multiaxial stress states can be calculated from the principal stresses as follows:

$$tan\theta = \frac{\sqrt{3}(\sigma_{II} - \sigma_{III})}{2\sigma_I - \sigma_{II} - \sigma_{III}} \tag{41}$$

where σ_I, σ_{II}, and σ_{III} represent the maximum, intermediate, and minimum principal stresses, respectively. Strength data under axisymmetric and multiaxial conditions are expressed as σ_I, σ_{II}, andσ_{III}, and by substituting these into Equation (40), the p, q, and θ for each stress state can be determined. Using these three parameters, a system of linear equations $A{\cdot}x = B$ is generated:

$$\begin{pmatrix} p_1 & q_1 sin\theta_1 & 1 \\ p_2 & q_2 sin\theta_2 & 1 \\ \vdots & \vdots & \vdots \\ p_n & q_n sin\theta_n & 1 \end{pmatrix} \begin{pmatrix} \frac{b_c}{V_0} \\ k \\ b_c \end{pmatrix} = \begin{pmatrix} q_1 cos\theta_1 \\ q_2 cos\theta_2 \\ \vdots \\ q_n cos\theta_n \end{pmatrix} \tag{42}$$

where A is the rectangular data matrix, x is the parameter vector, and B is the data vector.

Using the least squares method to solve Equation (42), b_c/V_0, k, and b are obtained. According to the expressions for the axisymmetric compression and extension lines in the p-q plane, φ_c and φ_e are calculated as follows:

$$\begin{cases} sin\varphi_c = \frac{3b_c}{6V_0 + b_c} \\ sin\varphi_e = \frac{3b_e}{6V_0 - b_e} \end{cases} \tag{43}$$

Thus, the material parameters φ_c, φ_e, and V_0 of the general formula for the strength criterion in stress space under the coupled influence of multiple failure mechanisms (Equation (31)) can be solved.

4. Experiments

4.1. Experimental Conditions

Experiments were conducted on Sichuan yellow sandstone and deep granite from the Jinchuan No. 2 mining area. High-pressure triaxial compression tests were performed using the TAW-2000 microcomputer-controlled electro-hydraulic servo rock triaxial testing machine and the portable self-sealing rock triaxial test high-pressure chamber pressurization system.

Standard rock mechanics specimens with dimensions of φ30 mm × 60 mm were prepared. The preparation of rock samples follows the ISRM-suggested method for the complete stress-strain curve under rigid experimental conditions [22]. For each set of confining and axial pressures, only one set of experimental results is selected. If there is a significant deviation in the results, the experiment will be repeated.

This conventional triaxial test relies on two independent loading systems and therefore cannot use the system settings to load axial and confining pressures simultaneously at the same rate. To ensure quasi-static conditions and reduce experimental error, an independent stepwise loading method for axial and confining pressures was adopted. The stepwise loading procedure is shown in Table 1. During the experiment, axial loading was initially controlled by stress, with a loading rate of 100 N/s. After reaching the hydrostatic pressure state, the radial pressure was maintained at a constant level throughout the experiment. Axial loading was then continued at the same rate until approximately 50% of the peak strength was reached, at which point it was switched to deformation control with a loading rate of 0.03 mm/min. During confining pressure loading, coarse control was first applied, with a loading rate of 1 MPa/s. When approaching 5 MPa of the hydrostatic pressure state, fine control was applied using a micro-adjustment handwheel with a loading rate of 0.2 MPa/s.

Table 1. Comparison table of the axial load and axial stress under step loading.

σ/MPa	P/kN	σ/MPa	P/kN	σ/MPa	P/kN
5	3.5	65	45.5	125	87.5
10	7	70	49	130	91
15	10.5	75	52.5	135	94.5
20	14	80	56	140	98
25	17.5	85	59.5	145	101.5
30	21	90	63	150	105
35	24.5	95	66.5	155	108.5
40	28	100	70	160	112
45	31.5	105	73.5	165	115.5
50	35	110	77	170	119
55	38.5	115	80.5	175	122.5
60	42	120	84	180	126

4.2. Experimental Results

High-confining-pressure conventional triaxial tests were conducted on Sichuan yellow sandstone and Jinchuan granite using the above stepwise loading scheme. For the Sichuan sandstone, 16 groups of different confining pressures ranging from 0 to 170 MPa were set, whereas for the Jinchuan granite, 13 groups of different confining pressures ranging from 0 to 160 MPa were set. The stress-strain curves for each rock sample during the tests are shown in Figures 9 and 10. The data for principal stress, mean stress, and deviatoric stress at failure are presented in Table 2. The failure modes of the samples are shown in Figures 11 and 12. Both the Sichuan yellow sandstone and Jinchuan granite exhibited similar failure mode trends with increasing confining pressure: under low confining pressure, the failure is primarily shear failure, with the angle of the failure plane decreasing as the confining pressure increases; whereas under higher confining pressure, the angle continues to decrease, gradually exhibiting a trend towards compressive-shear failure.

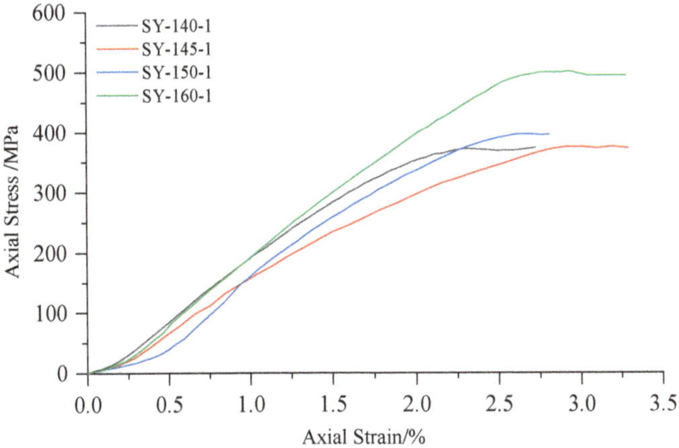

Figure 9. Stress-strain curves of Sichuan yellow sandstone in conventional triaxial compression tests.

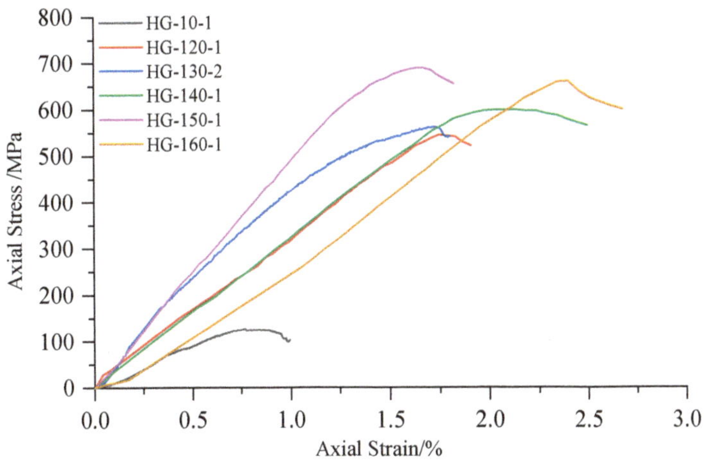

Figure 10. Stress-strain curves of Jinchuan granite in conventional triaxial compression tests.

Table 2. Summary of the conventional triaxial compression analysis of Sichuan yellow sandstone and Jinchuan granite.

Number	Wave Velocity/km/s	σ_2/MPa	σ_3/MPa	σ_1/MPa	p/MPa	q/MPa	θ_l/°
SY-0-1	1.43	0	0	49	16	49	−30
SY-10-1	1.58	10	10	107	42	97	−30
SY-20-1	1.69	20	20	173	71	153	−30
SY-30-1	1.42	30	30	222	94	192	−30
SY-40-1	1.54	40	40	219	100	179	−30
SY-50-1	1.43	50	50	267	122	217	−30
SY-60-2	1.5	60	60	274	131	214	−30
SY-65-1	1.52	65	65	322	151	257	−30
SY-70-1	1.46	70	70	361	167	291	−30
SY-90-1	1.46	90	90	398	193	308	−30
SY-110-2	1.72	110	110	377	199	267	−30
SY-140-2	1.43	140	140	430	237	290	−30
SY-145-1	1.61	145	145	477	256	332	−30
SY-150-1	1.46	150	150	460	253	310	−30

Table 2. *Cont.*

Number	Wave Velocity/km/s	σ_2/MPa	σ_3/MPa	σ_1/MPa	*p*/MPa	*q*/MPa	θ_l/°
SY-160-2	1.62	160	160	502	274	342	−30
SY-170-1	1.45	170	170	490	277	320	−30
HG-0-1	8.78	0	0	148	49	148	−30
HG-10-1	8.33	10	10	127	49	117	−30
HG-20-1	8.56	20	20	221	87	201	−30
HG-30-1	8.33	30	30	296	119	266	−30
HG-40-2	8.45	40	40	341	140	301	−30
HG-50-1	8.78	50	50	381	160	331	−30
HG-60-1	8.36	60	60	413	178	353	−30
HG-100-1	7.89	100	100	559	253	459	−30
HG-120-1	8.33	120	120	547	262	427	−30
HG-130-2	8.38	130	130	561	274	431	−30
HG-140-1	8.82	140	140	597	292	457	−30
HG-150-1	7.14	150	150	687	329	537	−30
HG-160-1	8.33	160	160	658	326	498	−30

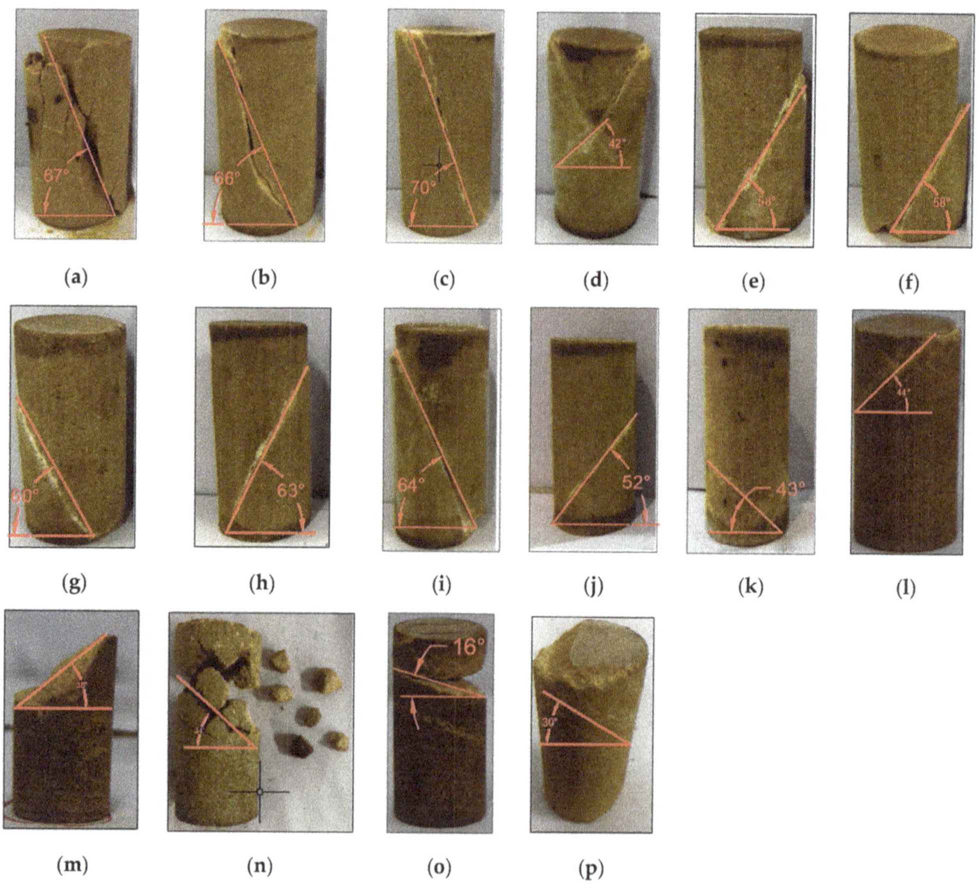

Figure 11. Failure modes of Sichuan yellow sandstone: (**a**) SY-0-1; (**b**) SY-10-1; (**c**) SY-20-1; (**d**) SY-30-1; (**e**) SY-40-1; (**f**) SY-50-1; (**g**) SY-60-2; (**h**) SY-65-1; (**i**) SY-70-1; (**j**) SY-90-1; (**k**) SY-110-2; (**l**) SY-140-2; (**m**) SY-145-1; (**n**) SY-150-1; (**o**) SY-160-1; (**p**) SY-170-1. (**a**–**p**) show the failure plane angles with gradually increasing confining pressure.

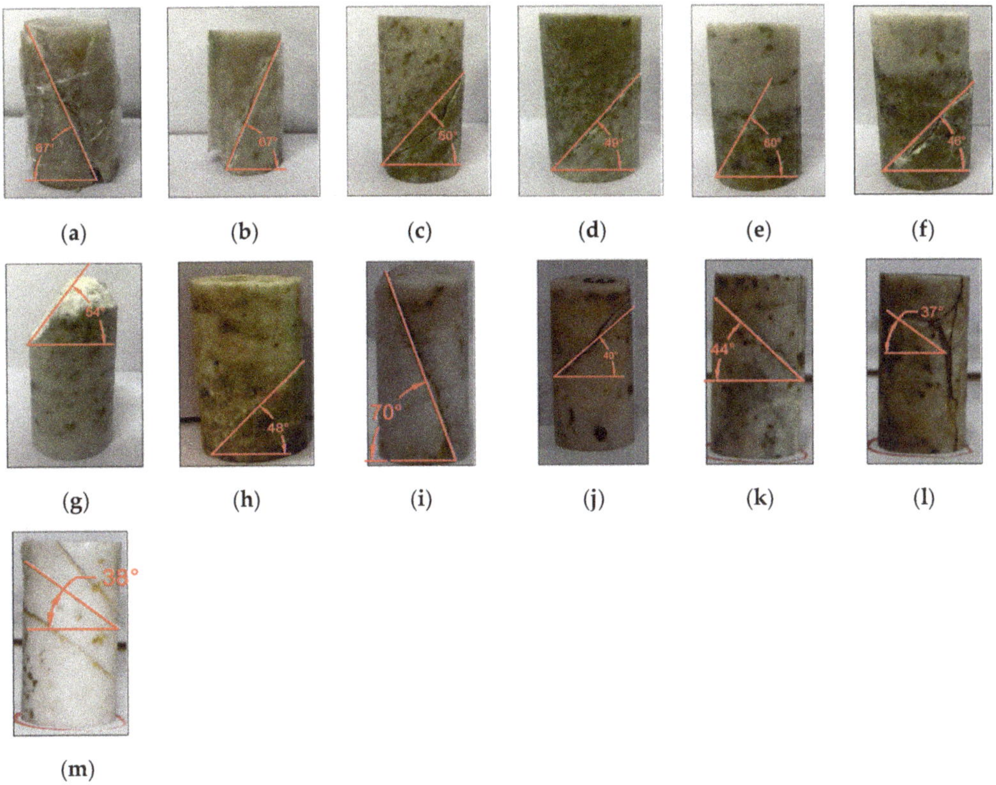

Figure 12. Failure modes of Jinchuan granite: (**a**) HG-0-1; (**b**) HG-10-1; (**c**) HG-20-1; (**d**) HG-30-1; (**e**) HG-40-2; (**f**) HG-50-1; (**g**) HG-60-1; (**h**) HG-100-1; (**i**) HG-120-1; (**j**) HG-130-2; (**k**) HG-140-1; (**l**) HG-150-1; (**m**) HG-160-1. (**a–m**) show the failure plane angles with gradually increasing confining pressure.

4.3. Applicability of the PMC Model for Strength with Multiple Failure Mechanisms

When $n = 1$, analysis is conducted using Berea sandstone experimental data (Bobich) (Table 3) [23]. The fitted failure surface in the p-q plane with the experimental data (conventional triaxial compression tests and conventional triaxial extension tests) is shown in Figure 13. The negative axis of q represents triaxial extension tests, and the positive axis of q represents triaxial compression tests. The PMC criterion fitting results closely match the experimental results, and the friction angle in conventional triaxial compression tests is smaller than in conventional triaxial extension tests. The difference between φ_c and φ_e illustrates the influence of the intermediate principal stress. Similarly, the study by Guo et al. explored the failure process of heterogeneous brittle rocks under uniaxial compression, highlighting the significant role of material heterogeneity in stress redistribution and crack formation [24]. Additionally, Wen et al. conducted qualitative and quantitative investigations on the effect of critical fissures on the failure process of rock specimens under plane strain compression, further emphasizing the importance of crack evolution in determining rock strength [25].

Table 3. PMC model fitting results.

Material	$\varphi_c/°$	$\varphi_e/°$	V_0/kPa
Berea Sandstone	37.7	44.2	21.7

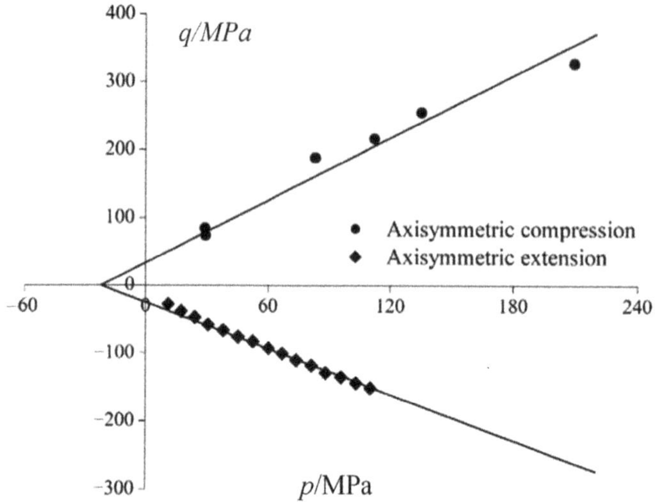

Figure 13. Experimental results obtained for the Berea sandstone. The negative axis of q represents triaxial extension tests, and the positive axis of q represents triaxial compression tests.

When n = 2, the experimental data for Jinchuan granite samples are analyzed. To ensure $V_0^{(1)} \geq V_0^{(2)}$, the region is divided into two segments: $p \geq 200$ MPa and $p \leq 200$ MPa. The six-parameter PMC model fitting results are shown in Table 4. The six-parameter PMC model for granite in the p-q plane is shown in Figure 14. The granite strength exhibits distinct piecewise linear characteristics, which can be fitted using the six-parameter PMC model. However, due to the lack of triaxial extension test data, this model is only applicable to the transitional stage of tensile-shear failure in rocks and cannot characterize the tensile failure mode of granite.

Table 4. Parameters used in the Jinchuan granite PMC model.

Failure Mechanism Number n	$\varphi_c/°$	$\varphi_e/°$	V_0/kPa
1	42.853	42.853	27.712
2	28.020	28.020	130.953

Note: Due to the lack of triaxial extension and true triaxial tests, only triaxial compression data are used for fitting; therefore, $\varphi_c = \varphi_e$.

When $n > 2$, the experimental data for Sichuan yellow sandstone samples are analyzed, and the fitting results are shown in Table 5. The fitted PMC failure surface with the experimental data in the p-q plane is shown in Figure 15. The strength of the yellow sandstone exhibits distinct piecewise strength characteristics, which conform to strength patterns under the influence of multiple failure mechanisms (Figure 15).

Table 5. Strength parameters of sandstone under the influence of multiple failure mechanisms.

Failure Mechanism Number n	V_0/MPa	$\varphi_c/°$	$\varphi_e/°$
1 (low-pressure failure segment)	14.66	40	40
2 (transition segment)	76.87	27	27
3 (high-pressure failure segment)	403.31	13	13

Note: Due to the lack of triaxial extension and true triaxial tests, only triaxial compression data are used for fitting; therefore, $\varphi_c = \varphi_e$.

Figure 14. Six-parameter PMC model for granite on the *p*-*q* plane. The six parameters represent the failure mechanism number *n* for *n* = 1 and *n* = 2 as φ_c, φ_e, and V_0. The piecewise linear relationship shown in the figure represents the results of the PMC six-parameter fitting. The diagram in the figure illustrates the cracking of Jinchuan granite, with the cracking plane angles gradually increasing from lower left to upper right as the confining pressure increases.

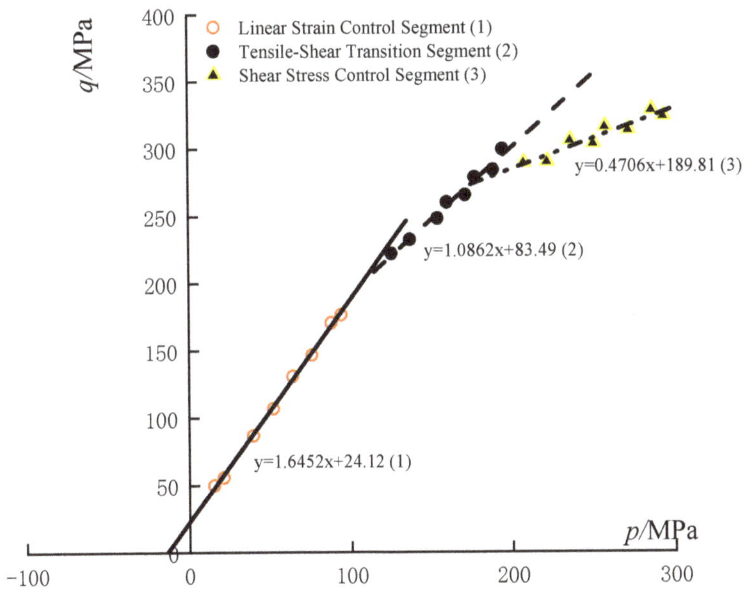

Figure 15. Fitting the failure strength of yellow sandstone with multiple mechanisms. The diagram shows the comparison between the PMC-fitted failure surface and the experimental data curves in the *p*-*q* plane.

5. Conclusions

The rock multiple failure mechanism theory model proposed in this paper addresses the limitations of considering only a single failure mechanism. The model introduces a unified expression through the PMC piecewise linear model to describe multiple failure modes, including tensile, splitting, and shear failures. It integrates established criteria such as the Mohr-Coulomb strength theory and the 12-sided PMC model and extends to include the asymmetric PMC-BPMC model providing a comprehensive tool for analyzing complex failure processes in rock materials. The model's applicability is supported by experimental validation using data from Jinchuan granite and Sichuan yellow sandstone.

(1) In light of the limitations of traditional rock strength theories, which consider only the impact of a single failure mechanism, and recognizing that tensile and splitting failures of rocks under low-stress conditions conform to the maximum elongation strain theory while shear-type failures occur at high-pressure levels, a multiple failure mechanism theory model for rocks has been proposed based on the PMC piecewise linear model. This model establishes a unified expression for multiple failure modes in the form of inequalities. Experimental verification has demonstrated that this criterion can effectively achieve nonlinear fitting for granite, sandstone, and other rock types.

(2) The rock multiple failure mechanism theory model's failure criterion can encompass the MC strength theory, the PMC strength model, and the 12-sided PMC model. It can also be extended to derive the asymmetric PMC-BPMC model, which considers isotropic yield. By integrating the principles of plastic flow rules and hardening laws, the model can be expanded to analyze and explain the mechanisms of pore collapse phenomena in rocks under high pressure, corresponding to the specialized "cap" yield model mentioned in the text.

(3) Using experimental data from Jinchuan granite and Sichuan yellow sandstone, the rock multiple failure mechanism theory model was fitted for $n = 1$ (considering only shear failure), $n = 2$ (considering the transition from shear failure to tensile failure), and $n > 2$ (coexistence of tensile and shear failure). The results indicate that the rock strength exhibits distinct segmented strength characteristics, consistent with the strength patterns under the influence of multiple failure mechanisms. This validates the applicability of the multiple failure mechanism theory model to rock materials. However, since this study only considers generalized shear stress, mean stress, and stress Lode angle parameters in the rock failure process, other minor stress influences and environmental factors were not discussed, presenting certain limitations.

Author Contributions: Conceptualization, S.L., Y.L., D.F. and L.Z. (Liang Zhao); Software, S.L. and Y.L.; Research, S.L. and Yuan; Abstract, S.L. and Y.L.; Validation, Y.L.; Formal analysis, S.L. and L.Z. (Liang Zhao); Data curation, S.L., L.Z. (Litian Zhang) and D.F.; Original draft, S.L. and L.Z. (Litian Zhang); Review and editing, Y.L.; Literature review, S.L. and Y.L.; Figures, Y.L. and S.L.; Tables, S.L. and Y.L.; Translation, S.L. and L.Z. (Litian Zhang); Funding, Y.L. All authors have read and agreed to the published version of the manuscript.

Funding: This research was funded by the National Key Research and Development Program of China (grant number 2022YFC2904102).

Institutional Review Board Statement: Not applicable.

Informed Consent Statement: Not applicable.

Data Availability Statement: The authors have not obtained permission to publish the data. Therefore, the data can be obtained from the corresponding author upon reasonable request.

Acknowledgments: Experimental equipment and technical support were provided by the University of Science and Technology Beijing. Data support was provided by Litian Zhang, a graduate from the University of Science and Technology Beijing.

Conflicts of Interest: The authors declare no conflicts of interest. The funders had no role in the design of the study; in the collection, analyses, or interpretation of data; in the writing of the manuscript; or in the decision to publish the results.

References

1. Labuz, J.F.; Zang, A. Mohr-Coulomb failure criterion. In *The ISRM Suggested Methods for Rock Characterization, Testing and Monitoring*; Springer International Publishing: Cham, Switzerland, 2014; pp. 227–231. [CrossRef]
2. Skrzypkowski, K.; Zagórski, K.; Zagórska, A. Determination of the Extent of the rock destruction Zones around a gasification Channel on the basis of Strength Tests of Sandstone and Claystone Samples Heated at High Temperatures up to 1200 C and Exposed to Water. *Energies* **2021**, *14*, 6464. [CrossRef]
3. Li, X.; Feng, F.; Li, D. Numerical simulation of rock failure under static and dynamic loading by splitting test of circular ring. *Eng. Fract. Mech.* **2018**, *188*, 184–201. [CrossRef]
4. Paul, B. Generalized pyramidal fracture and yield criteria. *Int. J. Solids Struct.* **1968**, *4*, 175–196. [CrossRef]
5. Paul, B. Macroscopic criteria for plastic flow and brittle fracture. *Fracture* **1968**, *2*, 313–496.
6. Meyer, J.P.; Labuz, J.F. Linear failure criteria with three principal stresses. *Int. J. Rock Mech. Min. Sci.* **2013**, *60*, 180–187. [CrossRef]
7. Makhnenko, R.; Labuz, J. Plane strain testing with passive restraint. *Rock Mech. Rock Eng.* **2014**, *47*, 2021–2029. [CrossRef]
8. Zeng, F.; Li, Y.; Labuz, J.F. Paul-Mohr-Coulomb failure criterion for geomaterials. *J. Geotech. Geoenviron. Eng.* **2018**, *144*, 06017018. [CrossRef]
9. Dehler, W.; Labuz, J.F. Stress path testing of an anisotropic sandstone. *J. Geotech. Geoenviron. Eng.* **2007**, *133*, 116–119. [CrossRef]
10. Stacey, T.R. A simple extension strain criterion for fracture of brittle rock. *Int. J. Rock Mech. Min. Sci. Geomech. Admin.* **1981**, *18*, 469–474. [CrossRef]
11. Huang, S. *Study on Mechanical Model of Brittle Rock under High Stress Condition and Its Engineering Applications*; Institute of Rock and Soil Mechanics, Graduate School of Chinese Academy of Sciences, Wuhan Institute of Geotechnical Mechanics: Wuhan, China, 2008.
12. Paul, B. A modification of the coulomb-mohr theory of fracture. *J. Appl. Mech.* **1961**, *28*, 259–268. [CrossRef]
13. Zhou, H.M.; Xiong, S.H.; Liu, X.H.; Yang, Z.H.; Kong, X.H. In-situ tension-shear tests and strength criterion studies on TGP shiplock slope rock mass. *Yanshilixue Yu Gongcheng Xuebao Chin. J. Rock Mech. Eng.* **2005**, *24*, 4418–4421.
14. Sun, J. *Theory and Practice of Underground Engineering Design*; Shanghai Science and Technology Publishing House: Shanghai, China, 1996.
15. Zhou, H.M.; Xu, P.; Sheng, Q.; Yu, Y. Application of new technology of rock mechanics test in TGP. *J. Yangtze River Sci. Res. Inst.* **2001**, *18*, 68.
16. Liu, F.Z. Study on mechanical property of rock in tension and tension shear state. *J. Yangtze River Sci. Res. Inst.* **1996**, *13*, 35.
17. Zhu, Z.L.; Li, J.L.; Wang, K.P. Experimental research on rock tension-shear strength and creep rupture related to the three gorges project. *J. Univ. Hydraul. Electr. Eng.* **1998**, *20*, 16–19.
18. Jianlin, L. Experimental study on rheological properties of rock in tension and shear. *J. Geotech. Eng.* **2000**, *22*, 299–303.
19. Li, P.; Yin, Y. Modification of Drucker-Prager criterion in tensile shear region. *Chin. J. Rock Mech. Eng.* **2010**, *29*, 3029–3033.
20. Folta, B.L. *Strength Testing Under Multi-Axial Stress States*; University of Minnesota: Minneapolis, MN, USA, 2016.
21. Labuz, J.F.; Zeng, F.; Makhnenko, R.; Li, Y. Brittle failure of rock: A review and general linear criterion. *J. Struct. Geol.* **2018**, *112*, 7–28. [CrossRef]
22. Fairhurst, C.E.; Hudson, J.A. Draft ISRM suggested method for the complete stress-strain curve for intact rock in uniaxial compression. *Int. J. Rock Mech. Min. Sci.* **1999**, *36*, 279–289.
23. Bobich, J.K. *Experimental Analysis of the Extension to Shear Fracture Transition in Berea Sandstone*; Texas A&M University: College Station, TX, USA, 2005.
24. Guo, S.; Qi, S.; Zou, Y.; Zheng, B. Numerical studies on the failure process of heterogeneous brittle rocks or rock-like materials under uniaxial compression. *Materials* **2017**, *10*, 378. [CrossRef] [PubMed]
25. Wen, J.; Tang, L.; Zhang, S.; Zhan, Q.; Wang, Y. Qualitative and quantitative investigations on the failure effect of critical fissures in rock specimens under plane strain compression. *Materials* **2023**, *16*, 611. [CrossRef] [PubMed]

applied sciences

MDPI

Article

Conducting Research to Identify Key Features and Critical Nodes in the Coalescence and Instability of Pre-Fabricated Jointed Rock

Buchu Zhang [1,2], Shichuan Zhang [1,2,*], Baotang Shen [1,2], Yangyang Li [1,2], Shilong Song [1] and Xuexian Han [1]

1 College of Energy and Mining Engineering, Shandong University of Science and Technology, Qingdao 266590, China; 15621166605@163.com (B.Z.); 18253151805@163.com (B.S.); lyy1987718@126.com (Y.L.); 15265119571@163.com (S.S.); 18265523191@163.com (X.H.)

2 Laboratory of Mine Disaster Prevention and Control, Shandong University of Science and Technology, Qingdao 266590, China

* Correspondence: zsc373260186@sdust.edu.cn

Abstract: The instability of jointed rock masses has been a persistent concern in China's underground geotechnical engineering, particularly regarding rock mass instability triggered by structural activation, such as faulting. This form of instability constitutes a significant type of dynamic geological hazard in the field of geotechnical engineering. Research on the mechanism of jointed rock mass instability typically concentrates on various characteristics associated with structural activation but frequently neglects the interplay between coalescence instability within the jointed zones and the intact zones, as well as the development and evolution of abrupt water channels. To delve into the coalescence instability characteristics between jointed and intact zones, this study conducted uniaxial compression tests on macro-scale pre-fabricated jointed sandstone. The research results show that the failure process of the specimen consists of a strong deviation linear stage, a sub-critical stage, and an unstable stage. The main failure process occurs during the sub-critical stage.

Keywords: jointed rock masses; coalescent failure; energy evolution; identification information

1. Introduction

In the surrounding rock of underground engineering projects, there are typically numerous geological discontinuities, such as fractures and faults, characterized by geological joints. These discontinuous and weak structural features are prone to the initiation of damage cracks, propagation at boundaries under external loads, and interconnection between joints. They directly influence the mechanical behavior and stability of the engineering rock mass. When subjected to mining disturbances, the behavior of the rock mass in jointed and intact zones displays distinct patterns of damage and instability, collectively impacting the overall mechanical characteristics of the excavation. Therefore, the study of coalescent fracture and instability evolution patterns between jointed and intact regions in jointed rock is of paramount importance.

Relevant scholars have conducted extensive research, exploring the failure and instability characteristics of jointed rock masses at both microscopic and macroscopic scales. These studies have primarily focused on aspects such as rock joint filling materials, dip angles, and connectivity. Regarding research on the mechanical properties and failure characteristics of jointed rocks with filled joints, scholars generally hold the following understandings: The shear strength of filled joints is determined by the combined effects of normal stress, filling thickness, and filling material saturation [1,2]. The degree of joint filling directly affects the mechanical performance and failure modes of rock joints [3]. The failure of jointed rock with filling initially occurs at the contact surface between joints and filling materials, followed by the failure of the filling medium [4]. The energy dissipation of rocks is inversely correlated with the thickness of joint filling and positively correlated with the strength of filling materials [5].

In the research concerning rock joint dip angles and connectivity, it has been observed that the pre-peak energy self-inhibition effect in jointed rock increases with an increasing joint dip angle, initially strengthening and subsequently weakening. This effect exhibits a positive correlation with joint length [6]. The sequence of failure in jointed rock undergoes changes as the joint dip angle increases, transitioning from filled-type weak joints to the rock matrix [7]. Joint dip angles of 45° and 90° exert the most significant influence on the mechanical properties of jointed rock, while the impact is minimal at a 15° dip angle [8]. Non-connected joints noticeably degrade the rock mass, with the strength of jointed rock positively associated with the joint dip angle. With increasing joint connectivity, the fracture toughness of jointed rock follows an approximately hyperbolic trend, decreasing [9]. An equivalent elastic model for rock masses featuring intersecting connected and non-connected joints has been developed using the equivalent elastic modeling approach [10].

In recent years, many coal mines in eastern China have reached depths exceeding one kilometer, resulting in extensive bidirectional jointing within the surrounding rock of tunnels, especially under high stress and intense mining conditions. This has contributed to increased complexity in the discontinuity and deformation damage of the surrounding rock. The process of crack propagation in rock specimens with two joints differs from that in specimens with a single joint, and the development of cracks is influenced by factors such as joint spacing, dip angle, and location. Consequently, numerous scholars have delved into the mechanisms of crack propagation through the interaction of two joints.

In the case of layered jointed rock masses, as the joint angle increases, the impact of joint surfaces on failure characteristics decreases [11]. In non-connected jointed rock masses during unloading, the maximum shear stress occurs on both sides of the joint, followed by the center of the rock bridge, while the joint surface experiences the minimum shear stress [12]. The numerical values of elastic strain energy and dissipation energy for double-jointed rock specimens exhibit significantly smaller abrupt increases or decreases compared to single-jointed specimens [13]. Under dynamic loading, specimens with double cracks at different angles display an X-shaped distribution of the primary fracture [14]. In jointed rock masses, coplanar joint tips predominantly undergo pure shear failure, while adjacent joint tips experience both tension and shear failures. Non-coplanar joints primarily exhibit a combination of shear and tension failures [15]. The deformation and damage process of prefabricated jointed rock masses under freeze–thaw cycles can be divided into four stages: compaction, elastic deformation, pre-peak instability, and post-peak instability damage [16]. Additionally, the mechanical properties of defective red sandstone are weakened due to the pore pressure from free water [17]. During stress, defective rocks primarily exhibit shear cracking, and at low crack inclinations ($a < 45$ degrees), there is a notable formation of concentrated zones [18].

The previously mentioned research has extensively investigated various aspects of jointed rock, encompassing its failure modes, deformation characteristics, and the influence of mechanical parameters. Most scholars have directed their research efforts toward jointed rock masses, with a predominant focus on damage mechanics and fracture mechanics. They have explored the effects of various spatial distributions, such as angles and widths, as well as the filling characteristics of joints on the mechanical properties of rock masses. Additionally, they have delved into patterns of macro and micro-scale crack propagation. Nonetheless, there is a notable scarcity of research on the deformation and evolution processes of jointed rock masses with multiple joints, particularly in terms of analyzing their mutual interactions. A deeper, more comprehensive analysis in this realm is clearly warranted.

This study entails the design of red sandstone specimens with dual joints, positioned at a 45-degree dip angle, varying in length. Cement slurry is used to fill these dual joints, allowing for the differentiation between the jointed and intact regions. Subsequently, laboratory compression tests are conducted to investigate the slip, propagation of secondary cracks, and distribution patterns of deformation fields in the dual-jointed rock during the loading process. Furthermore, numerical simulations of jointed rock are executed to clarify the fracture and instability characteristics influenced by joint response between jointed and intact regions in 45-degree dual-jointed rock. The study also encompasses the identification of physical data at critical nodes throughout the loading process.

2. Experimental Research

2.1. Preparation of Pre-Fabricated Jointed Red Sandstone

In accordance with international standards for rock mechanics testing, we meticulously processed the red sandstone into standardized rectangular rock specimens measuring 50 mm × 50 mm × 100 mm (length × width × height). To replicate the presence of natural joints in the rock, we utilized a rock-cutting machine to create varying-sized cracks along both sides of the red sandstone samples, subsequently filling these fissures with cement slurry. Within a controlled indoor environment, the cement slurry was meticulously prepared by blending 325-grade cement and water, maintaining a cement-to-water mass ratio of 2:1. The outcomes of rock mechanics testing revealed that the compressive strength of the standard cement sample, featuring a water-to-cement ratio of 0.5, measured 32 MPa, while the tensile strength reached 5.5 MPa, and the density stood at 3.0 g/cm^3. The pristine red sandstone specimens demonstrated a uniaxial compressive strength of 50.7 MPa, representing an 18.7 MPa advantage over the standard cement sample (with a water-to-cement ratio of 0.5). Consequently, the dissimilarity in strength could be effectively simulated using cement slurry to replicate structural characteristics.

2.2. Design of Experimental Plan

This experiment was designed with four sets of jointed rock specimens, and the experimental setup is detailed in Table 1 and Figure 1. Initially, a cutting machine was used to create incisions in the red sandstone, ensuring an average width of 5 mm for each cut and an elliptical shape at the ends of the incisions. Subsequently, a cement slurry was injected into these incisions using a syringe, and waterproof tape was used to secure it to the surface of the red sandstone. After the cement slurry filling was completed, the prepared jointed rock specimens (as shown in Figure 2) were placed in an incubator for 28 days. Following the curing period, uniaxial compression tests were conducted on the specimens under servo-controlled displacement loading, with an axial displacement rate of 0.001 mm/s.

Table 1. Size design plan for jointed red sandstone.

Numbering	D/mm	H/mm	α/°	h/mm	l/mm	l/D	Remark
1-1	50	100	45	35	35	0.7	
1-2	50	100	45	25	25	0.5	
1-3	50	100	45	15	15	0.3	
1-4	50	100	45	0	0	0	Penetration

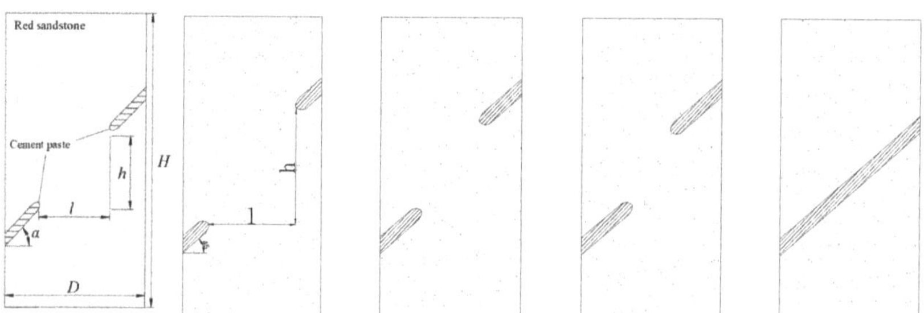

Figure 1. Design scheme for jointed rock mass.

Figure 2. Completed red sandstone jointed rock mass production.

2.3. Test Equipment and System

The experimental testing system comprises the test loading system, acoustic emission device, static strain detection equipment, and image acquisition system. The experiments were conducted using an AG-X250 testing machine (The machine is manufactured by Qingdao Qiankunxing Intelligent Technology Co., Ltd., Qingdao, China) for uniaxial loading, with a loading rate of 0.001 mm/s, and the sensitivity of the testing machine was set to 80%. Simultaneously, a Donghua (Jingjiang, China) DH-3816N static strain testing instrument (as shown in Figure 3a) was used for data acquisition during the experimental process, with a sampling interval of 60 points per minute and a strain sensitivity coefficient ranging from 1.0 to 3.0. BX120-3AA strain gauges (Beijing Yiyang Yingzhen Testing Technology Co., Ltd., Beijing, China) were selected as strain sensors, with a base length × width of 6.6 × 3.6 mm and a strain limit of 2000 μm/m. Two strain gauges were placed on the surface of each red sandstone specimen, and the sensors were sequentially labeled as S1 and S2 from top to bottom. S1 was oriented perpendicular to the defect trend, while S2 was parallel to the defect trend. The positions of the strain gauges on the intact red sandstone and the four sets of defective rock specimens were identical, as illustrated in Figure 3b, with a physical representation shown in Figure 3c, the device installation location is shown in Figure 4. During the experiment, the MISTRAS series PCI acoustic emission system produced by the American Physical Acoustics Company (Beijing, China) was used to monitor acoustic emission events in the specimens during uniaxial loading. Additionally, a SONY 4K high-speed camera (Sony Corporation, Tokyo, Japan) was employed to capture the entire process of surface crack propagation. Before the experiment, the timing synchronization of the test loading control system, acoustic emission device, and image acquisition equipment was adjusted to ensure their correspondence, The schematic diagram of the testing system is shown in Figure 5.

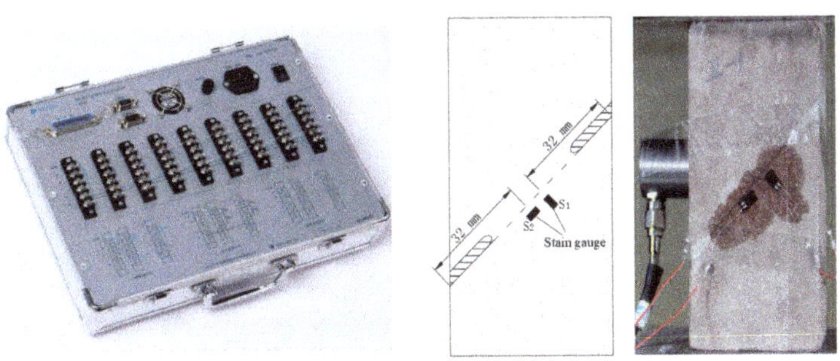

(a) (b) (c)

Figure 3. DH-3816N static strain testing system (**a**); location of strain gauge arrangement (**b**); physical diagram of sensor layout (**c**).

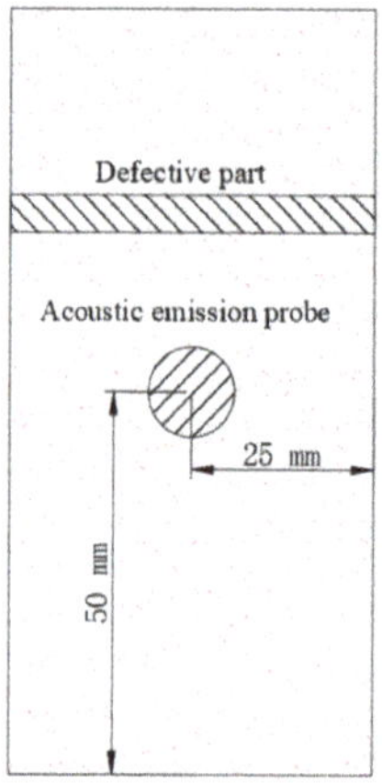

Figure 4. Acoustic emission layout.

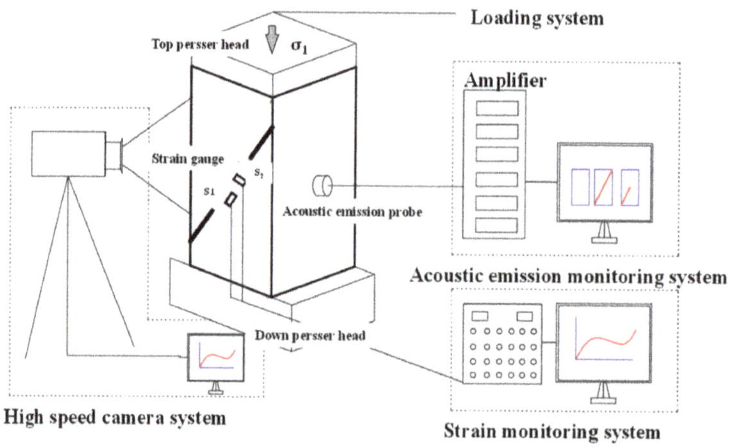

Figure 5. Schematic diagram of the test system.

3. Test Results and Discussion

3.1. Analysis of Strength Variation and Failure Characteristics in Jointed Red Sandstone

The red sandstone specimens used in the experiments include regions with filled joints, and when compared to intact red sandstone, these samples exhibit varying degrees of changes in their mechanical characteristics and deformation properties due to the influence of joints. An analysis of the uniaxial compression test results for these samples reveals noticeable differences in axial stress variations and failure modes among specimens with different joint sizes.

Figure 6 depicts the trends in axial stress over time under uniaxial loading for red sandstone with different joint sizes. The experimental results indicate that, in comparison to red sandstone without joints, the compressive strength of the jointed red sandstone samples No. 1-1 to 1-4 decreased by 9.86%, 45.76%, 40.24%, and 51.68%, respectively. The presence of filled joints results in a reduction in the rock's load-bearing capacity. Additionally, the increase in joint size to some extent contributes to the decreased load-bearing capacity of the rock. Sample No. 1-4 exhibits the most significant decrease in compressive strength.

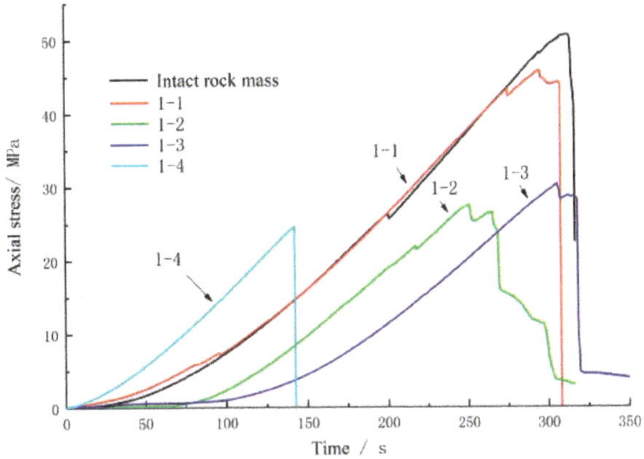

Figure 6. Axial stress vs. time curve of jointed rock mass.

In the jointed red sandstone, tensile failure is the primary mode of failure, accompanied by localized shear failure in the jointed regions, as depicted in Figure 7. Samples No. 1-1 to No. 1-3 all display tension cracks that penetrate the samples at the filled joint ends, propagating perpendicular to the loading direction. In the case of No. 1-4, the filled joint extends across the entire sample, and under the influence of applied stress, the jointed portion of the sample experiences sliding shear failure along the section. The failure modes of the samples indicate that, due to size effects, No. 1-2 initially develops cracks on the specimen's end face during loading, resulting in significant reductions in peak strength due to localized splitting and tensile failure.

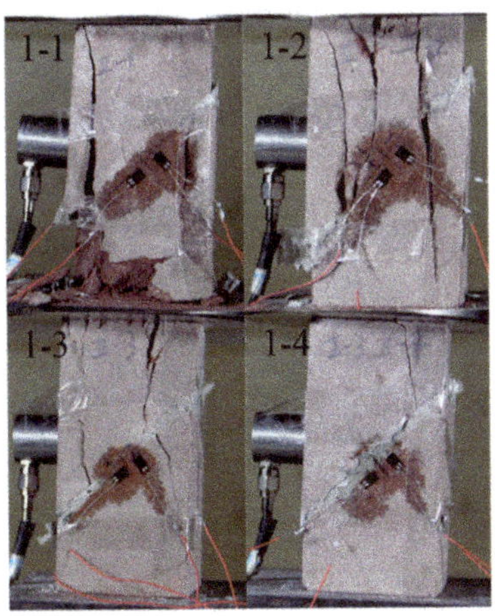

Figure 7. Failure forms of jointed rock masses.

3.2. *Analysis of Cooperative Failure Characteristics in Jointed Rock Mass during the Instability Process*

3.2.1. Characteristics of Accelerated Interaction Zones during the Instability Process

In conventional uniaxial compression tests, the stress–strain curve for rock can be divided into four segments: the compaction phase (from a to b), elastic phase (from b to c), plastic phase (from c to d), and post-peak phase (from d to e), as illustrated in Figure 8.

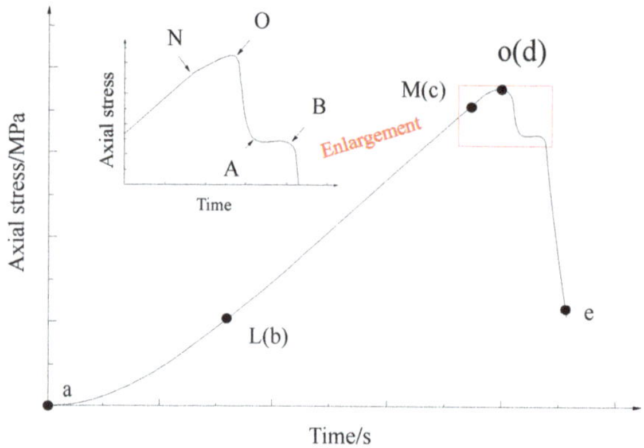

Figure 8. Stress–time curve and instability stage division.

Based on prior research [19,20], it is established that fracture interaction predominantly occurs during the strong non-linear deviation stage and the instability stage. The jointed rock mass exhibits varying physical field information, such as stress and strain, during these different stages. The emergence of cooperation during this phase is attributed to the interaction between strong and weak segments. Typically, before the applied force reaches a certain threshold, the various regional units within the jointed rock mass undergo spontaneous and irregular independent movements with minimal mutual influence. However, as it enters the strong non-linear deviation stage and the instability stage, the correlation between strong and weak segments becomes more pronounced than the independent movement of individual units. This correlation becomes the primary driving force behind cooperative motion among different regions. Consequently, interaction demonstrates zone-specific characteristics.

The study integrates methodologies from prior research to classify the strong non-linear deviation stage, sub-instability stage, and instability stage. Figure 8 illustrates the stress–time curve derived from the stress–strain curve obtained via the loading test machine. Critical stress values at key moments are identified with letters.

The LM (b-c) segment signifies the linear deformation stage, characterized by elastic deformation in the rock, with the elastic model remaining essentially constant during this phase.

The MO (c-d) segment marks the departure from linearity, where the stress–strain curve exhibits an upward concave trend. This phase involves microcracking and plastic deformation in the rock, accompanied by changes in the stress field. The NO subsegment represents a significant departure from linearity, featuring a slower rate of stress increase.

The OB segment represents the sub-instability stage, featuring a post-peak drop in the curve as internal rock fractures expand and connect. Within this stage, the OA subsegment represents a static sub-instability stage, with stress changes transitioning to a plateau, decline, or stable fluctuations in the stress–time curve. The AB subsegment represents a dynamic sub-instability stage, characterized by a rapid decline in the curve.

The sub-instability stage OB lies between the stress peak point and the rapid instability point. During this stage, the stress–strain relationship may exhibit non-linear fluctuations. This phase is pivotal as jointed rock masses approach rapid instability, serving as a period of strain accumulation leading to strain release.

3.2.2. Analysis of Cooperative Failure Characteristics in Jointed Red Sandstone

The stress–time trends during the instability stage and the corresponding sample failure modes are illustrated in Figure 9, with crucial points labeled using letters in the stress–time curves. Samples No. 1-1 to 1-4 feature jointed regions of varying sizes, each displaying distinctive failure modes during the loading process.

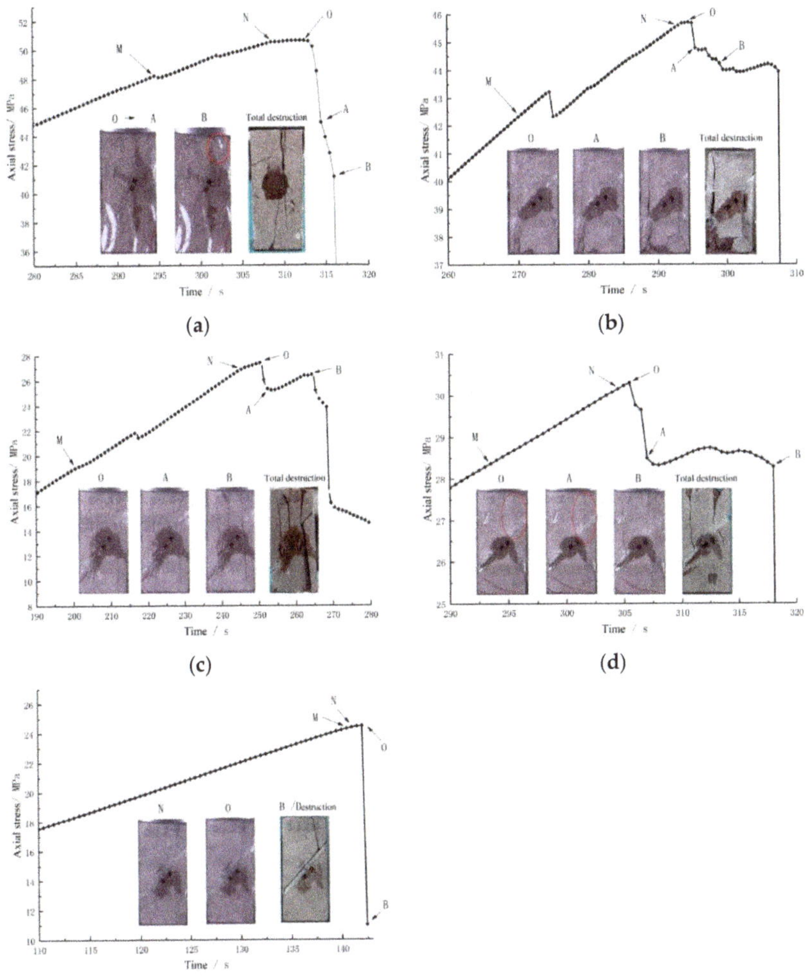

Figure 9. The change of stress at the key points with time and the corresponding failure mode. (**a**) Complete red sandstone; (**b**) No. 1-1 (l/D = 0.7); (**c**) No. 1-2 (l/D = 0.5); (**d**) No. 1-3 (l/D = 0.3); (**e**) No. 1-4 (l/D = 0).

In the case of l/D = 0.7 (No. 1-1, Figure 9b), initial cracks emerge at the tips of the lower jointed section of the sample and steadily propagate along the axial stress direction. During the O-A stage, the stress curve exhibits a linear decline, and axial stress experiences a sudden drop. Nevertheless, no failure occurs in this stage, as no fresh cracks develop on the sample surface, and existing cracks do not propagate further. Beyond point A in the stress curve, axial stress remains stable and fluctuates until point B, when tensile cracks connect. It is clear that after point B, material failure escalates, and tensile failure becomes the primary mode of specimen failure.

For l/D = 0.5 (No. 1-2, Figure 9c), following the attainment of peak strength, initial cracks manifest first at the tips of both the upper and lower jointed sections of the sample. These cracks extend along the axial stress direction after point A. In the A-B stage, the sample undergoes crack propagation and increased damage. Beyond point B, the sample rapidly dissipates a substantial amount of energy and experiences abrupt failure, leading to a sharp reduction in load-bearing capacity.

In the case of l/D = 0.3 (No. 1-3, Figure 9d), during the OB stage, the stress trends and failure modes of sample No. 1-3 resemble those of No. 1-2. The primary failure mode for samples No. 1-2 and No. 1-3 is tensile failure, with rock damage initially occurring at the ends of the jointed regions.

For l/D = 0.3 (No. 1-4, Figure 9e), where the joint completely traverses the red sandstone, its failure mode differs from the other samples. Shear failure predominates in this case, with the sample undergoing sliding shear failure along the separation planes of the jointed and intact sections. Localized shear failure gradually progresses to shear failure across the entire cross-section. After loading the sample to point O, numerous cracks develop and extend within the jointed portion until point B, at which the entire sample experiences sliding.

3.3. *Sub-Instability Stage Zoning Characteristics in Pre-Fabricated Jointed Red Sandstone*

Figure 10 depicts the percentage of the OA segment and AB segment within the sub-instability stage. It is noteworthy that as the joint size increases, the proportion of the static sub-instability stage in relation to the entire sub-instability stage gradually diminishes, while the dynamic sub-instability stage gains prominence. When considering the observed failure modes of jointed red sandstone at different times, it becomes apparent that larger joint sizes lead to a prolonged dissipation of accumulated elastic energy within the red sandstone, resulting in an extended period of rock failure.

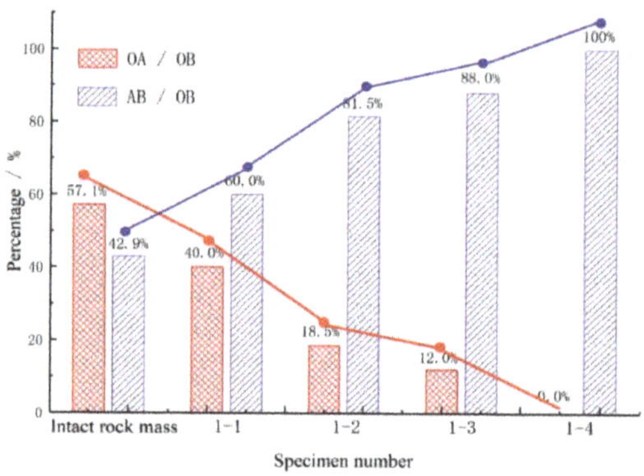

Figure 10. The percentage of OA segment and AB segment in the sub-instability stage.

Upon analyzing data from multiple experimental sets, it becomes evident that jointed rock masses of varying sizes exhibit distinct levels of interaction between the jointed and intact regions during different stress stages in the instability process. The instability process of jointed rock masses can be subdivided into three pivotal stages: the strong departure from the linearity stage (NO), the static sub-stability stage (OA), and the dynamic sub-stability stage (AB). During the strong departure from the linearity stage, the jointed region experiences initial failure, marked by the initiation and propagation of cracks in this area. At this stage, the intact region of the jointed rock mass remains largely undamaged, and the interaction between the jointed and intact regions is relatively weak. The intact region continues to bear the majority of the load. In the sub-instability stage, cracks in the jointed region rapidly extend and interconnect, causing damage to the intact region. During this stage, the interaction between the jointed and intact regions reaches its peak. Upon entering the fully unstable stage, the jointed rock mass experiences complete failure.

3.4. Analysis of Strain Variation in the Intact Region of Rock Mass under the Influence of Joints

(1) Division of Characteristic Strain Change Stages in Jointed Rock Mass. Based on the observed strain change characteristics in Figure 11a,c,e,g, it becomes evident that red sandstone of varying joint sizes exhibits a similar pattern of strain changes in relation to the intact region. These trends can be categorized into three distinct stages:

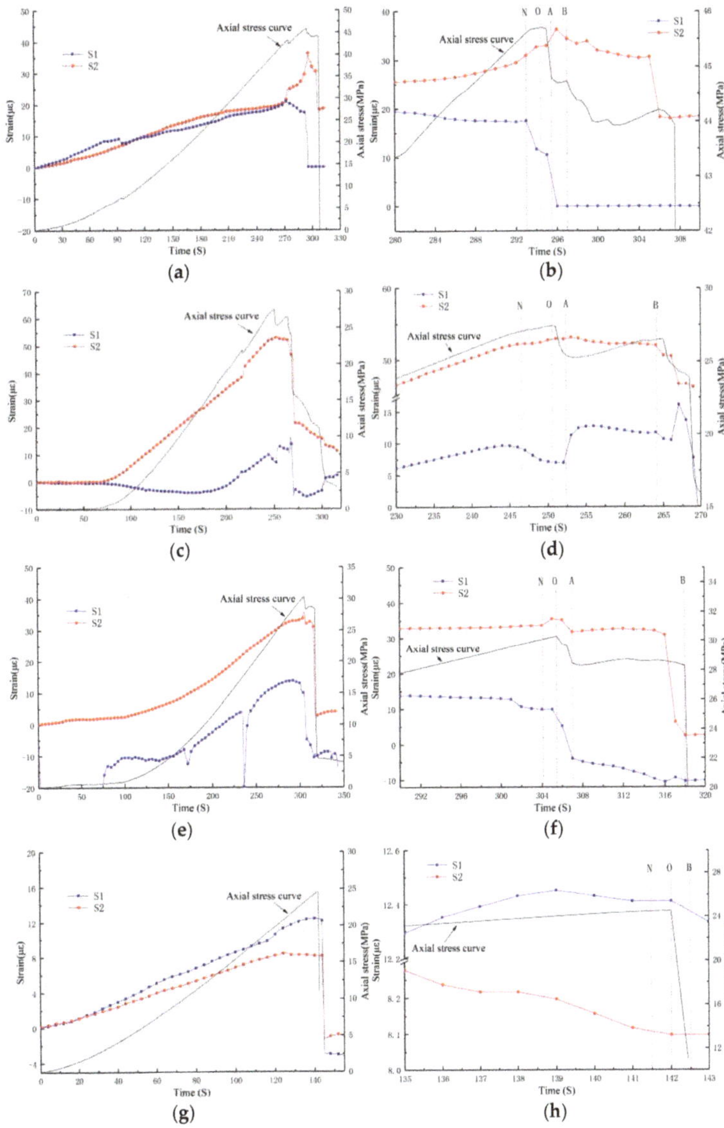

Figure 11. Strain variation characteristics of 11 1-1~1-4 specimens. (**a**) The strain change of 1-1 specimen in the whole process; (**b**) the strain change of 1-1 specimen in sub-instability stage; (**c**) strain changes during the overall process of 1-2 specimens; (**d**) strain variation during sub instability stage of 1-2 specimens; (**e**) strain changes during the overall process of 1-3 specimens; (**f**) strain changes during the sub-instability stage of 1-3 specimens; (**g**) strain changes during the overall process of 1-4 specimens; (**h**) the strain change of 1-4 specimen in the sub-instability stage.

① Compaction stage, elastic stage, and early plastic stage: As the applied stress increases, the strain in the jointed region relative to the intact region steadily rises. This is accompanied by localized fluctuations in axial stress.

② Late plastic stage (strong departure from the linearity stage) to sub-instability stage: During this phase, the strain in the jointed region relative to the intact region transitions from gradual fluctuations to abrupt and intense variations. This leads to significant oscillations in strain monitoring data, marking a notable departure from the previous stable changes. Consequently, this stage plays a critical role in identifying when the rock mass enters a state of complete instability.

③ Instability stage: In this stage, the test specimen has undergone complete failure. The strain in the jointed region relative to the intact region stabilizes at a specific value or reverts to zero, indicating the complete failure of the specimen.

(2) Sub-Instability Stage Strain Transient Characteristics in Jointed Rock Mass

Sample 1-4, influenced by shear failure, did not exhibit a typical sub-instability stage, and its strain changes differed significantly from those of the other specimens.

In Sample 1-1, Strain S1 displayed the following pattern: an accelerated increase during the NO stage (+2 με), followed by relative stability during the OA stage, and a sudden sharp increase followed by a sharp decrease during the AB stage (+5 με to −4 με). On the other hand, Strain S2 showed a sharp decrease during the NO stage (−7 με), followed by relative stability during the OA stage, and a rapid decrease to zero during the AB stage (−10 με), as depicted in Figure 11b. Consequently, during the strong departure from the linearity stage (NO), the jointed rock mass initially emits instability signals, such as a sudden increase or decrease in strain, although the magnitude of these signals is relatively low. In the static sub-instability stage (OA), the jointed rock mass maintains a phase of stability. In the dynamic sub-instability stage (AB), the jointed rock mass releases intense instability signals, with strain changes reaching as high as 50%. This stage serves as a precursor to the complete instability of the jointed rock mass.

In Samples 1-2 and 1-3, Strains S1 and S2 exhibited a similar trend to that of Sample 1-1 during the N-B stage, characterized by significant strain changes in the NO stage and rapid changes in the AB stage. However, the extent of strain variation in Samples 1-2 and 1-3 was lower compared to Sample 1-1. This phenomenon can be attributed to differences in the sizes of the jointed portions. In fixed-sized red sandstone samples, as the proportion of the jointed portion increases, its influence on the intact portion of the red sandstone becomes more pronounced. Under the same external loading conditions, when the proportion of the jointed portion is relatively small (as in Sample 1-1), the strain energy released due to the jointed portion's failure is lower, resulting in a relatively higher accumulation of strain energy in the intact portion of the red sandstone. Consequently, the physical signals released during the instability process (with strain gauges located in the intact portion) become more pronounced. Conversely, when the joint proportion is larger (as shown in Figure 11b), the jointed portion concentrates stress to a greater extent, causing a higher release of strain energy, The specific areas of mutual influence are shown in Figure 12. Consequently, the physical signals released during the instability process in the intact portion become less prominent.

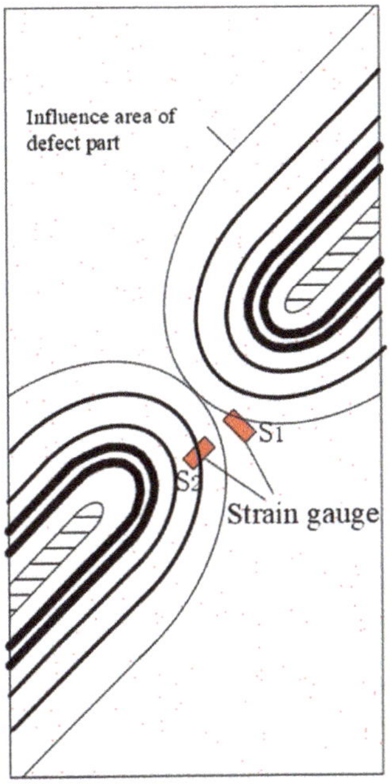

Figure 12. Synergistic influence area.

3.5. Analysis of Acoustic Emission Characteristics during the Instability Process of Jointed Red Sandstone

Figure 13 illustrates the acoustic emission signal characteristics during the sub-instability stage for the four test specimens. From the acoustic emission monitoring data, it is evident that Sample 1-4, which experienced sudden shear instability, did not generate a significant number of AE events during the instability stage. However, at 137 s in Sample 1-4, the sensor detected an amplitude of 97 dB and an energy of 29,000, which corresponds to the moment when an internal shear crack parallel to the joint orientation occurred within the jointed portion, as observed at the test site.

From Figure 13a–f, it can be observed that Samples 1-1, 1-2, and 1-3 did not produce significant acoustic emission signals during the strong departure from the linearity stage (NO). However, during the strong departure from the linearity stage (NO), they began to generate noticeable acoustic emission events, with amplitudes around 85 dB and energy levels below 5000.

During the static sub-instability stage (OA), these specimens consistently produced intense acoustic emission signals. Notably, Sample 1-2 exhibited a peak in acoustic emission signals with an amplitude of up to 97 dB and energy levels as high as 16,000.

In the dynamic sub-instability stage (AB), as the specimens gradually lost stability and cracks initiated and propagated, they produced a continuous stream of acoustic emission signals. Analyzing the acoustic emission characteristics of the jointed rock mass reveals that as it enters the strong departure from the linearity stage and sub-instability stage, it consistently generates acoustic emission signals. The amplitudes and energy levels of these signals initially peak during the loading process. The continuous generation of these acoustic emission signals signifies that the jointed rock mass is approaching failure. Therefore, it can be inferred that the onset of the sub-instability stage marks the beginning of cooperative failure in the jointed rock mass.

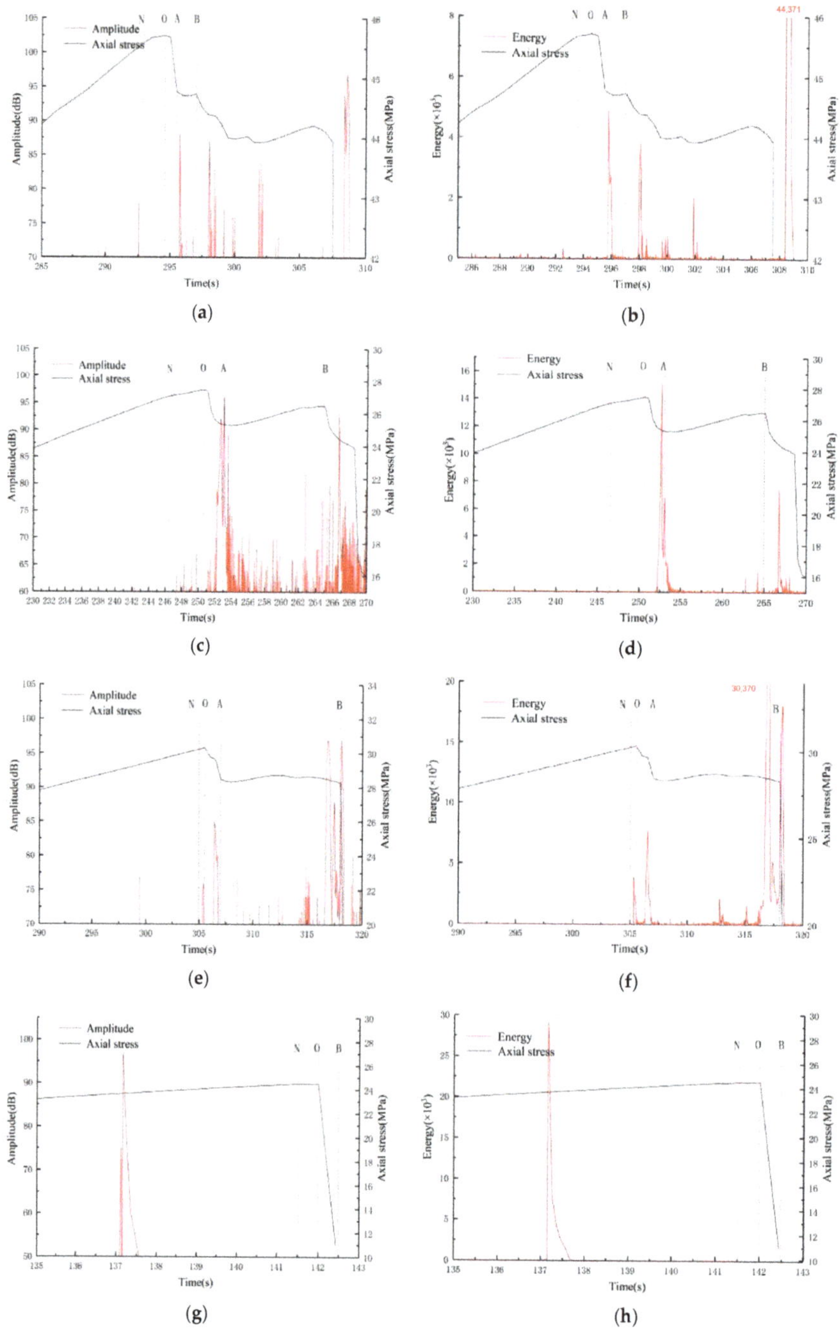

Figure 13. Analysis of acoustic emission results of 1-1~1-4 specimens. (**a**) Time–amplitude–stress of 1-1 specimen; (**b**) time–energy–stress of 1-1 specimen; (**c**) time–amplitude–stress of 1-2 specimen; (**d**) time–energy–stress of 1-2 specimen; (**e**) time–amplitude–stress of 1-3 specimen; (**f**) time–energy–stress of 1-3 specimen; (**g**) time–amplitude–stress of 1-4 specimen; (**h**) time–energy–stress of 1-4 specimen.

4. Detailed Mechanical Response of Jointed Red Sandstone Model Specimens

In the preceding sections, we explored the crack propagation and failure patterns of a set of prefabricated jointed rock specimens through uniaxial compression tests. The mechanical behavior under uniaxial compression primarily involves a mixed tensile–shear failure, so using the PFC2D 5.0 numerical simulation software, a two-dimensional fracture mechanics model is established employing the distinct element method (DEM). This modeling encompasses tensile failure induced by tensile strain, and subsequently, the evolution of cracks under uniaxial stress conditions.

4.1. Validation and Application of Fractals

To demonstrate the crack propagation mechanisms discussed in PFC2D, Sample No. 1-3 was modeled as an example. In this case, crack initiation and propagation were tested through uniaxial compression experiments. The geometric structure of the rock model matched that of Sample No. 1-3 in the laboratory experiments, with vertical and horizontal stresses of approximately 20 MPa and 0 MPa, respectively. Based on experimental results with intact red sandstone, the rock had a strength of UCS = 50.7 MPa; cohesion c = 3.6 MPa, internal friction angle, tensile strength = 4.0 MPa; Young's modulus E = 3.0 GPa; and Poisson's ratio ν = 0.24. Under basic conditions, initial and residual crack apertures were assumed to be 10, with crack shear and normal stiffness at 100 GPa/m. The simulated rock crack propagation of Sample No. 1-3 under uniaxial compression is depicted in Figure 14. During the simulation loading process, small cracks were initially observed at the tip of the end face and propagated in the vertical direction. In the later stages of the simulation, multiple sets of cracks appeared in the rock bridge, and the merging and penetration of these cracks resulted in the failure of rock bridge strength. Based on the numerical simulation results, it can be concluded that crack initiation and propagation are similar to the laboratory test results.

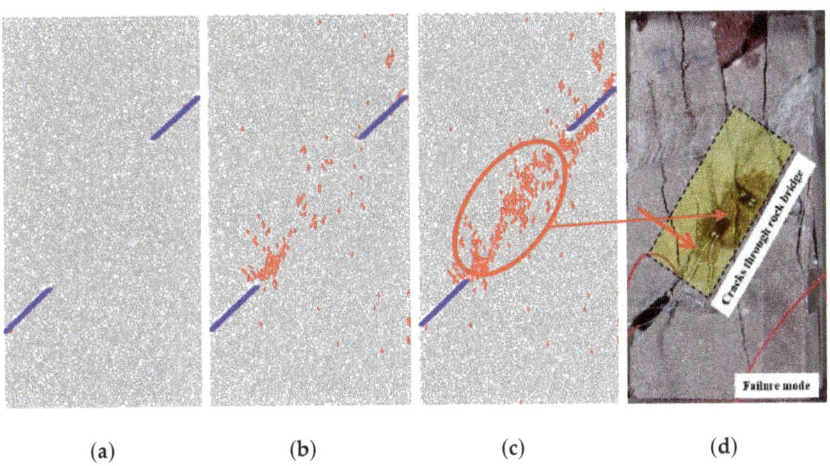

Figure 14. Validation model of rock crack propagation under uniaxial compression. (**a**) Initial stage of simulation; (**b**) mid-simulation; (**c**) the end of simulation; (**d**) laboratory test.

As shown in Figure 15, based on the computed results, the peak strength data closely match the laboratory results, and the stress–strain curve trends are similar to each other (Figure 15a). In this section, the failure mode of the jointed rock mass is analyzed using the safety factor (Fos). From Figure 15b, we can clearly see that the numerical model effectively simulates the instability and failure process of the rock structure. Under the influence of shear stress, damage to the structure primarily concentrates around the line connecting the two pre-existing fractures, propagating from the crack tips towards the center of the model.

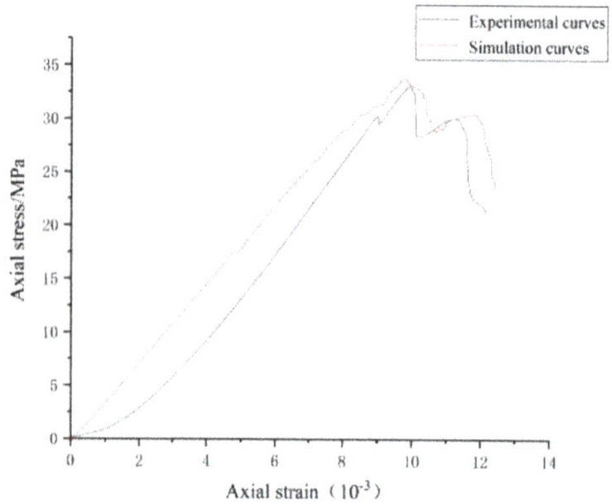

(a)

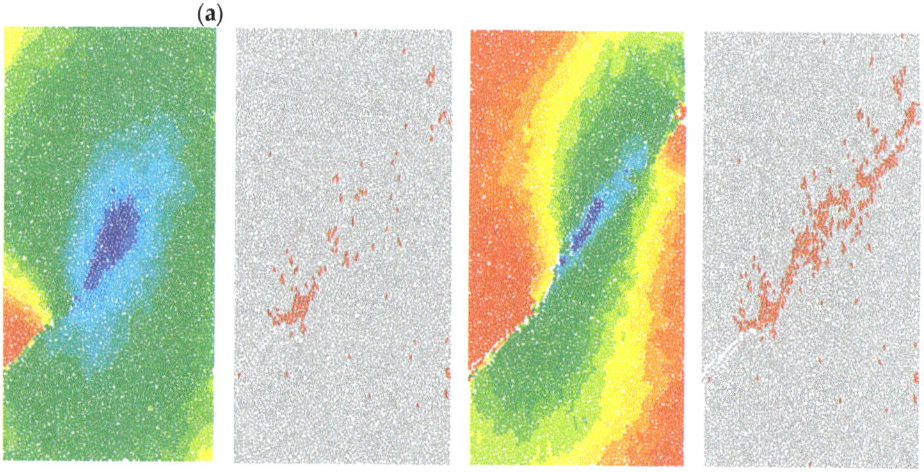

(b)

Figure 15. Verification of uniaxial compression test. (**a**) Axial stress–strain curves in step-by-step calculations and laboratory tests; (**b**) the results of displacement and crack propagation in numerical simulation.

4.2. Modeling and Simulation Results

In the numerical simulation, the specimen model has dimensions of 50 mm in length and 100 mm in height. The pre-existing fractures in the model are inclined at an angle of 45 degrees, similar to those in Sample No. 1-3. The aperture of these pre-existing fractures measures 5 mm. We have outlined three distinct phases corresponding to different stages, which can be roughly categorized based on the impact of uniaxial loading-induced vertical stress on this jointed rock mass.

Initial loading phase (strong departure from the linearity stage): During this phase, the particle displacement direction at the ends is nearly parallel to the loading direction, influenced by the vertical stresses applied at both ends. The particle displacement direction gradually shifts towards both sides of the model's long axis, with larger angular deviations observed closer to the short axis. At the short axis, particle displacement direction is nearly perpendicular to the model's long axis (Figure 16-NO).

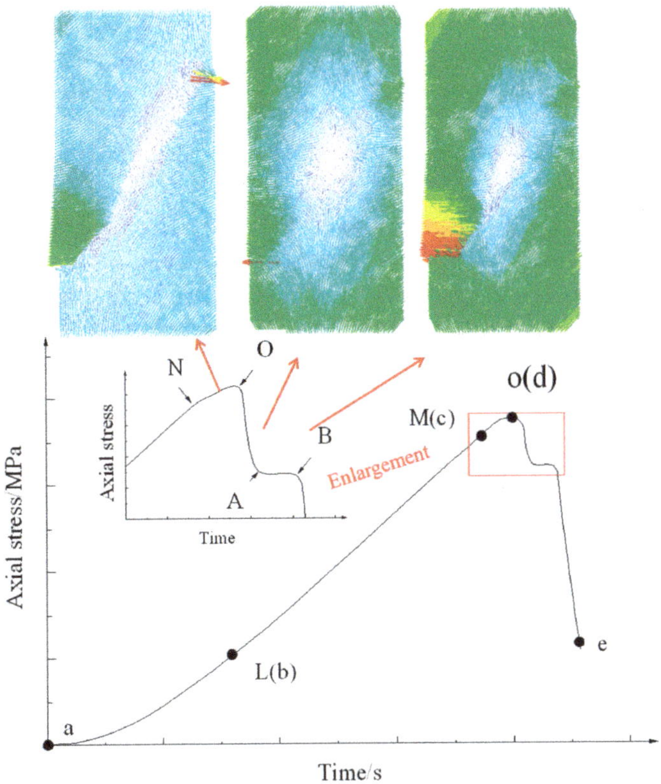

Figure 16. Particle displacement pattern of the model.

Mid-loading phase (sub-instability stage): In this stage, influenced by the pre-existing fractures, there is a sudden change in particle displacement direction near these fractures. At the locations of these fractures, the particle displacement direction is almost perpendicular to the model's long axis. This phenomenon occurs due to the opposite displacement directions on the upper and lower surfaces of the pre-existing fractures, resulting in a 45-degree angle between these surfaces and the loading direction. Consequently, the resultant force is perpendicular to the loading direction. During this phase, differences in particle displacement around the tips of the pre-existing fractures become apparent, with significant variations in particle displacement magnitude and opposing directions on either side of the fracture lines. The model begins to exhibit shear failure tendencies (Figure 16-OA).

Late-loading phase (instability stage): At this stage, particle displacement directions on either side of the connecting lines between the pre-existing fractures are distinctly opposite, resembling a normal fault. Particles on the upper side move downward, while those on the lower side move upward, with particle displacement directions essentially parallel to the direction of the pre-existing fractures. During this phase, the primary mode of failure in the model is shear failure (Figure 16-AB).

Figure 17 depicts the force chains and crack distribution throughout the loading process of the layered rock. Darker colors and wider lines indicate higher stress concentrations. This process can be roughly categorized into three stages:

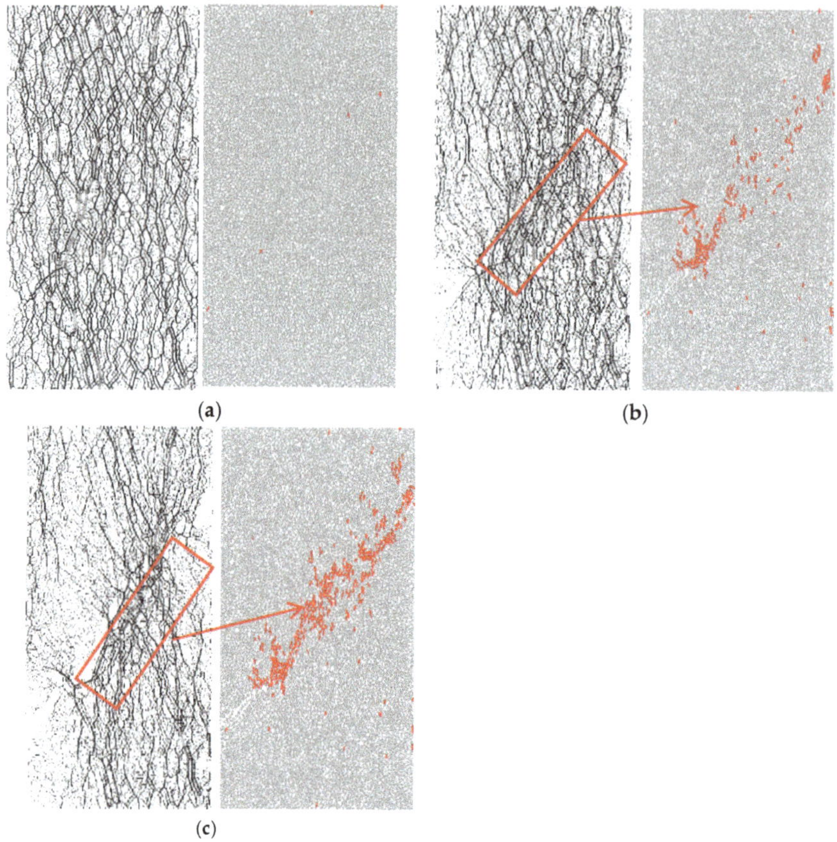

Figure 17. Distribution of force chain and crack in loading process of jointed rock. (**a**) initial loading; (**b**) mid loading; (**c**) end of loading.

Initial loading stage: During this phase, the force chains are distributed relatively evenly within the model.

Mid-loading stage: As pre-existing cracks come into play, stress concentration notably diminishes within the region covered by the horizontal projection of these cracks. Stress is redirected along the model's longitudinal axis towards the interior. Concurrently, stress concentration increases in areas not covered by the horizontal projection of the pre-existing cracks, with the most significant stress concentration occurring around the connecting lines of these two pre-existing cracks.

Late-loading stage: During this phase, the force chains around the connecting lines of the two pre-existing cracks rupture, ultimately leading to the model's destruction. It becomes apparent that there are tensile stresses in the vertical direction and tensile failure cracks surrounding them. Consequently, it can be inferred that the model's failure is predominantly a result of a mixed mode involving both shear and tensile failure.

5. Conclusions

Through indoor rock mechanics experiments, we have explored the interplay between robust and weaker segments within jointed rock masses throughout the failure process. By gathering an extensive array of physical field data, encompassing stress, strain, and acoustic emissions, we have undertaken a comprehensive analysis of the transformation of jointed rock masses from individual behavior to collective failure just before reaching instability.

(1) Under the influence of stress, the primary load-bearing components within the jointed rock mass are the comparatively sturdy ones. Initially, the less robust segments within the rock mass undergo weakening and evolve into regions of strain release. As these points of strain release and interconnect and the release areas expand, the overall strain accumulation intensifies due to interactions, eventually leading to the rapid failure zones.

(2) During the strongly non-linear phase, jointed rock masses emit initial indicators of instability, such as abrupt increases or decreases in strain (or continuous acoustic emission signals). However, the extent of signals released during this phase is relatively limited. In the sub-instability phase, the rock mass emits pronounced signals of instability, with strain changes reaching levels of up to 50% (with amplitude and energy reaching their first peak). The continuous generation of physical signals during this phase signifies that the jointed rock mass is on the brink of failure. Therefore, it can be inferred that the entry into the sub-instability phase marks the onset of cooperative failure.

(3) The entry into the sub-instability phase symbolizes the initiation of cooperative failure. During the strongly non-linear phase, isolated regions of strain accumulation and release points initially emerge within the weaker segments and influence the interfaces between the jointed and intact regions. In the static sub-instability phase, interactions commence, and the regions of strain accumulation within the jointed areas multiply and rapidly expand, impacting the accumulation of strain in the intact regions. As we progress into the dynamic sub-instability phase, there is mutual connection and interaction between the strain release points in the jointed and intact regions, resulting in an intensified cooperative process that reaches its peak intensity. Subsequently, the rock mass undergoes a transition into complete instability.

Author Contributions: Conceptualization, B.Z. and S.Z.; methodology, B.S.; software, X.H. and B.S.; validation, S.S.; formal analysis, Y.L.; investigation, B.Z.; resources, B.Z.; data curation, X.H. and Y.L.; writing—original draft preparation, B.Z.; writing—review and editing, B.Z.; visualization, S.Z. and S.S.; supervision, B.S.; project administration, S.Z.; funding acquisition, S.Z. All authors have read and agreed to the published version of the manuscript.

Funding: This research was funded by the Open Fund of State Key Laboratory of Water Resource Protection and Utilization in Coal Mining (No. GJNY-20-113-19), National Natural Science Foundation of China (No. 52004147, 51974173).

Institutional Review Board Statement: Not applicable.

Informed Consent Statement: Not applicable.

Data Availability Statement: The original contributions presented in the study are included in the article, further inquiries can be directed to the corresponding author.

Conflicts of Interest: The authors declare no conflicts of interest.

References

1. Jiao, F.; Xu, J.; Guo, B.; Peng, S.; Yan, F.; Chen, Y. Shear test on the influence of filling thickness on the shear strength of rock joints. *J. Min. Saf. Eng.* **2022**, *39*, 405–412.
2. Indraratna, B.; Premadasa, W.; Brown, E.T.; Gens, A.; Heitor, A. Shear strength of rock joints influenced by compacted infill. *Int. J. Rock Mech. Min.* **2014**, *70*, 296–307. [CrossRef]
3. She, C.X.; Sun, F.T. Study of the Peak Shear Strength of a Cement-Filled Hard Rock Joint. *Rock Mech. Rock Eng.* **2018**, *51*, 713–728. [CrossRef]
4. Xiao, W.; Yu, H.; Zhu, Z.; Li, Y.; Liu, W. Experimental study on the dilatancy characteristics of thin-bedded filling rock joints. *Chin. J. Geotech. Eng.* **2020**, *42*, 1499–1508.
5. Chai, S.; Wang, H.; Jing, Y.; Jia, N. Experimental study on dynamic compression characteristics of cumulative damage of filled jointed rock. *Chin. J. Rock Mech. Eng.* **2020**, *39*, 2025–2037.
6. Wang, G.; Cao, T.; Wen, X.; Sun, F.; Zhang, L. Evolution law of pre-peak energy self-inhibition of single-joint sandstone under uniaxial compression. *J. China Coal Soc.* **2021**, *46* (Suppl. S1), 211–221.

7. Ma, Q.; Su, Q.; Ma, D.; Yuan, P. Experimental study on SHPB dynamic mechanical failure characteristics of sandstone in deep roadway with different joint dip angles. *Chin. J. Rock Mech. Eng.* **2020**, *39*, 1104–1116.
8. Jin, A.; Lu, T.; Wang, B.; Chen, S.; Zhang, J. Experimental study on dip angle characteristics of key joints in rock mass based on improved mechanical equivalence. *Chin. J. Rock Mech. Eng.* **2023**, *42*, 76–87.
9. Liu, T.; Yang, R.; Ding, L.; Li, X.; Zeng, L. Study on mechanical properties and meso-cracking mechanism of granite with non-interpenetrated joints. *Chin. J. Rock Mech. Eng.* **2023**, *42*, 1070–1082.
10. Deng, Z.; Wu, J.; Shang, J.; Xie, L. The equivalent elastic model and strength characteristics of rock mass with transfixion-non-transfixion cross joints. *J. China Coal Soc.* **2018**, *43*, 3098–3106.
11. Li, S.; Lin, H.; Lin, Q.; Wang, Y.; Zhao, Y.; Hu, H. Mechanical behavior and failure characteristics of double-layer composite rock-like specimens with coplanar double joints under uniaxial loading. *Trans. Nonferrous Met. Soc. China* **2023**, *33*, 2815–2831. [CrossRef]
12. Zhao, Y.; Liu, J.; Jin, A.; Sun, H.; Wang, B.; Wei, Y. Study on failure characteristics of non-interpenetrated jointed rock mass under loading and unloading conditions. *J. Cent. South Univ. (Nat. Sci. Ed.)* **2020**, *51*, 1893–1901.
13. Wang, G.; Zhang, L.; Xu, M.; Liang, Z.; Ran, L. Study on energy evolution mechanism of damage and failure of non-interpenetrated jointed rock mass under uniaxial compression. *Chin. J. Geotech. Eng.* **2019**, *41*, 639–647.
14. Wang, Q.; Xia, K.; Wu, B.; Xu, Y.; Liu, F. Experimental study on dynamic failure of prefabricated parallel double jointed rock-like material plate. *J. Tianjin Univ. (Nat. Sci. Eng.)* **2019**, *52*, 1099–1108.
15. Zare, S.; Karimi-Nasab, S.; Jalalifar, H. Analysis and determination of the behavioral mechanism of rock bridges using experimental and numerical modeling of non-persistent rock joints. *Int. J. Rock Mech. Min.* **2021**, *141*, 104714. [CrossRef]
16. Wang, J.; Li, J.; Zhou, K.; Jiang, C.; Shen, Y.; Jia, H. Fracture mechanical properties of sandstone with pre-fabricated cracks under freeze-thaw cycles. *Theor. Appl. Fract. Mech.* **2024**, *131*, 104444. [CrossRef]
17. Yin, T.; Yin, J.; Wu, Y.; Yang, Z.; Liu, X.; Zhuang, D. Water saturation effects on the mechanical characteristics and fracture evolution of sandstone containing pre-existing flaws. *Theor. Appl. Fract. Mech.* **2022**, *122*, 103605. [CrossRef]
18. Yuan, H.; Xiao, T.; She, H.; Huang, M. Crack propagation law of rock with single fissure based on PFC2D. *Front. Earth Sci.* **2023**, *10*, 977054. [CrossRef]
19. Jin, M.A.; Guo, Y.S. Accelerated synergism prior to fault instability: Evidence from laboratory experiments and an earthquake case. *Seismol. Geol.* **2014**, *36*, 547–561.
20. Jin, M.A.; Sherman, S.I.; Guo, Y.S. Identification of meta-instable stress state based on experimental study of evolution of the temperature field during stick-slip instability on a 5° bending fault. *Sci. China Earth Sci.* **2012**, *55*, 869–881. [CrossRef]

applied sciences

MDPI

Article

Mechanical Responses and Fracture Evolution of Marble Samples Containing Stepped Fissures under Increasing-Amplitude Cyclic Loading

Yongchun Yu [1], Yu Wang [2,*], Xuefeng Yi [2] and Zhenzhen Chen [2]

1 Shangri-La Yunkuang Hongniu Mining Co., Ltd., Shangri-La 674400, China; yychun_ykhm@126.com
2 Department of Civil Engineering, School of Civil & Resource Engineering, University of Science & Technology Beijing, Beijing 100083, China; yxfeng_ustb@sina.com (X.Y.); chenzz_ustb@126.com (Z.C.)
* Correspondence: wyzhou@ustb.edu.cn

Abstract: This work aims to reveal the effect of rock bridge length (RBL), i.e., 10, 20, 30, or 40 mm, on the fatigue mechanical responses and fracture evolution of marble samples containing stepped fissures under multilevel cyclic loading paths. Comprehensive investigations were conducted on fatigue strength, deformation, damping evolution, and damage propagation. The test results demonstrate that fatigue strength, volumetric deformation, and fatigue lifetime increase as rock bridge length increases. The energy dissipation reflected by the damping ratio indicates that much energy is consumed to drive crack propagation, especially for rock with larger rock bridge segments at the final cyclic loading stage (CLS). An index of strain incremental rate is proposed to predict rock failure development. It is found that volumetric strain rate is a better early warning sign than axial strain rate. Warning time decreases with increasing rock bridge length; it is suggested that rock with large segments has good ability to resist external fatigue loading.

Keywords: volumetric deformation; rock bridge segment; fracture evolution; damage modelling; strain incremental rate

1. Introduction

As a typical geomaterial, rock mass is characterized by non-homogeneity, discontinuity, and anisotropy. Natural flaws, e.g., joints, weak planes, faults, cleavage, schistosity, etc., lead to non-linear geomechanical behaviors in naturally fractured rock mass [1–5]. The long-term stability of rock mass is obviously influenced by these discontinuities; notably, the structural deterioration of rock mass is accelerated when exposed to cyclic or fatigue loading [6–9]. Therefore, investigations into the mechanical responses of flawed rock mass are crucial to maintain the operation of underground rock infrastructures.

Laboratory mechanical tests like uniaxial compression or triaxial compression are often used to study the cyclic or fatigue mechanical properties of rock. These properties are affected not only by the geometric structure of flaws, but also by several loading parameters such as amplitude, frequency, mean stress, and the number of fatigue cycles [10–13]. Fatigue loading tests are categorized into constant-amplitude and variable-amplitude tests. Previous studies have primarily explored the influence of loading parameters—such as waveform, maximum stress, mean stress, amplitude, frequency, and cycle number—on the strength and deformation characteristics of rocks. The sinusoidal signal is widely acknowledged as closely resembling the dynamic waveforms generated by blast vibrations, earthquakes, or rock bursts. Therefore, sinusoidal stress waveforms are commonly employed in fatigue mechanical tests [14,15]. Variable-amplitude cyclic loading, involving multiple levels, is considered more representative of the actual stress conditions in engineered rock masses than constant-amplitude loading tests. Field monitoring data also confirm that rock masses experience multi-level fatigue loading [16–18]. In underground rock engineering, blasting- or mining-induced earthquakes impose non-constant-amplitude fatigue loading on the surrounding rock structures. Numerous studies have investigated intact or flawed rocks subjected to variable-amplitude or variable-frequency fatigue loading, revealing insights into rock deformation, damage propagation, and failure patterns from both mesoscopic and macroscopic perspectives [11,19,20]. Nevertheless, the impact of rock bridge segments on rock fatigue response remains uncertain. Non-connected rock bridge segments within the rock mass serves as high-energy storage zones that impede fissure connection and associated instability events.

In summary, there is a paucity of studies investigating the impact of rock bridge length (RBL) on volumetric deformation, damping characteristics, and damage evolution. Therefore, this study conducted a series of increasing-amplitude cyclic loading tests on marble samples with RBLs of 10 mm, 20 mm, 30 mm, and 40 mm. Fatigue mechanical properties, including damping effects, fatigue deformation, strain incremental rate, damage evolution, and fracture development, were systematically analyzed. The findings enhance the understanding of how rock bridge segments influence the fracture evolution of flawed rock and offer new insights into their role in resisting external fatigue disturbances.

2. Materials and Methods

2.1. Rock Sample Preparation

The rock material was obtained from a roadway of a lead–zinc polymetallic mine in Baoshan city, Yunnan Province, China. The lead–zinc polymetallic deposit, located in the northern part of Baoshan in the Sanjiang region, is a skarn-type polymetallic deposit hosted within Early Paleozoic carbonate rocks. The geographical coordinates are approximately 99°10′12″–99°11′31″ E and 25°33′46″–25°24′51″ N, with the mining area spanning 3.005 km^2, situated directly north of the Baoshan urban area at a straight-line distance of 40 km. The Baoshan block is delineated to the west by the Lushi suture zone, to the east by the Changning–Menglian suture zone, to the north by the intersection of the Nujiang–Ruili fault and the Lancangjiang fault, and to the south by the Dian–Burma–Thai–Malay block. Overall, it exhibits a wedge-shaped configuration that is narrower in the north and broader in the south, interposed between the Tengchong–Ruili block and the Lanping–Simao basin [21,22]. The block has undergone a geological history characterized by the formation, evolution, and closure of the Tethys Ocean, followed by the continental–continental collision orogeny between the Eurasian and Indian plates. This unique and complex tectonic environment, coupled with favorable metallogenic geological conditions, has created an advantageous setting for the concentration and mineralization of various useful metallic elements [23].

During the field investigation, as shown in Figure 1a, multiple intermittent joints were observed on the rock mass surface. Due to blast vibrations and excavation disturbance, it was found that the discontinuous joints were connected, increasing the risk of geo-disaster from spalling and collapse occurring at the roadway. The connection of the intermittent joints poses a huge threat to the stability of the roadway, and could affect normal mine production. As a result, the influence of intermittent joints on rock fracture exposed to dynamic loading needs to be thoroughly investigated. At the laboratory scale, marble samples were prepared with double-stepped fissures using a water jet (Figure 1b) to mimic the rock bridge structure. The water jet had a weak disturbance effect on the intact marble samples, ensuring the preparation quality of the tested marble samples. The combination of high-pressure water and garnet abrasive, emitted from a nozzle with a diameter of 0.75 mm, can create a defect with an aperture measuring 1 mm. The inclination of the fissures was determined from the field measurement of rock joints, and the fissure angle (α) was set to 30°. The rock bridge approach angle (β) was set to 75°. Four kinds of marble samples with different rock bridge lengths were prepared, i.e., 10 mm, 20 mm, 30 mm, and 40 mm. Figure 1c shows typical marble samples containing double-stepped fissures with different rock bridge lengths of 10, 20, 30, and 40 mm.

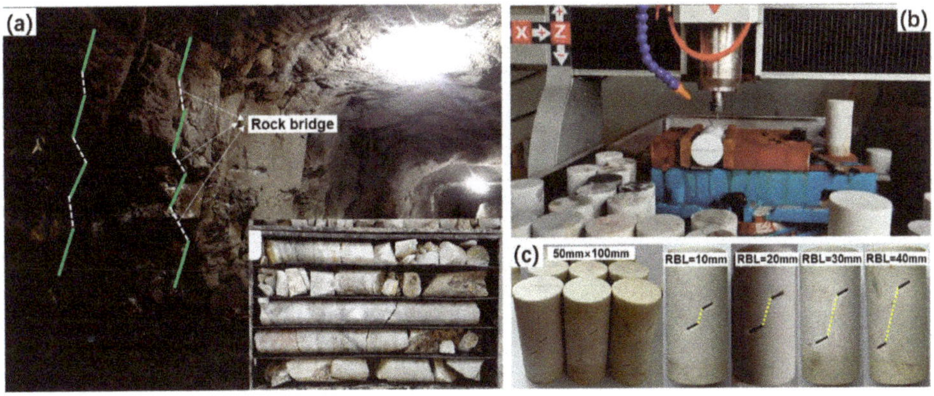

Figure 1. Rock structure characteristics of the lead–zinc mine. (**a**) Rock mass in a roadway. (**b**) Fissure prefabrication in intact marble sample. (**c**) Typical marble samples with two fissures of different rock bridge lengths.

2.2. Experimental Setup

A servo-controlled fatigue mechanical apparatus (RTFS 1000) was employed to carry out the uniaxial increasing-amplitude cyclic loading test, as shown in Figure 2. The host frame has a stiffness of 8 GN/m and the maximum load capacity is 1000 kN. The testing machine has the capacity to apply complicated disturbance loading, e.g., sinusoidal, square, and triangular cyclic loading, on rock or rock-like specimens. The testing machine can achieve a dynamic loading frequency ranging from 0 to 10 Hz, with a maximum loading amplitude of 1.5 mm. Concurrently, a central computer was used to monitor the deformation and strength parameters, including axial strain, axial stress, and transverse strain, at the same sampling frequency. Axial strain is measured using two grating transducers accurate to 0.1 μm. Radial deformation is measured by a linear variable differential transformer (LVDT), positioned at the marble specimen's center with a precision of 0.1 μm.

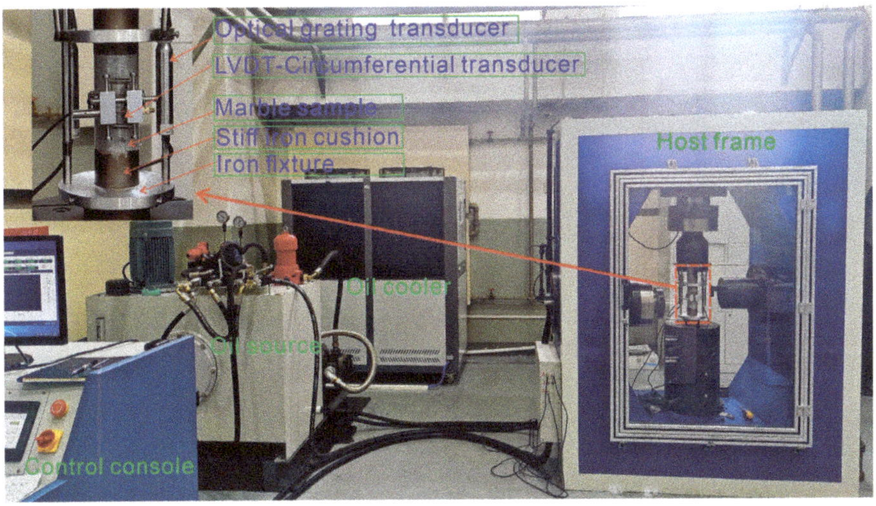

Figure 2. The apparatus for the rock samples containing stepped fissures. Axial and radial deformation were measured using optical grating and LVDT installed at the longitudinal and circumferential segments, respectively.

2.3. Testing Procedures

In this study, fatigue loading tests with multiple levels were conducted on marble samples containing double fissures. Initially, each sample was loaded to a stress of 10 MPa at a constant displacement rate of 0.06 mm/min (equivalent to 1.0×10^{-5} s^{-1}). Subsequently, cyclic dynamic loading was applied based on the stress disturbance characteristics obtained from blast vibration measurements, utilizing a dynamic loading frequency of 0.5 Hz. This resulted in each cycle of loading and unloading being completed within 2 s.

During the fatigue loading process, the first cyclic dynamic loading stage involved applying a stress-controlled sinusoidal cyclic load with a stress amplitude of 5 MPa. In each subsequent cyclic loading stage (CLS), the stress amplitude was incremented by 5 MPa, also under sinusoidal cyclic loading control mode. This cycling of stress continued until the sample eventually failed. For each dynamic cyclic loading stage, the granite sample underwent 30 stress cycles per loading level. The testing protocol for marble is detailed in Table 1, where σ_{min} is the fixed minimum cyclic stress (10 MPa), ${\sigma_{max}}^i$ is the maximum cyclic stress of the first CLS, σ_a is the increasing stress amplitude between two consecutive CLSs, ${\sigma_{max}}^f$ is the ultimate failure stress corresponding to rock failure, CLS is the cyclic loading stage, and CDF is the cyclic loading frequency. The differences in rock bridge length and rock failure at different CLSs and fatigue strengths are shown in Table 1 and Figure 3.

Table 1. The loading schemes for marble samples exposed to cyclic loads in this work.

Specimen ID	σ_{min} (MPa)	${\sigma_{max}}^i$ (MPa)	σ_a (MPa)	${\sigma_{max}}^f$ (MPa)	CLS (/)	Cycles in Each CLS (Cycle)	CDF (Hz)
RTSF -1,2,3	10	15	5	55	9	30	0.5
RTSF -4,5,6	10	15	5	65	11	30	0.5
RTSF -7,8,9	10	15	5	70	12	30	0.5
RTSF -10,11,12	10	15	5	80	14	30	0.5

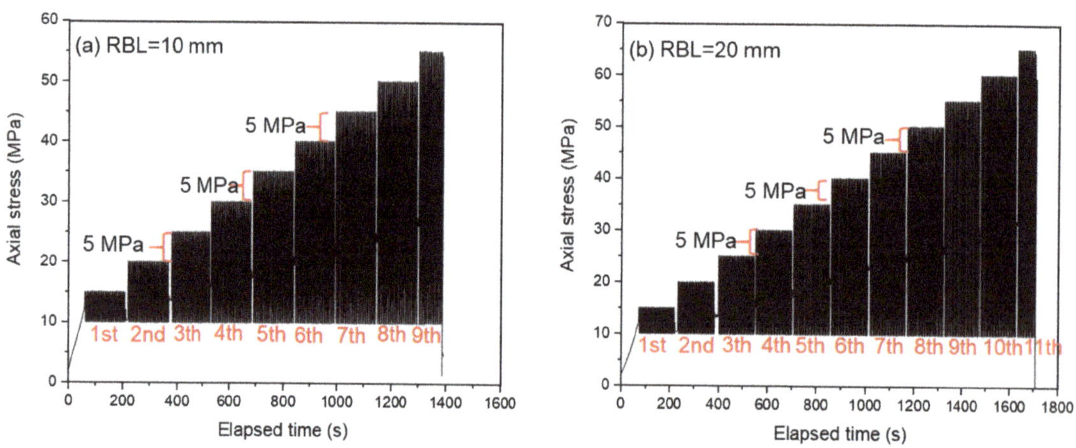

Figure 3. *Cont.*

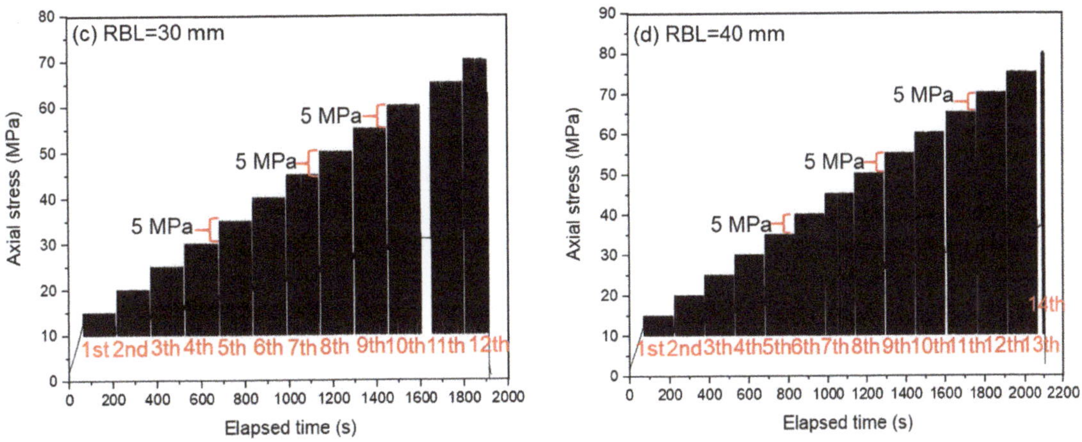

Figure 3. Plots of axial stress (σ) and elapsed time (s) for rock containing stepped fissures with rock bridge lengths of 10, 20, 30, and 40 mm.

3. Testing Results

3.1. Typical Stress–Strain Responses

As depicted in Figure 4a–d, the samples were subjected to uniaxial increasing-amplitude fatigue loading paths. Influenced by rock bridge length, the marble samples exhibit varying fatigue loading stages and cycles before failure. The fatigue lifetime increases with longer rock bridge lengths, measuring 258, 316, 353, and 394 cycles, respectively, for samples with rock bridge lengths of 10 mm, 20 mm, 30 mm, and 40 mm. Irreversible plastic deformation causes the loading curve to deviate from the unloading curve, resulting in the formation of a hysteresis loop whose shape changes over time. After measuring axial strain (ε_a) and lateral strain (ε_r), volumetric strain (ε_v) can be calculated using the formula $\varepsilon_v = \varepsilon_a - 2\varepsilon_r$.

Volumetric strain in rock provides a more comprehensive reflection of its deformation characteristics by integrating both axial and lateral deformations. The shape of the hysteresis loop on the volumetric stress–strain curve varies across different stages of fatigue loading. Changes in the sparse and dense characteristics of volumetric strain exhibit distinct trends. The shape of the hysteresis loop is intricately linked to the behaviors of crack initiation, propagation, and coalescence. Before reaching the last fatigue loading stage, the hysteresis loop consistently shifts from sparse to dense at the onset of each loading stage. This shift indicates greater plastic deformation occurring with each increase in stress amplitude. However, during the final loading stage, the hysteresis loop gradually becomes sparser until the granite samples eventually fail. This progression suggests that significantly more energy is expended to form the crack coalescence pattern, resulting in greater energy dissipation during the last cyclic stage compared to the preceding loading stages.

3.2. Deformation Characteristics Analysis

It can be seen that the stress–strain curves are influenced by rock bridge length. Looking beyond the stress–strain curves, the relationship between strain and cycle number at various CLSs is revealed, as plotted in Figure 5. The existence of fissures and the initiation, propagation, and connection of the rock bridge have a great influence on strain development. The evolution curves of axial, radial, and volumetric strain present fluctuating trends. In Figure 5a–d, the graphs depict the maximum axial strain per cycle versus cycle number. It is observed that axial strain increases rapidly during the initial loading stages of each cyclic loading sequence (CLS), after which it stabilizes. However, as the CLS approaches failure, the axial strain increases rapidly once again. This pattern is consistent across all tested rocks with different rock bridge lengths (RBLs), showing a similar evolution of axial strain where the incremental rate accelerates continuously until rock failure. The evolution of radial strain with cycle number, as illustrated in Figure 5e–h, shows an initial slow increase followed by a rapid pattern as the cycle number increases. The measured radial strain indicates expansion deformation in the middle part. Radial expansion reflects damage evolution, crack propagation, and coalescence. Influenced by rock bridge length (RBL), radial expansion is significantly affected by the presence of rock bridge segments, with spalling and the collapse of voids enhancing radial deformation development. The progression of radial strain accelerates and reaches its maximum in the final few cyclic loading stages. The evolution of volumetric strain with cycle number is depicted in Figure 5i–l. Volumetric strain serves as a reliable predictor of rock sample fracture and instability. The pattern of the volumetric strain curve closely resembles that of the radial strain curve, suggesting that radial deformation controls volumetric strain and significantly influences overall rock deformation. The curve illustrates a transition in rock behavior from compression-dominated to expansion-dominated deformation. Inflection points on the volumetric strain curve signify unstable crack propagation. Specifically, these inflection points occur at the first, second, third, and fifth CLSs for rock samples with RBLs of 10 mm, 20 mm, 30 mm, and 40 mm, respectively. These points mark the critical stages where crack propagation becomes unstable.

3.3. Dynamic Parameter Analysis

In Figure 5, dynamic elastic modulus (E_d) and damping ratio (D_r) are utilized to characterize the dynamic properties of the cyclic stress hysteresis curve for the marble samples across every fatigue loading stage [24,25]. These parameters offer insights into how the material behaves dynamically under cyclic loading conditions, providing information on its stiffness (E_d) and energy-dissipation capability (D_r). The definition of these two parameters can be found in the literature (Wang et al. [26]). It can be seen that D_r displays different patterns for the tested marble samples, and the curve shape is affected by RBL (Figure 6a–d). This result indicates that accumulative damage and crack propagation behaviors are influenced by rock bridge length; moreover, the structural deterioration during fatigue loading is influenced by rock bridge length. For the sample with an RBL of 10 mm, the Dr shows a pattern of first increasing then decreasing, and then increasing again. For the sample with an RBL of 20 mm, it first decreases and then increases until failure. For the sample with an RBL of 30 mm, it increases slowly at the first several CLSs and then increases quickly until failure. For the sample with an RBL of 40 mm, it first decreases and then increases until failure. The differential change trend of D_r reflects the fracture mechanism, and it is influenced by the rock bridge between the two fissures. In the fatigue loading stage, D_r presents a fluctuation trend, indicating the propagation and interlocking of cracks.

Concerning the evolution of E_d, Figure 7a–d present a decreasing trend with CLS, indicating that increasing load cycles leads to a decrease in E_d. However, E_d first presents an increasing and then deceasing trend at different CLSs. The analysis of dynamic elastic modulus (E_d) confirms the earlier findings from the cyclic stress–strain curves, demonstrating that greater plastic strain leads to a larger hysteresis loop and a reduced dynamic elastic modulus (E_d). However, Ed exhibits a sharp decline towards sample failure across all tested samples.

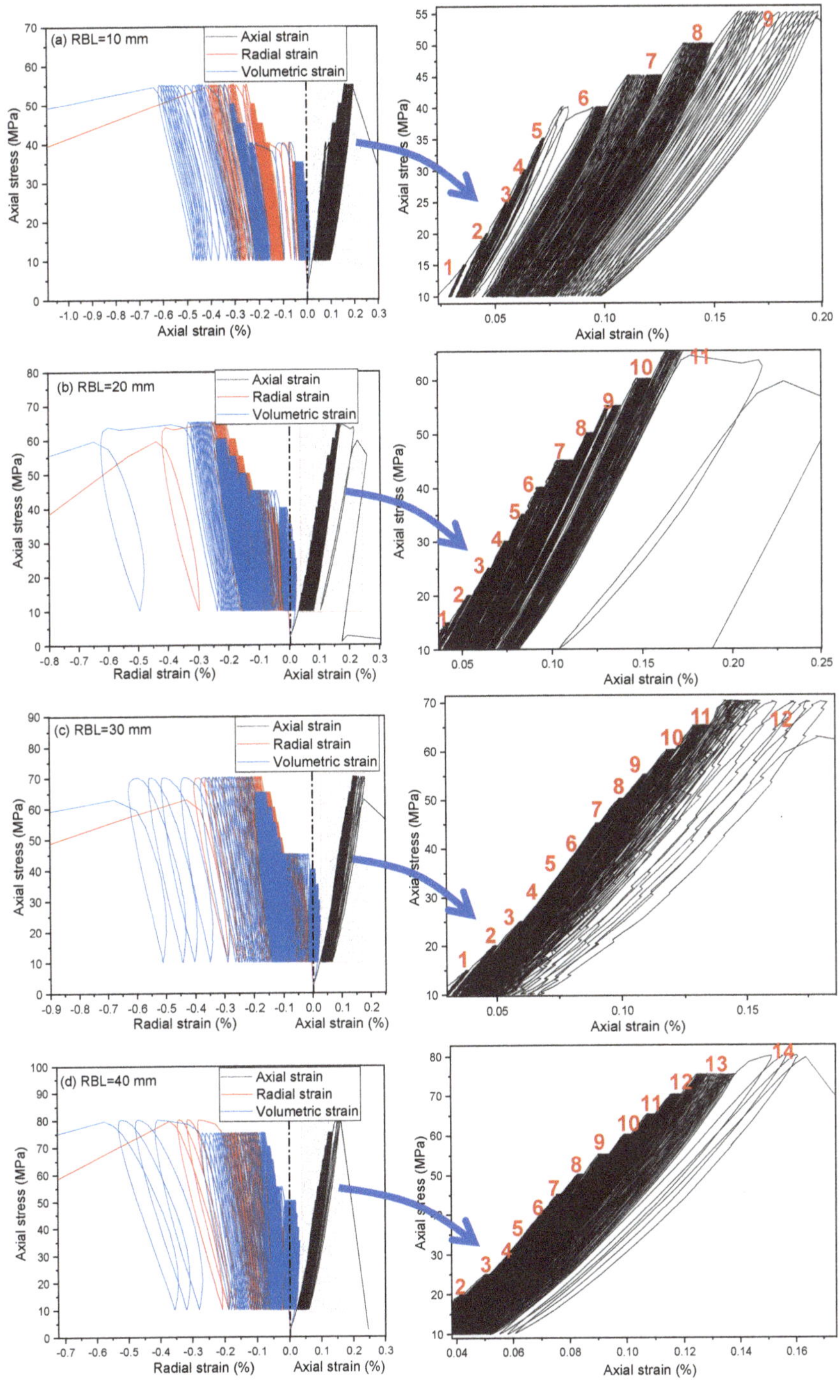

Figure 4. Axial stress versus strain for marble specimens subjected to cyclic loading. (**a**–**d**) Rock bridge lengths are 10 mm, 20 mm, 30 mm, and 40 mm, respectively. The numbers in the figures indicate the cyclic loading stage.

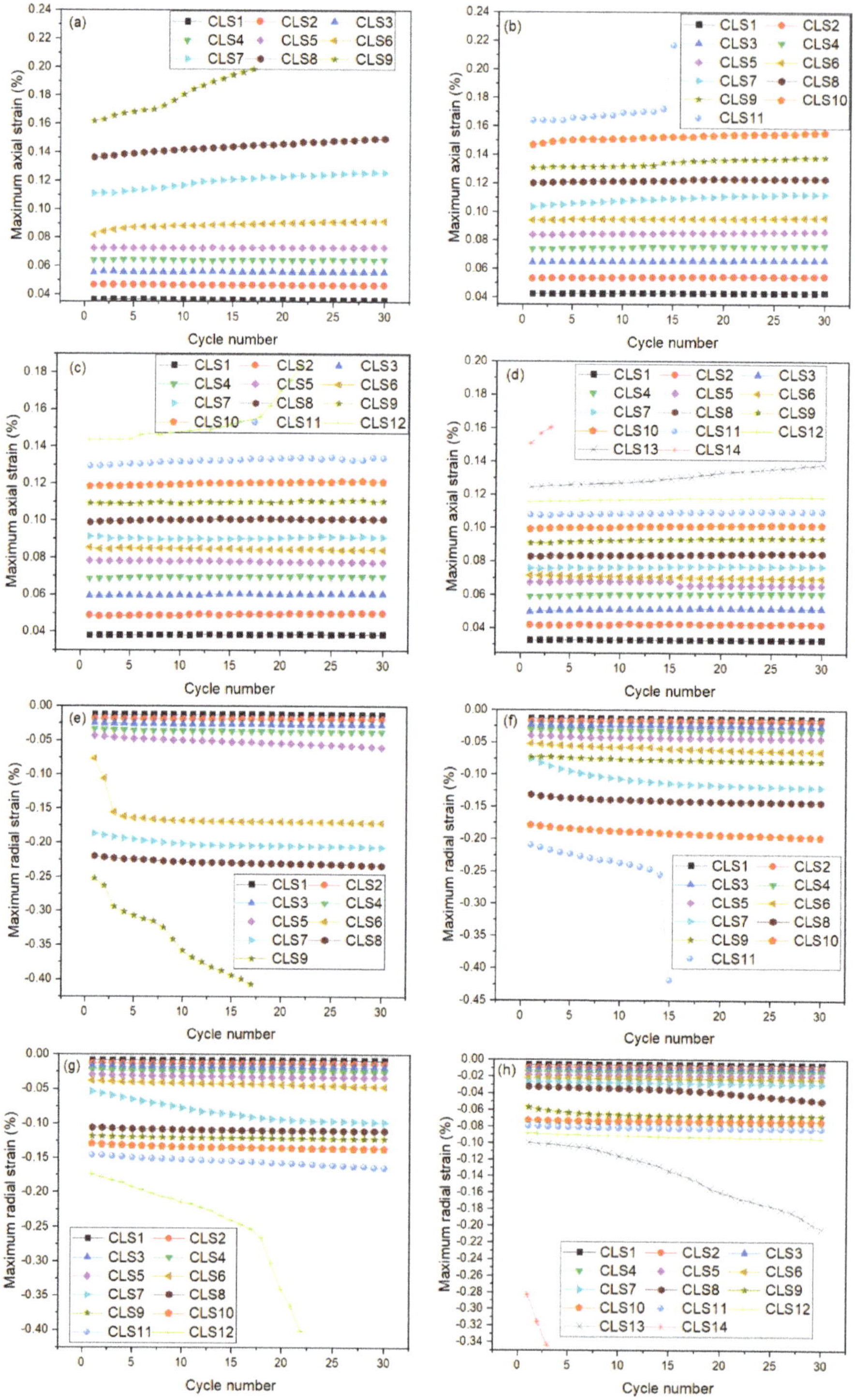

Figure 5. *Cont.*

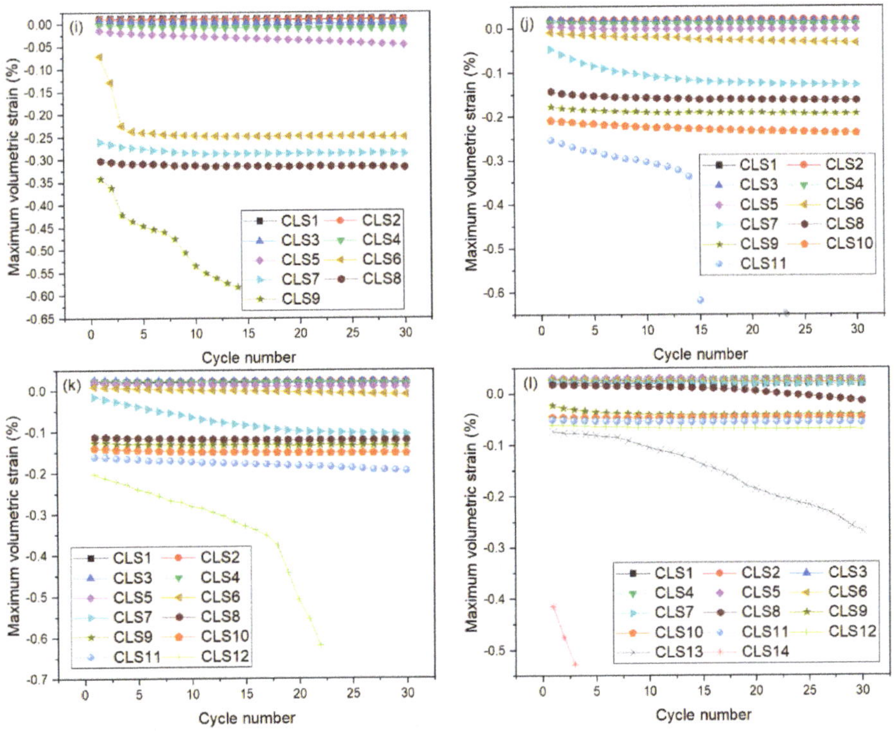

Figure 5. Description of maximum axial, radial, and volumetric strain with cycle number at different cyclic loading stages. ((**a**–**d**): The maximum axial strain versus cycle number; (**e**–**f**): The maximum radial strain versus cycle number; (**g**–**l**): The maximum volumetric strain versus cycle number).

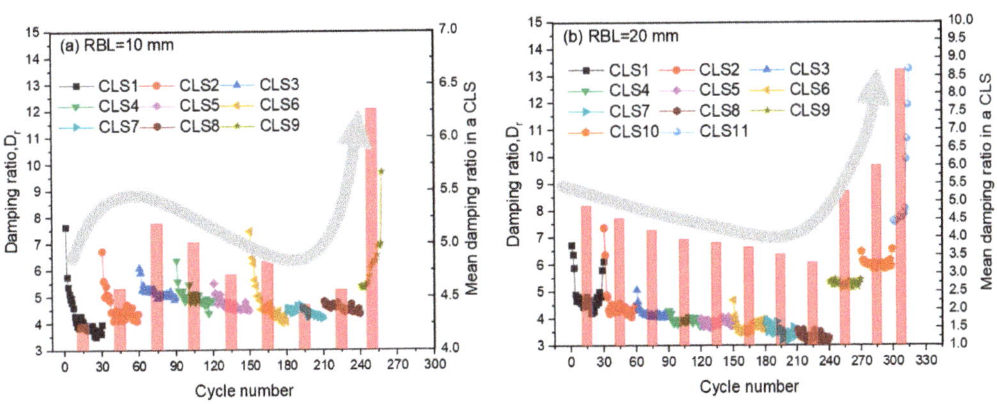

Figure 6. *Cont.*

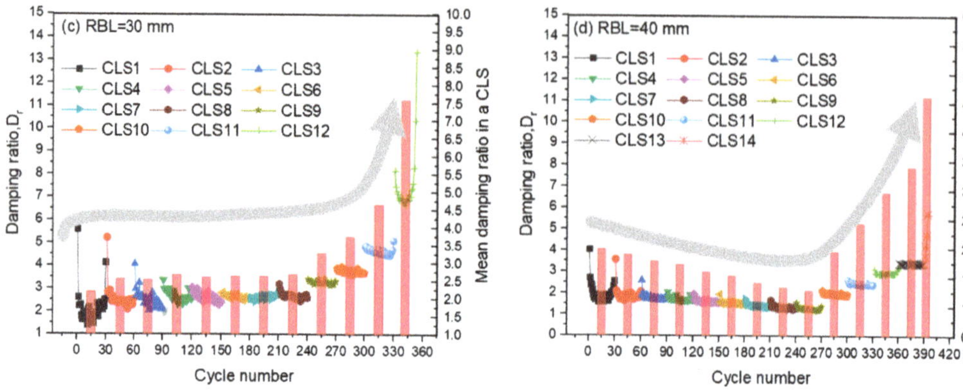

Figure 6. Evolution of damping ratio with cycle number at different loading stages. (**a**–**d**) Marble samples containing stepped fissures with rock bridge lengths of 10, 20, 30, and 40 mm, respectively.

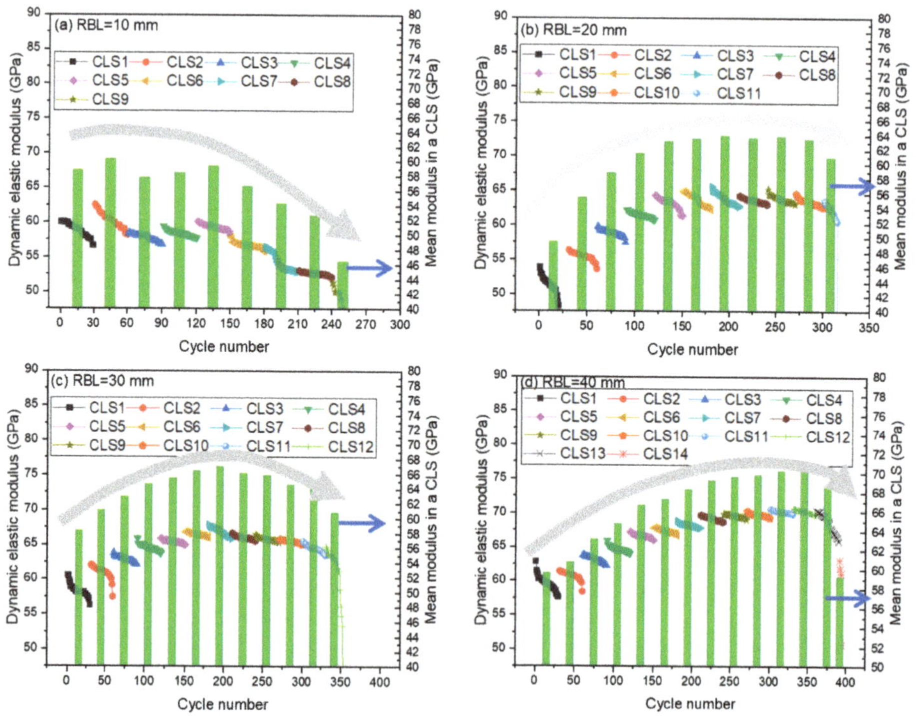

Figure 7. The evolution of dynamic elastic modulus (E_d) with cycle number at different loading stages for marble samples containing stepped fissures with rock bridge lengths of 10 mm (**a**), 20 mm (**b**), 30 mm (**c**), and 40 mm (**d**).

3.4. Strain Incremental Rate Analysis

This section discusses the impact of RBL on rock failure by analyzing the incremental rates of axial and volumetric strain. Typically, strain rate serves as a metric to quantify the rate of damage incurred by a rock during deformation. The consensus exists that irrespective of whether a rock experiences creep, static, or cyclic loading [27–29], the strain rate consistently correlates positively with the damage ratio. In this context, the strain incremental rate refers to the incremental change in strain (axial, radial, or volumetric) at the upper stress point of each cyclic loading stage (CLS), as depicted in Figure 8. This rate is calculated by dividing the difference in strain values between the different cycles by the number of cycles elapsed: $(\varepsilon_{an} - \varepsilon_{a1})/(N - 1)$, where ε_{an} and ε_{a1} represent the axial strain of the Nth and 1st cycles, respectively, at the upper stress level. The cycle number (N) in each CLS is 30 in this work. This study focuses on analyzing the axial strain rate (ε_{ar}) and volumetric strain rate (ε_{vr}), while the circumferential LVDT device records local deformation as radial strain. The calculation of incremental axial and volumetric strain rates is detailed in Tables 2 and 3. The gradual failure of marble is observed to be influenced by RBL, and sudden increases in axial and volumetric strain are attributed to the formation of rock bridges. Regions highlighted in green, yellow, and red denote varying levels of risk. It is noted that the risk levels categorized by axial and volumetric strain rates exhibit distinct patterns. As the volumetric strain rate is a comprehensive reflection of axial and radial deformation, the early warning using the volumetric rate is favorable for instability prediction. From Table 3, it is shown that warning predictions are found at the 5th, 8th, 10th and 11th levels; these are earlier than the warnings using axial strain rate (i.e., the 7th, 11th, 11th, and 13th levels; Table 2).

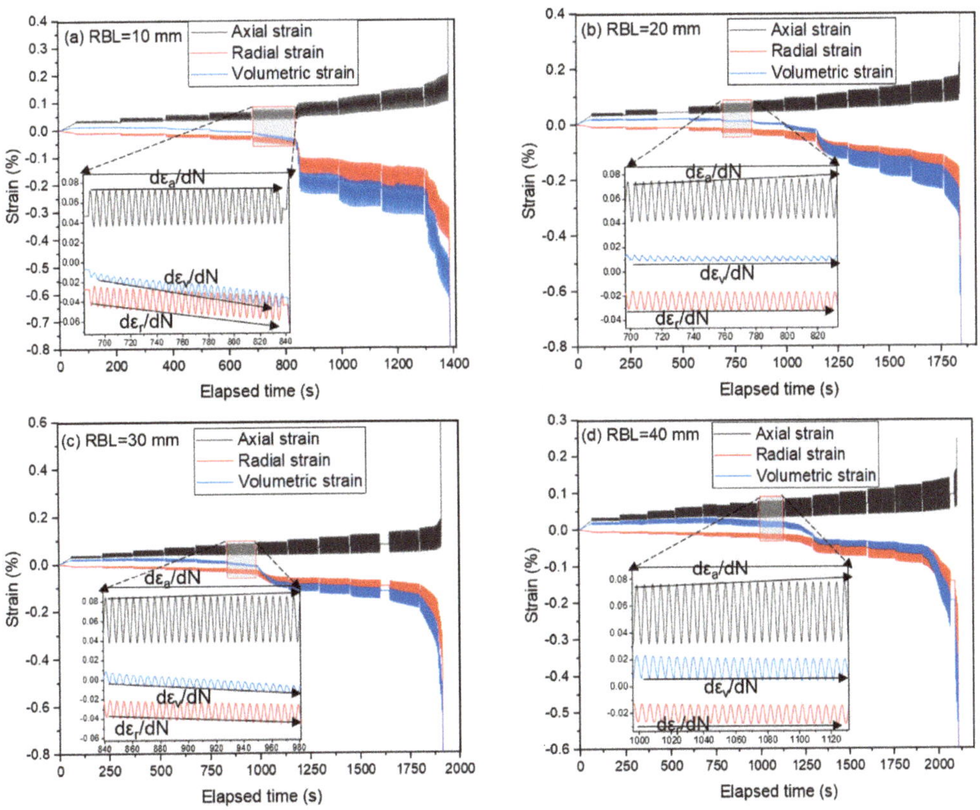

Figure 8. Definition of axial, radial, and volumetric strain incremental rates. (**a–d**) Marble samples containing stepped fissures with RBLs of 10, 20, 30, and 40 mm, respectively.

Table 2. Axial strain rates for the tested rock samples at different CLSs.

CLS	Maximum Stress (MPa)	Axial Strain Rate [%/N − 1]			
		RBL = 10 mm	RBL = 20 mm	RBL = 30 mm	RBL = 40 mm
1	15	6.666×10^{-5}	1.333×10^{-5}	3×10^{-5}	1×10^{-5}
2	20	1.533×10^{-4}	3.033×10^{-4}	3.866×10^{-4}	3.146×10^{-4}
3	35	2.1×10^{-4}	4.666×10^{-4}	3.5×10^{-4}	3.146×10^{-4}
4	40	3.933×10^{-4}	5.4×10^{-4}	3.133×10^{-4}	3.04×10^{-4}
5	45	4.8×10^{-4}	6.466×10^{-4}	2.466×10^{-4}	1.6×10^{-4}
6	50	6.266×10^{-4}	3.233×10^{-4}	2.266×10^{-4}	1.466×10^{-4}
7	55	0.00114 *	5.4×10^{-4}	2.3×10^{-4}	2.346×10^{-4}
8	60	0.00133	3.633×10^{-4}	3.2×10^{-4}	2.533×10^{-4}
9	65	0.00163	5.033×10^{-4}	3.233×10^{-4}	3.093×10^{-4}
10	70		5.733×10^{-4}	3.533×10^{-4}	2.56×10^{-4}
11	75		0.00119 *	0.00121 *	2.746×10^{-4}
12	80			0.00154	2.96×10^{-4}
13	85				0.00106 *
14					0.00165

Note: The green region denotes safety, the yellow region signifies uncertainty, and the red region indicates danger. The * underscores the notable volume dilatancy observed in the rock sample, marking a shift from compression dominance to expansion dominance.

Table 3. Volumetric strain rates for the tested rock samples at different CLSs.

CLS	Maximum Stress (MPa)	Volumetric Strain Rate [%/N − 1]			
		RBL = 10 mm	RBL = 20 mm	RBL = 30 mm	RBL = 40 mm
1	15	1×10^{-6}	1×10^{-6}	6.667×10^{-6}	1.666×10^{-6}
2	20	4.6×10^{-5}	3.333×10^{-6}	1.04×10^{-4}	1.213×10^{-4}
3	35	2.5×10^{-4}	9.733×10^{-5}	1.4×10^{-5}	1.346×10^{-4}
4	40	4.066×10^{-4}	2.68×10^{-4}	1.266×10^{-4}	7.733×10^{-5}
5	45	0.00122 *	3.893×10^{-4}	3.133×10^{-4}	8.667×10^{-5}
6	50	0.00677	4.032×10^{-4}	6.667×10^{-4}	1.4×10^{-4}
7	55	0.00117	4.653×10^{-4}	3.16×10^{-4}	1.32×10^{-4}
8	60	0.00107	0.00234 *	5.533×10^{-4}	1.14×10^{-4}
9	65	0.01166	0.00105	4.3×10^{-4}	8.773×10^{-4}
10	70	Failure	0.00301	0.00103 *	1.907×10^{-4}
11	75	--	0.01285	0.00143	2.987×10^{-4}
12	80	--	Failure	0.01294	0.00329
13	85	--	--	Failure	0.00667 *
14		--	--	--	0.00861
		--	--	--	Failure

Note: The green region denotes safety, the yellow region signifies uncertainty, and the red region indicates danger. The * underscores the notable volume dilatancy observed in the rock sample, marking a shift from compression dominance to expansion dominance.

3.5. Modelling of Damage Evolution

In the flawed rock samples depicted in Figure 5a–d, the axial strain occasionally exhibits a decreasing trend, indicating sporadic deformation, as recorded by two optical gratings. Conversely, the LVDT system, positioned at the rock bridge segment, effectively measures the deformation of the rock bridge, particularly the radial strain. Furthermore, the shape of the hysteresis loop on the radial stress–strain curve is more distinct compared to that on the axial stress curve. As seen in Figure 5e–h, the development of radial strain at each cyclic loading stage is more pronounced than that of axial strain. The rapid increase in radial deformation results from crack coalescence, leading to significant damage. Therefore, in this study, radial strain is deemed more suitable for describing the propagation of rock damage. Radial deformation is utilized to model the evolution of rock damage and establish the relationship between damage factors and cycles. A damage variable defined using the maximum radial strain, D_{rs}, is expressed as follows:

$$D_{rs} = \frac{\varepsilon_{max}^{n} - \varepsilon_{max}^{0}}{\varepsilon_{max}^{f} - \varepsilon_{max}^{0}} \tag{1}$$

where ε_{max}^{0}, ε_{max}^{n}, and ε_{max}^{f} are the initial maximum strain, instantaneous maximum strain after n cycles, and ultimate maximum strain, respectively.

Using Equation (1), we depict the evolution of this damage variable against cycle number in Figure 9. The characteristics of damage propagation are consistent with the analysis of radial deformation. Initially, damage within a CLS is relatively minor during the early loading stages. Subsequently, damage escalates as the CLSs increase, and the incremental rate accelerates accordingly. Towards the end of the loading sequence, an inverted "S"-shaped pattern of damage propagation becomes evident in the last several CLSs.

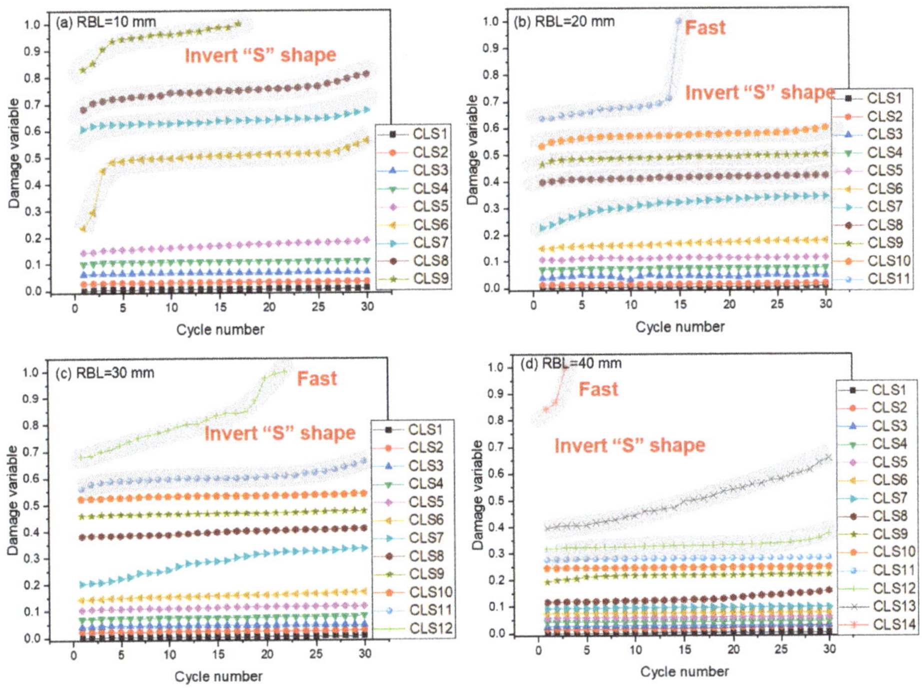

Figure 9. Damage variable versus cycle number at different loading stages. (**a–d**) Rock samples with RBLs of 10, 20, 30, and 40 mm, respectively.

Based on the findings presented in Figure 9, we establish a self-defined damage evolution model that incorporates radial strain and deformation, as illustrated in Figure 10. The form of this proposed model is represented as follows:

$$D = 1 - (1 - (n/N_f)^{a})^{b} \tag{2}$$

The damage variable D results from irreversible plastic deformation. Initially, rock damage is zero before loading (N = 0) and reaches one when failure occurs completely (N = N_f). Here, n denotes the number of loading cycles and N_f signifies the fatigue lifetime. Parameters a and b are associated with material characteristics. Figure 10 displays the fitting results using Origin 2021 software for the marble samples with RBLs of 10 mm,

20 mm, 30 mm, and 40 mm. These results, along with corresponding correlation coefficients, affirm the applicability of the proposed model in describing the evolution of rock damage.

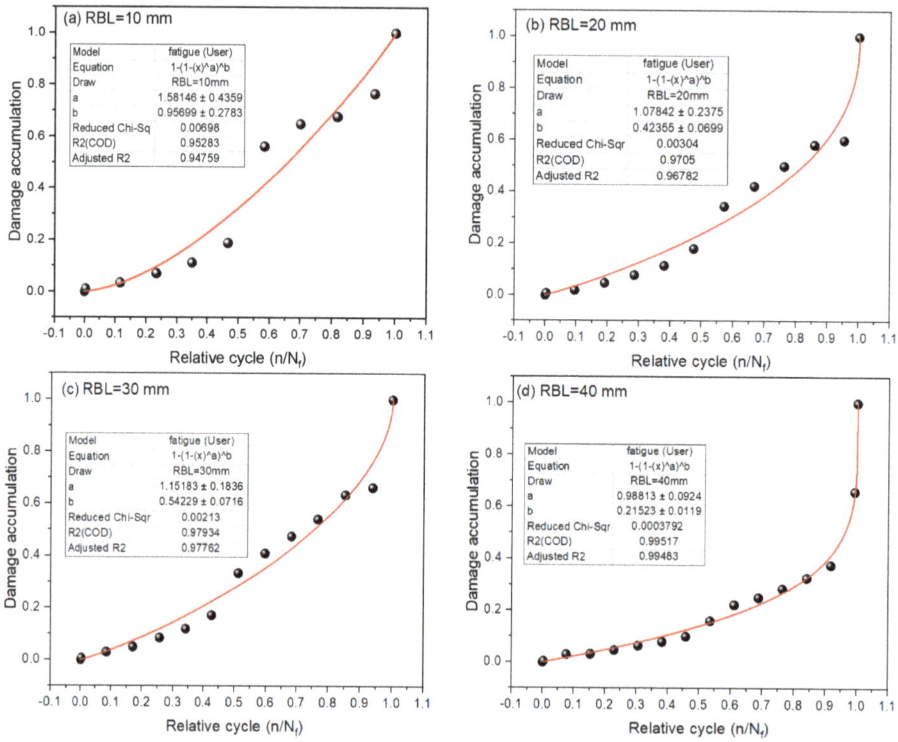

Figure 10. Modelling of damage evolution equation for typical fissure-containing marble samples. (**a**–**d**) Rock samples with RBLs of 10, 20, 30, and 40 mm, respectively.

4. Summary and Conclusions

This study utilized marble samples containing double-stepped fissures of varying rock bridge length to conduct fatigue loading tests with increasing amplitude. The experimental findings systematically illustrate how the length of the rock bridge (RBL) affects the fatigue mechanical properties of brittle rock material, encompassing stress–strain behaviors, fatigue strain progression, damping characteristics, dynamic elastic modulus, and the evolution of damage. The principal findings are synthesized as follows:

(1) The presence of pre-existing rock bridge segments significantly impacts crack propagation and coalescence. As the length of the rock bridge increases, both rock fatigue deformation and fatigue lifetime decrease. Notably, the sample with the longest rock bridge segment exhibited the lowest volumetric deformation, suggesting that damage propagation was least pronounced in this case.

(2) Both damping ratio and dynamic elastic modulus are influenced by the presence of a rock bridge segment. The damping ratio reveals significant energy dissipation, particularly during the final stages of cyclic loading, highlighting substantial energy consumption driving crack propagation.

(3) A strain incremental rate index is introduced to forecast the progression of rock failure. It is observed that volumetric strain rate serves as an effective early warning indicator compared to axial strain rate. The warning time diminishes as rock bridge length

increases, suggesting that rocks with larger segments exhibit greater resistance to external fatigue loading.

Author Contributions: Conceptualization, Y.Y.; methodology, X.Y.; software, Z.C.; writing—review and editing, Y.W.; funding acquisition, Y.W. All authors have read and agreed to the published version of the manuscript.

Funding: This article is funded by the National Natural Science Foundation of China (52174069), the Beijing Natural Science Foundation (8202033), and the Henan Key Laboratory for Green and Efficient Mining & Comprehensive Utilization of Mineral Resources (Henan Polytechnic University) (KCF2203).

Institutional Review Board Statement: Not applicable.

Informed Consent Statement: Not applicable.

Data Availability Statement: The original contributions presented in the study are included in the article; further inquiries can be directed to the corresponding author.

Conflicts of Interest: Author Yongchun Yu was employed by the company Shangri-La Yunkuang Hongniu Mining Co., Ltd. The remaining authors declare that the re-search was conducted in the absence of any commercial or financial relationships that could be construed as a potential conflict of interest.

References

1. Gong, B.; Wang, S.; Sloan, S.W.; Sheng, D.; Tang, C.A. Modelling rock failure with a novel continuous to discontinuous method. *Rock Mech. Rock Eng.* **2019**, *52*, 3183–3195. [CrossRef]
2. Bobet, A.; Fakhimi, A.; Johnson, S.; Morris, J.; Tonon, F.; Yeung, M.R. Numerical models in discontinuous media: Review of advances for rock mechanics applications. *J. Geotech. Geoenvironmental Eng.* **2009**, *135*, 1547–1561. [CrossRef]
3. Lisjak, A.; Grasselli, G. A review of discrete modeling techniques for fracturing processes in discontinuous rock masses. *J. Rock Mech. Geotech. Eng.* **2014**, *6*, 301–314. [CrossRef]
4. Shirani Faradonbeh, R.; Taheri, A.; Karakus, M. Fatigue failure characteristics of sandstone under different confining pressures. *Rock Mech. Rock Eng.* **2022**, *55*, 1227–1252. [CrossRef]
5. Zheng, Q.; Liu, E.; Yu, D.; Liu, M. Fatigue and damage properties of non-consecutive jointed mudstone samples subjected to cyclic triaxial loading. *Bull. Eng. Geol. Environ.* **2020**, *79*, 2467–2481. [CrossRef]
6. Liu, Y.; Dai, F.; Dong, L.; Xu, N.; Feng, P. Experimental investigation on the fatigue mechanical properties of intermittently jointed rock models under cyclic uniaxial compression with different loading parameters. *Rock Mech. Rock Eng.* **2018**, *51*, 47–68. [CrossRef]
7. Jafari, M.K.; Pellet, F.; Boulon, M.; Hosseini, K.A. Experimental study of mechanical behaviour of rock joints under cyclic loading. *Rock Mech. Rock Eng.* **2004**, *37*, 3–23. [CrossRef]
8. Cerfontaine, B.; Collin, F. Cyclic and fatigue behaviour of rock materials: Review, interpretation and research perspectives. *Rock Mech. Rock Eng.* **2018**, *51*, 391–414. [CrossRef]
9. Fathi, A.; Moradian, Z.; Rivard, P.; Ballivy, G. Shear mechanism of rock joints under pre-peak cyclic loading condition. *Int. J. Rock Mech. Min. Sci.* **2016**, *83*, 197–210. [CrossRef]
10. Li, K.; Yang, S.; Liu, C.; Chen, Y.; Zhang, G.; Ma, Q. Mechanical response and microscopic damage mechanism of pre-flawed sandstone subjected to monotonic and multilevel cyclic loading: A laboratory-scale investigation. *Int. J. Min. Sci. Technol.* **2023**, *33*, 1487–1510. [CrossRef]
11. Moghaddam, R.H.; Golshani, A. Fatigue behavior investigation of artificial rock under cyclic loading by using discrete element method. *Eng. Fail. Anal.* **2024**, *160*, 108105. [CrossRef]
12. Brown, E.T.; Hudson, J.A. Fatigue failure characteristics of some models of jointed rock. *Earthq. Eng. Struct. Dyn.* **1973**, *2*, 379–386. [CrossRef]
13. Zhang, T.; Xu, W.; Wang, H.; Wang, R.; Yan, L.; Shi, A. Anisotropic mechanical characteristics and energy evolution of artificial columnar jointed rock masses subjected to multi-level cyclic loading. *Int. J. Fatigue* **2022**, *165*, 107215. [CrossRef]
14. Vaneghi, R.G.; Thoeni, K.; Dyskin, A.V.; Sharifzadeh, M.; Sarmadivaleh, M. Fatigue damage response of typical crystalline and granular rocks to uniaxial cyclic compression. *Int. J. Fatigue* **2020**, *138*, 105667. [CrossRef]
15. Erarslan, N.; Williams, D.J. Mechanism of rock fatigue damage in terms of fracturing modes. *Int. J. Fatigue* **2012**, *43*, 76–89. [CrossRef]
16. Wang, Y.; Li, C.H.; Han, J.Q. On the effect of stress amplitude on fracture and energy evolution of pre-flawed granite under uniaxial increasing-amplitude fatigue loads. *Eng. Fract. Mech.* **2020**, *240*, 107366. [CrossRef]
17. Karakus, M.; Akdag, S.; Bruning, T. Rock fatigue damage assessment by acoustic emission. In Proceedings of the International Conference on Geo-Mechanics, Geo-Energy and Geo-Resources (IC3G), Melbourne, Australia, 28–29 September 2016; pp. 9–82.

18. Shu, L.Y.; Liu, Z.S.; Wang, K.; Zhu, N.N.; Yang, J. Characteristics and classification of microseismic signals in heading face of coal mine: Implication for coal and gas outburst warning. *Rock Mech. Rock Eng.* **2022**, *55*, 6905–6919. [CrossRef]
19. Xu, B.; Liu, X.; Zhou, X.; Xie, Y.; Suliman, L.; Liu, X.; Lin, G.; Huang, J. Investigation on the macro-meso fatigue damage mechanism of rock joints with multiscale asperities under pre-peak cyclic shear loading. *Soil Dyn. Earthq. Eng.* **2021**, *151*, 106958. [CrossRef]
20. Singh, M.; Khalkho, P. Modelling of Fatigue Damage Evolution of Two Natural Rocks Under Cyclic Loading. *Int. J. Geosynth. Ground Eng.* **2022**, *8*, 8. [CrossRef]
21. Tao, Y.; Hu, R.Z.; Zhu, F.L.; Ma, Y.S.; Ye, L.; Cheng, Z.T. Ore-forming age and the geodynamic background of the Hetaoping lead-zinc deposit in Baoshan, Yunnan. *Acta Petrol. Sin.* **2010**, *26*, 1760–1772.
22. Deng, M.G.; Zhao, J.X.; Liu, F.X.; Yu, H.J.; Sun, B.D.; Liu, F.; Li, S.B. Discussion on sources of metallogenic fluids and materials of the Shuitoushan Pb-Zn deposit in Zhenkang, western Yunnan: Evidence from H, O, S and Pb isotopes. *Acta Petrol. Sin.* **2017**, *33*, 2001–2017.
23. Li, W.; Yin, G.; Lu, Y.; Wang, Y.; Yu, H.; Cao, X.; Zhang, S. Delineation and significance of the Hongshan-Shudu serpentinite mélange zone within the "Sanjiang" volcanic-magmatic arc in Southwest China. *Acta Petrol. Sin.* **2010**, *26*, 1661–1671.
24. Wang, Y.; Wei, X.M.; Li, C.H. Dynamic behavior of soil and rock mixture using cyclic triaxial tests and X-ray computed tomography. *Arab. J. Geosci.* **2019**, *12*, 229. [CrossRef]
25. Lamas-Lopez, F. Field and Laboratory Investigation on the Dynamic Behavior of Conventional Railway Track-Bed Materials in the Context of Traffic Upgrade. Ph.D. Thesis, Ecole Nationale des Ponts et Chaussees, Universite Paris-Est, Paris, France, 2016.
26. Wang, Y.; Li, C.; Han, J.; Wang, H. Mechanical behaviours of granite containing two flaws under uniaxial increasing-amplitude fatigue loading conditions: An insight into fracture evolution analyses. *Fatigue Fract. Eng. Mater. Struct.* **2020**, *43*, 2055–2070. [CrossRef]
27. Wang, Y.; Yang, H.N.; Han, J.Q.; Zhu, C. Effect of rock bridge length on fracture and damage modelling in granite containing hole and fissures under cyclic uniaxial increasing-amplitude decreasing-frequency (CUIADF) loads. *Int. J. Fatigue* **2022**, *158*, 106741. [CrossRef]
28. Brantut, N.; Heap, M.J.; Meredith, P.G.; Baud, P. Time-dependent cracking and brittle creep in crustal rocks: A review. *J. Struct. Geol.* **2013**, *52*, 17–43. [CrossRef]
29. Li, X.; Konietzky, H.; Li, X.; Wang, Y. Failure pattern of brittle rock governed by initial microcrack characteristics. *Acta Geotech* **2019**, *14*, 1437–1457. [CrossRef]

materials

MDPI

Article

Mechanical Properties of Rock-like Materials Under Disturbance Loads at Different Lateral Pressures

Yonghong Liu, Fujun Zhao *, Qiuhong Wu and Zhouyuan Ye

School of Resource & Environment and Safety Engineering, Hunan University of Science and Technology, Xiangtan 411201, China; lyhong0607@163.com (Y.L.); qiuhong.wu@hnust.edu.cn (Q.W.); 1010044@hnust.edu.cn (Z.Y.)

* Correspondence: fjunzhao@mail.hnust.edu.cn; Tel.: +86-15616745687

Abstract: Underground surrounding rock engineering displays unique mechanical properties after being subjected to disturbance loads. In this study, the self-developed CX-8568 impact-disturbance surrounding rock test system was utilized to conduct dynamic tests on gypsum specimens subjected to different lateral pressures. The results show that the presence of lateral pressure enhances the specimen's ability to withstand disturbance loads, which shows higher lateral pressure results in a greater number of disturbance cycles required for specimen failure. Lateral pressure inhibits both the transverse and axial deformation of the specimen, leading to an increase in the elastic modulus and average cyclic disturbance times as lateral pressure rises. When the lateral pressure is held constant, the residual plastic strain of the specimen increases continuously with the number of cyclic disturbance cycles, while the elastic modulus of the specimen decreases steadily as the cyclic disturbance cycles increase. The application of disturbance loads causes significant spalling and damage to the free surface of the specimen under varying lateral pressures. At low lateral pressures, the specimen primarily experiences tensile splitting, whereas at high lateral pressures, shear failure occurs at the ends of the specimen, while tensile failure is observed in the middle. Through this study, we can more clearly understand the mechanical properties and failure characteristics of rock under disturbed load and provide theoretical guidance for the stability of rock engineering.

Keywords: disturbance loads; lateral pressure; mechanical properties; failure mechanism

1. Introduction

In the fields of coal mining, tunnel excavation, and chamber excavation, the surrounding rock is frequently subjected to disturbance loads [1–4], which may originate from various sources including stress waves generated by excavation blasting and mechanical rock drilling, as well as mining disturbances during adjacent tunnel excavation. These disturbance loads have the potential to deteriorate rock mechanical properties and trigger secondary disasters [5–7]. Consequently, examining the mechanical behavior and failure characteristics of surrounding rock under such disturbance loads is essential for gaining insights into the mechanisms underlying rock damage, deterioration, and instability.

Numerous researchers have conducted extensive investigations on the mechanical properties of rocks under disturbance loads, encompassing strength characteristics, deformation features, and failure characteristics. In the realm of uniaxial cyclic loading and unloading [8–10], Zhang et al. [11] performed a series of cyclic loading and unloading tests on red sandstone to explore the relationship between elastic strain, viscoelastic strain, and viscoplastic strain during cyclic loading. Gong et al. [12] proposed a calculation method for LURR at different loading–unloading points based on elastic modulus and developed a rock burst criterion incorporating the lag time ratio index. Zhou et al. [13] utilized X-ray, CT imaging, and 3D reconstruction techniques to quantitatively describe the impact of micropores on rock failure characteristics through customized digital microstructures.

Regarding triaxial cyclic loading and unloading [14–16], Wang et al. [17] designed four gradient cyclic loading and unloading tests of true triaxial fluid–solid coupling of coal and analyzed the deformation, seepage, damage, strength, and failure characteristics of coal samples. Tan et al. [18] conducted triaxial cyclic loading and unloading tests on sandstone under six confining pressures using the MTS-815 rock mechanics testing system, revealing that the elastic modulus of the rock samples followed a pattern of cyclic hardening, stability, and cyclic softening. Ning et al. [19] conducted triaxial cyclic loading and unloading tests on granite at different levels and obtained linear models for deformation modulus, radial strain, and residual strain during cyclic loading. Scholars have also delved into cyclic impact tests [20–22]. Dai et al. [23] carried out cyclic impact tests on granite using the Hopkinson rod experimental system, investigating the dynamic response and damage evolution characteristics of rock samples under cyclic impact at different axial pressures. Luo et al. [24] quantified the fracture interface using a 3D contour scanner and analyzed the dynamic mechanical properties and damage failure patterns of coal and rock with different bedding angles under triaxial cyclic impact tests. Fan et al. [25] studied the dynamic stress–strain relationship of water-cooled granite at different temperatures and cycles, discussing the effects of temperature and the number of cycles on dynamic stress, elastic modulus, maximum strain, and strain rate.

While progress has been made in uniaxial and triaxial tests, there is a significant gap in understanding rock behavior under biaxial cyclic loading and unloading, especially in practical underground engineering, where the transition from a triaxial stress field ($\sigma_1 \geq \sigma_2 \geq \sigma_3 > 0$) to a biaxial stress field ($\sigma_3 = 0$) occurs [26–29]. To address this gap, this study employs a self-developed impact-disturbance testing system for surrounding rock to conduct tests on gypsum specimens under varying lateral pressures. The investigation aims to examine the influence of disturbance loads on crack evolution and failure characteristics of specimens, with the goal of providing guidance for the stability of underground rock engineering.

2. Experimental Methodology

2.1. Specimen Preparation

In this paper, gypsum materials are selected for experiments due to their advantages of innocuity, short molding time, and mass production feasibility [30,31]. The mixing ratio of gypsum specimen is m_{gypsum}:m_{water} at 1:0.5, and the specimen dimensions are 150 mm × 150 mm × 150 mm. To ensure the stability of the specimens' strength, they were standardly cured for 28 days. Table 1 presents the basic mechanical parameters of the standard specimens.

Table 1. Basic mechanical parameters of standard specimens.

Material Type	m_{gypsum}:m_{water}	Compressive Strength/MPa	Internal Friction Angle/(°)	Elastic Modulus/GPa	Poisson's Ratio	Density/g·cm^{-3}
Gypsum	1:0.5	12.53	44.3	2.66	0.21	2.75

2.2. Testing System and Loading Procedure

The experiment employed a self-developed CX-8568 impact-disturbance surrounding rock test system, as shown in Figure 1a. This system is capable of performing single or combined loading in three directions, including static loading and unloading, dynamic disturbance, and impact testing on large-sized specimens. In this study, stress in the Z-axis direction is denoted as σ_1, and stress in the X-axis direction as σ_2, with the amplitude of the disturbance loads being σ_s. The strain is measured by the built-in spring displacement meter. The specimens were placed on the x–y plane of the test rig, as shown in Figure 1b. Lubricant was applied to the specimens' surface to minimize friction between the pressure plate and the specimens.

Figure 1. CX-8568 impact-disturbance surrounding rock test system and test procedure. (**a**) CX-8568 impact-disturbance surrounding rock test system. (**b**) Details of compression of test pieces under different lateral reductions. (**c**) Simplified diagram of the specimen under compression. (**d**) Loading path of disturbed load of the sample under different lateral pressures.

Based on the uniaxial compressive strength of standard specimens, four lateral pressures σ_2 of 0 MPa, 2 MPa, 4 MPa, and 6 MPa were selected to conduct cyclic disturbance tests on specimens. The procedure includes the following:

1. Conducting biaxial static compression tests on specimens under varying lateral pressures at a loading rate of 1 kN/s to determine the biaxial compressive strength σ_{bcs} for each lateral pressure.
2. Using 80% of the σ_{bcs} under each lateral pressure as the average stress σ_1 for cyclic disturbance loads, and 20% of σ_{bcs} as the amplitude σ_s of the disturbance loads, performing cyclic disturbance tests until specimen failure. The specific steps for the cyclic disturbance test involve applying loads along the X-axis and Z-axis at a rate of 1 kN/s, maintaining the preset lateral pressure σ_2 on the X-axis, loading the Z-axis to the preset average cyclic disturbance stress σ_1, and then subjecting the specimen to disturbance loads until failure with a sine-wave disturbance waveform at a frequency of 2 Hz. The failure process is recorded using a camera. Figure 1c,d are the compression diagram and stress path diagram of the specimen, respectively.

3. Experimental Results and Analysis

3.1. Cyclic Disturbance Test Results

Table 2 presents the cycle counts to failure for gypsum specimens under various lateral pressures when subjected to disturbance loads. Figure 2 is the relationship between the average cycle counts and the lateral pressure. Observations from Table 2 indicate that gypsum specimens exhibit the fewest cycle counts to failure at a lateral pressure of 0 MPa. As lateral pressure increases, there is a corresponding increase in the average cycle counts to failure, suggesting that lateral pressure can enhance the rock's capacity to withstand fatigue loads. In Figure 2, for each increment of 2 MPa in lateral pressure, the multiplicative increase in the average cycle counts to failure for the gypsum specimens shows a successive decline, with factors of 2.2, 1.8, and 1.1, respectively. This trend mirrors that observed in the literature, underscoring that the rock's fatigue resistance does not improve linearly with increasing lateral pressure; instead [27,32], it plateaus at a certain threshold. This observation implies that there is a definitive upper limit to the enhancing effect of lateral pressure on the rock's fatigue resistance.

Table 2. Cyclic times of disturbance loads on gypsum specimens with different lateral pressures.

Specimen Number	Lateral Stress σ_2/MPa	Axial Stress σ_1/MPa	Disturbance Amplitude σ_S/MPa	Disturbance Frequency *f*/Hz	Cycle Counts/*n*	Average Cycle Counts/$\overline{n}$
R0-1 R0-2 R0-3	0	10.21	2.55	2	3 2 3	3
R2-1 R2-2 R2-3	2	11.02	2.76	2	6 8 4	6
R4-1 R4-2 R4-3	4	12.10	3.02	2	9 13 10	11
R6-1 R6-2 R6-3	6	12.80	3.20	2	11 10 15	12

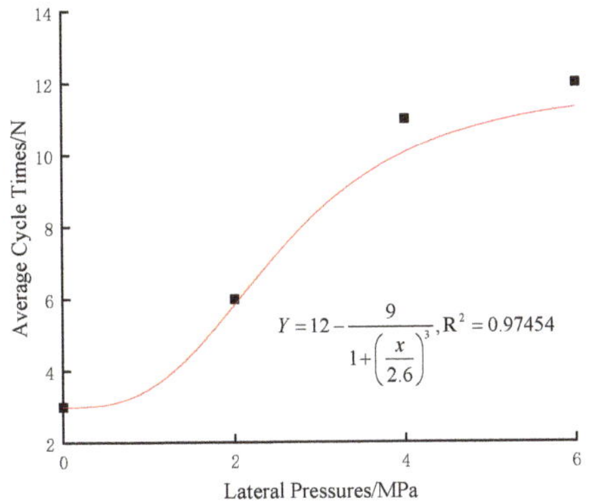

Figure 2. Curve of the average cycle counts of the specimen with lateral pressure.

3.2. Mechanical Properties of Specimens

To investigate the mechanical characteristics of gypsum specimens under disturbance loads, a representative set of results was analyzed for each of the four lateral pressures applied: 0 MPa, 2 MPa, 4 MPa, and 6 MPa. It is important to note that in this study, tensile stress and strain are defined as negative values. The specimens selected for this analysis were R0-1, R2-2, R4-1, and R6-3.

3.2.1. Stress–Strain Curve

Figure 3 presents the stress–strain curves for specimens R0-1, R2-2, R4-1, and R6-3 under different lateral pressures. In Figure 3a, specimen R0-1 undergoes a series of stages as stress increases: an initial compression and elasticity phase, a yield stage where strain increases rapidly, and a disturbance failure stage. Initially, as σ_1 increases, the specimen's internal cracks are compressed, resulting in a clear concave shape in the stress–strain curve. The relationship between stress and strain is linear elastic before entering the yield stage. At 80% of the biaxial compressive strength, the specimen begins to exhibit perturbation. After the third cycle of disturbance, a sudden increase in axial strain indicates specimen failure.

Specimens R2-2, R4-1, and R6-3 also experience these four stages, but their stress–strain curves display different characteristics with increasing lateral pressure. Key observations include the following:

- The presence of lateral pressure causes the specimens to be in a compressed state in the transverse direction during the initial loading stage, with the initial strain on the X-axis being greater than zero. Higher lateral pressure results in a larger area under the trans-verse stress–strain curve, indicating more energy stored in the transverse direction.
- At the initial stage of static load application, the internal cracks of the specimens compact rapidly due to biaxial pressure, reducing the proportion of the axial compaction stage as lateral pressure increases.
- In the late stage of static load application, the proportion of the axial yield stage increases, suggesting that increasing lateral pressure genuinely enhances the bearing capacity of the rocks.
- In the stage of disturbance load, with the increase in disturbance times, the hysteretic curve changes from dense to sparse when the lateral pressure is 2 MPa, while the hysteretic curve grows more evenly when the lateral pressure is 4 MPa and 6 MPa, and the hysteretic curve is denser when the lateral pressure is 6 MPa, which indicates that the actual strain growth of the specimen slows down in a single cycle after the lateral pressure increases.

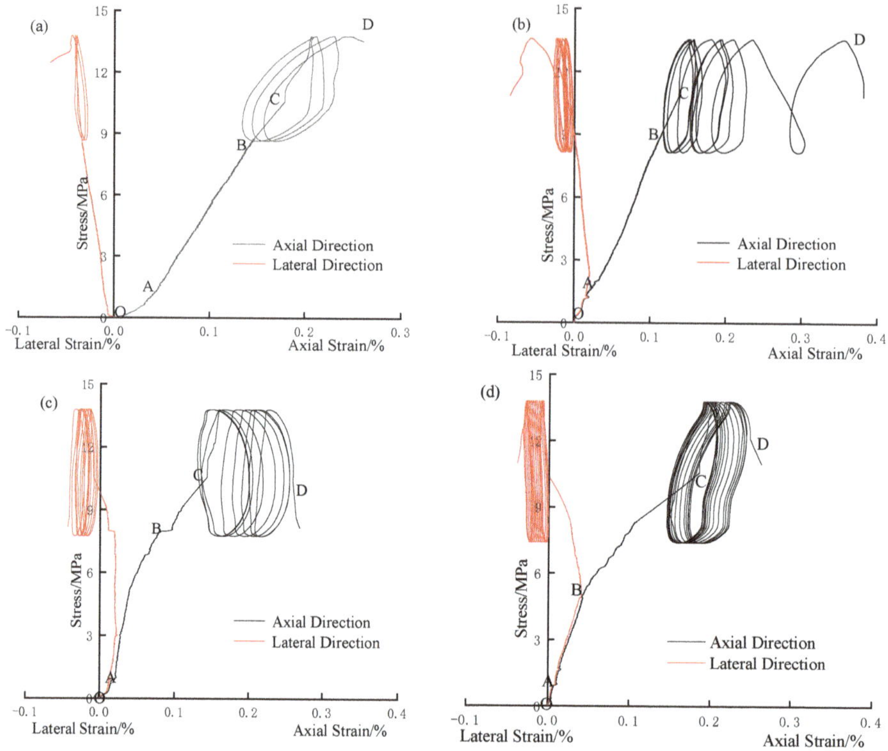

Figure 3. Stress–strain curves of specimens with different lateral pressures. (**a**) Stress–strain curves R0-1. (**b**) Stress–strain curves of R2-2. (**c**) Stress–strain curves of R4-1. (**d**) Stress–strain curves of R6-3.

3.2.2. Deformation Characteristics

Gypsum, as an elastic–plastic material, experiences both elastic and plastic deformation in response to cyclic disturbance. Elastic deformation returns to its original state after each cycle, while plastic deformation results in residual strain. Therefore, under the action of disturbance loads, the total axial strain ε_Z of the gypsum specimen is the sum of its axial elastic strain ε_E and axial plastic strain ε_P [19,33].

Figure 4 illustrates the axial plastic strain, axial maximum strain, and lateral maximum strain of the specimens during cyclic disturbance. It shows that the residual plastic strain of gypsum specimens increases with the number of cyclic disturbances under different lateral pressures, indicating that disturbance loads at an 80% σ_{bcs} stress level cause irreversible damage to the specimens. Additionally, the growth rates of axial residual plastic strain and axial maximum strain for specimens R2-2 and R4-1 become gradually higher than those for R6-3 as the number of cyclic disturbances increases. However, the growth rates of transverse maximum strain are similar, suggesting that plastic failure of the specimens is mainly influenced by axial strain.

In the initial phase of cyclic disturbance (Figure 4a), the axial residual plastic strain for all specimens initially decreases with the increase in lateral pressure and then increases. For instance, the residual plastic strain of R0-1 is 0.00134 at the first cyclic disturbance and increases at rates of 1.08 and 1.09 for the second and third disturbances, respectively. This indicates a notably high failure rate for the specimen under disturbance loads when the lateral pressure is 0 MPa. In contrast, the residual plastic strains of R2-2 and R4-1 are 0.00116 and 0.00118, respectively, at the first disturbance, both lower than that of R0-1.

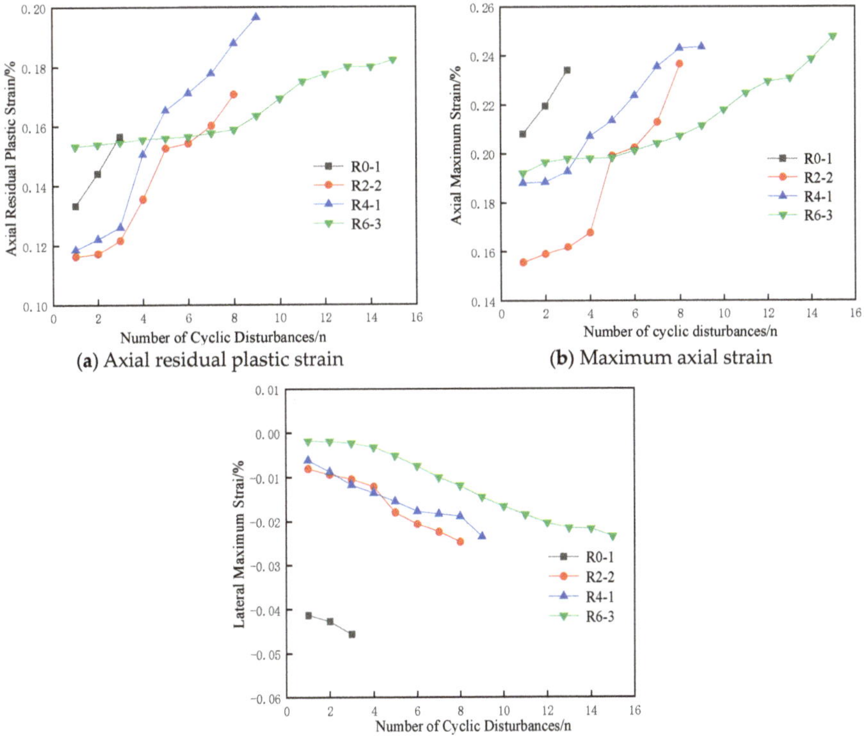

(**a**) Axial residual plastic strain (**b**) Maximum axial strain

(**c**) Maximum lateral strain

Figure 4. Axial residual plastic strain, axial maximum strain, and maximum lateral strain of specimens with different lateral pressures.

The axial and lateral maximum strains of R2-2 and R4-1, shown in Figure 4b,c, follow a similar pattern, suggesting that the presence of lateral pressure can reduce axial and lateral deformation of the specimens, thereby affecting their strength. When the lateral pressure is increased to 6 MPa, the maximum axial and lateral deformation of R6-3 during the first cyclic disturbance are still lower than that of R0-1. However, the maximum residual plastic strain is higher than that of R0-1, indicating that while lateral pressure can restrict deformation, the disturbance loads at this stress level remain the primary factor influencing specimen deformation.

3.2.3. Strength Characteristics

Elastic modulus is a fundamental parameter that quantifies a material's resistance to externa deformation [34,35]. In biaxial cyclic disturbance tests, the rate at which specimens' resistance to deformation changes varies due to the application of different lateral pressures. Use the rate of change of two adjacent elastic moduli to represent the magnitude of the deformation resistance.

Figure 5 presents the curve of the elastic modulus and the change rate of the elastic modulus of the specimens during cyclic disturbance. As shown in Figure 5a, under the same lateral pressure, the specimens' resistance to deformation decreases with an increase in the number of cyclic disturbances, which is reflected by the decrease in the elastic modulus. Additionally, for the same number of cyclic disturbances, a higher lateral pressure results in a greater elastic modulus, indicating a stronger ability to resist deformation.

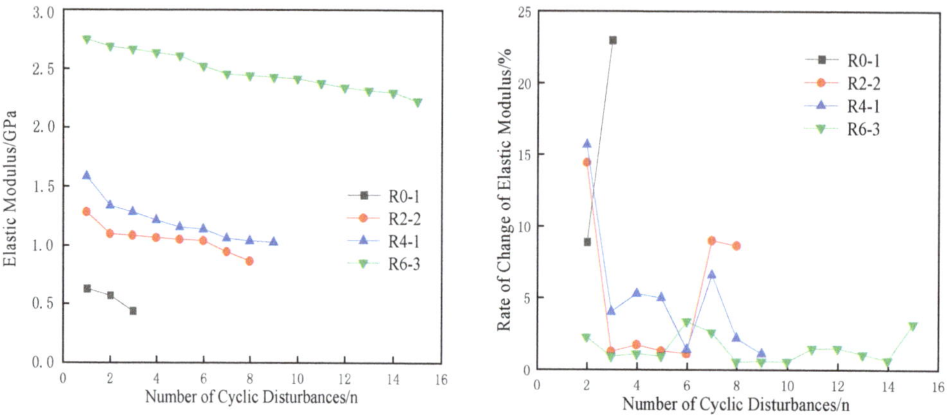

(a) Elastic modulus (b) Rate of change in elastic modulus

Figure 5. Elastic modulus and rate of change in elastic modulus of specimens with different lateral pressures.

In Figure 5b, it is evident that the decrease rate of the elastic modulus of the specimens under different lateral pressures increases rapidly near the end of cyclic disturbance. For instance, at a lateral pressure of 0 MPa, the change rate of the elastic modulus for specimen R0-1 increases rapidly, signaling an impending fracture. At lateral pressures of 2 MPa and 4 MPa, the elastic modulus of the specimens decreases rapidly during the initial cyclic disturbances, suggesting that lateral pressure enhances the specimens' ability to resist deformation. At a lateral pressure of 6 MPa, the elastic modulus of the specimen changes moderately until the final cyclic disturbance.

3.3. Crack Evolution Process and Failure Characteristics of Gypsum Specimens

Camera recordings of the compression deformation process have enabled detailed observations of crack appearance, development, and penetration on the specimen surfaces, facilitating the analysis of crack evolution and failure characteristics under different lateral pressures. To illustrate the crack propagation process, especially with the presence of strain gauges, a corresponding sketch has been created as shown in Figure 6.

3.3.1. Crack Evolution Process

In Figure 6a, specimen R0-1 initially exhibits a small tensile crack on the left side. At 80 s, this crack extends to the upper and lower sides, and a tensile–shear crack forms in the middle and upper parts, approximately parallel to the axial direction. By 84 s, the left side is nearly fractured, and the right side is extruded as the disturbance loads commence. At 86 s, the tensile–shear crack in the middle expands rapidly, accompanied by two parallel cracks, leading to specimen destruction as both sides detach.

In Figure 6c, specimen R4-2 first shows a short tensile crack in the lower right corner. After 60 s, the crack bulges in the upper-right corner. At 70 s, tensile cracks on the right side connect, and an oblique downward shear crack appears in the middle and upper parts. The disturbance loads then begin. At 74 s, the right area is severely extruded and peeled off, and the shear crack spreads, connecting with another in the middle and lower part, ultimately causing specimen destruction.

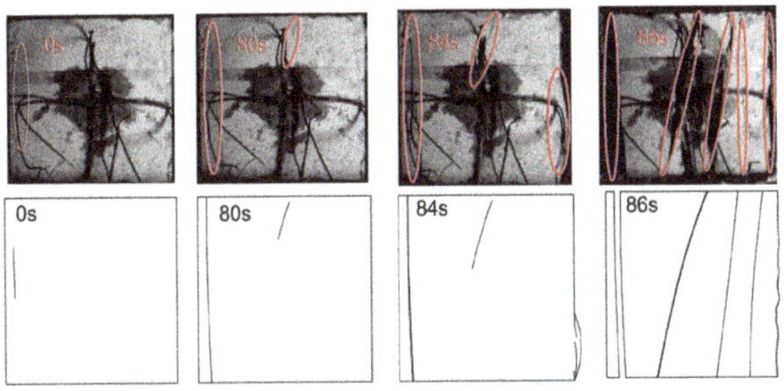

(**a**) R0-1

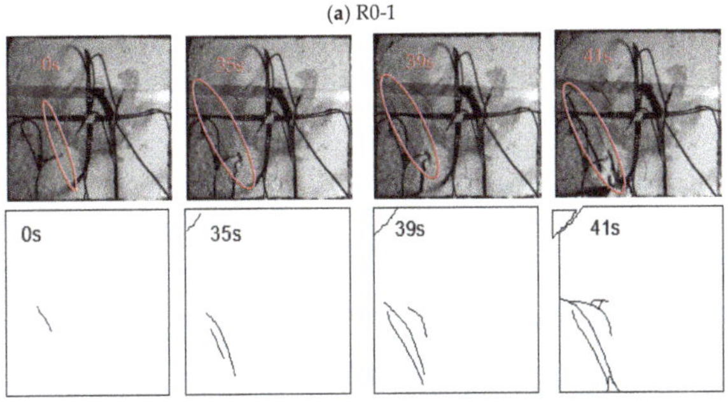

(**b**) R2-2

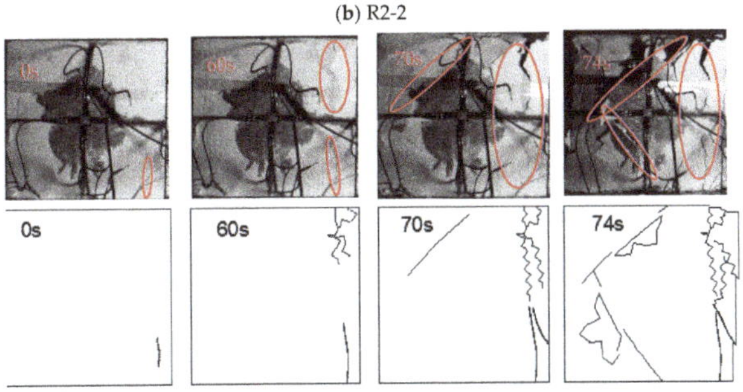

(**c**) R4-1

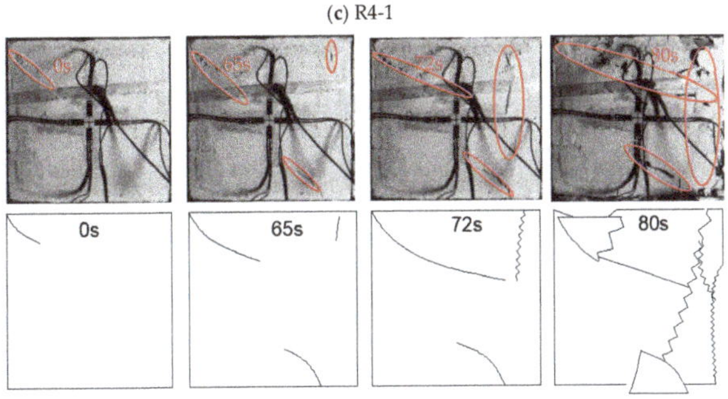

(**d**) R6-3

Figure 6. Crack evolution process and failure characteristics of gypsum specimens.

Comparing the crack propagation processes under different lateral pressures reveals several trends: under lower lateral pressures, specimens primarily develop tensile or tensile–shear cracks parallel to the axial direction. In contrast, higher lateral pressures are more likely to produce shear cracks at the ends and tensile cracks in the middle. After the disturbance loads are applied, the spalling of the free surface and accelerated crack propagation are pronounced, even at higher lateral pressures. Additionally, the sound emitted during specimen failure is crisp at 0 MPa lateral pressure but becomes dull and quieter as the lateral pressure increases.

3.3.2. Failure Characteristics of Specimens

Figure 7 is the failure characteristics of the specimens under disturbance loads at various lateral pressures. When the lateral pressure is 0 MPa, the primary failure mode of the specimen is tensile splitting, and the fracture surface of the specimen is clear without any powder residue after crushing. However, as the lateral pressure increases to 2 MPa, it becomes apparent from the side view that the specimen experiences shear failure. As the lateral pressure continues to increase, the visibility of the main crack's failure diminishes progressively. There is a significant increase in the number of small cracks in the tensile failure area in the middle, and their distribution becomes chaotic. Additionally, fractures in the shear failure area at the end exhibit a stepped appearance and contain more residue. It is worth noting that the expansion of the specimen's free surface becomes more pronounced as the lateral pressure increases, and the fragmentation after failure becomes smaller.

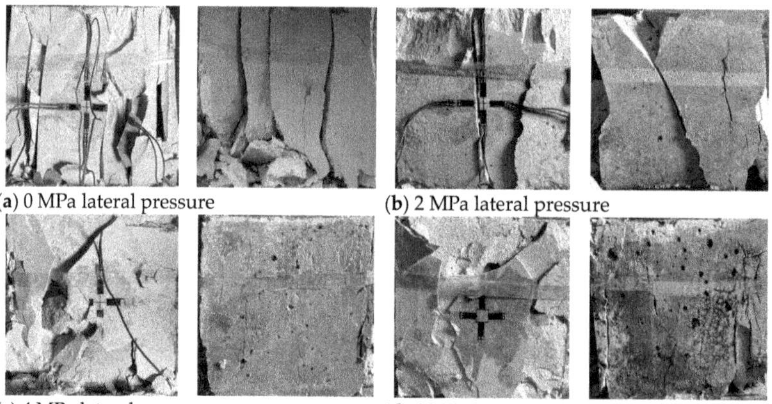

(**a**) 0 MPa lateral pressure (**b**) 2 MPa lateral pressure

(**c**) 4 MPa lateral pressure (**d**) 6 MPa lateral pressure

Figure 7. Failure characteristics of gypsum specimens with different lateral pressures.

4. Discussion

4.1. Characteristics of Damage Variables of Specimens Under Different Lateral Pressures

In Figure 4, the maximum axial strain and maximum lateral strain of the specimen increase with an increasing number of cycles, indicating that the disturbance load causes irreversible damage to the specimen. Xiao et al. [36] define the damage variable for the i-th cycle as the ratio of the axial strain in this cycle to the axial strain at ultimate failure. Figure 8 is the relationship between the specimen's damage variable and the relative cycle number under different lateral pressures, where the relative cycle number is the ratio of the i-th cycle to the total number of cycles. The discussion will exclude specimen R0-1 since it was destroyed after only three cycles of disturbance. As seen from Figure 8, the damage variables of specimens R2-2, R4-1, and R6-3 are mainly divided into two stages with the increase in relative period: initially, they experience slow growth until reaching the middle stage and then accelerate towards failure. Compared with the three stages of typical uniaxial cyclic disturbance damage variables [37], there are some notable differences. Firstly, the disappearance of the initial rapid growth stage can be attributed to the presence of lateral pressure, which reduces the axial deformation of the specimens during the cyclic disturbance stage and leads to a gradual increase in damage during the initial phase. Secondly, the later growth stage slows down due to the application of lateral pressure, which increases the compressive strength and elastic modulus of the specimen [38], thereby enhancing its ability to resist deformation. Consequently, the growth of damage in the later stages of failure is decelerated.

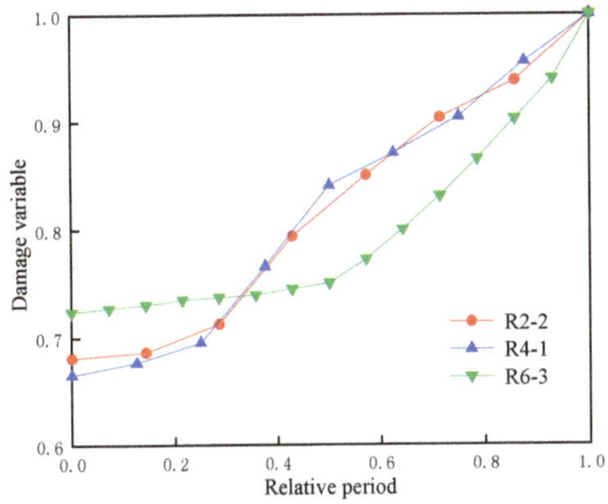

Figure 8. Relationship between damage variables of specimens under different side pressures with relative period.

4.2. Failure Mechanism of Rock Under Disturbed Load Under Different Lateral Pressures

The failure mechanisms of rock under varying lateral pressures are depicted in Figure 9. In this test, the applied disturbance loads consist of a constant static load and a cyclic loading and unloading at a constant frequency.

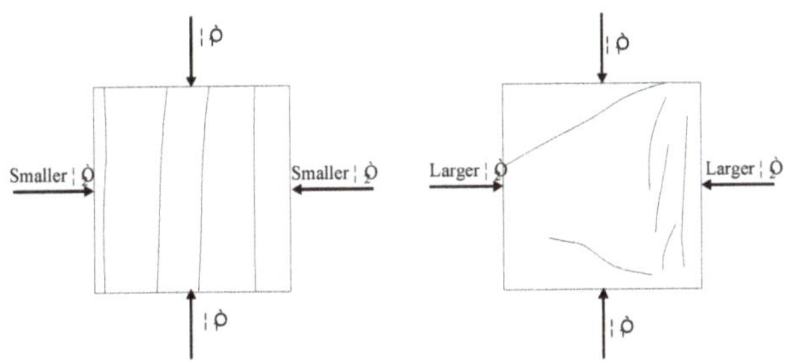

(**a**) Smaller lateral pressure conditions (**b**) Larger lateral pressure conditions

Figure 9. Failure mechanism of specimens with different lateral pressure.

Under smaller lateral pressure conditions (as shown in Figure 9a), the friction between the internal cracks of the rock specimen is minimal during the axial stress σ_1 loading process. Consequently, the rock tends to expand and deform in the directions of the free surfaces σ_2 and σ_3. If cyclic loading and unloading continue at a certain frequency, the lateral unconstrained nature of the rock prevents the effective recovery of deformation during unloading, leading to an accelerated deformation rate. Therefore, under low lateral pressure, the rock specimen is susceptible to rapid destruction after the application of disturbance loads, primarily due to tensile failure.

Conversely, under larger lateral pressure conditions (as shown in Figure 9b), the friction between the end of the rock specimen and the press, denoted by σ_3, increases during the axial stress loading. The combined effect of σ_1, σ_2, and σ_3 creates an approximate triaxial compression state ($\sigma_1 > \sigma_2 > \sigma_3 > 0$) at the end of the specimen, leading to shear failure. Meanwhile, the middle part of the specimen, constrained by continuous extrusion, can only expand towards the free surface, resulting in tensile cracks. Under larger lateral pressure, the specimen is prone to shear cracks at the ends and tensile cracks in the middle. The significant friction between internal cracks under high lateral pressure somewhat mitigates the impact of cyclic loading and unloading on deformation. Moreover, the high frequency of cyclic loading and unloading increases the internal friction of the rock specimen, causing more pronounced deformation in the middle tensile area compared to the end shear area, leading to more severe damage in the middle area.

Based on the analysis of the failure mechanisms under disturbance loads and various lateral pressures, it is clear that rock can experience tensile failure under both low and high lateral pressures. The failure rate in the tensile failure area increases rapidly under the influence of disturbance loads. To prevent the impact of disturbance loads on underground rock engineering, measures such as appropriately increasing the density of anchors in the affected area to control the destructive influence of lateral tensile deformation of the surrounding rock are essential.

5. Conclusions

The experimental investigation into the mechanical and failure characteristics of gypsum specimens under disturbance loads at varying lateral pressures has yielded several key findings:

- The lateral pressure significantly enhances the ability of the specimen to resist the disturbance load. With the increase in lateral pressure, the number of failure cycles increases, the compaction stage decreases, the yield stage expands, and the hysteresis curve of the failure stage under the disturbance load is denser.
- Under a constant number of cyclic disturbances, the increase in lateral pressure reduces the ability of lateral and axial deformation, while increasing the elastic modulus and the average number of cycles of failure. Under constant lateral pressure, the increase in cyclic disturbances times leads to the continuous increase in residual plastic strain and the decrease in elastic modulus.
- The application of disturbance loads leads to more pronounced free-surface spalling and peeling of gypsum specimens, which accelerates the failure rate under different lateral pressures.
- The lateral pressure level significantly affects the crack evolution and failure behavior of gypsum specimens. The lower lateral pressures mainly lead to tensile cracking and tensile splitting failure, while higher lateral pressures leading to the initial cracking are not obvious. The failure characteristics of the specimen are shear fracture at both ends and tensile fracture in the middle.

The results obtained in this study can provide some useful considerations and suggestions on the stability of underground rock engineering.

Author Contributions: Conceptualization, Y.L.; methodology, Y.L.; software, Y.L. and Z.Y.; validation, Z.Y.; formal analysis, Y.L. and Z.Y.; investigation, Z.Y.; resources, F.Z. and Q.W.; data curation, Y.L.; writing—original draft preparation, Y.L.; writing—review and editing, F.Z. and Q.W.; visualization, Y.L.; supervision, F.Z.; project administration, F.Z. and Z.Y.; funding acquisition, F.Z. and Q.W. All authors have read and agreed to the published version of the manuscript.

Funding: The work was supported by the National Natural Science Foundation of China (Grant Nos. 51674116 and 51904101) and Natural Science Foundation of Hunan Province (Grant Nos. 2020JJ4311 and 2020JJ5188).

Institutional Review Board Statement: Ethical review and approval were waived for this study not involving humans or animals. Approval for the study was not required in accordance with local/national legislation.

Informed Consent Statement: Informed consent was obtained from all subjects involved in the study.

Data Availability Statement: The data presented in this study are available on request from the corresponding author. The data are not publicly available due to we tested in the laboratory.

Conflicts of Interest: The authors declare no conflicts of interest.

References

1. Geranmayeh Vaneghi, R.; Thoeni, K.; Dyskin Arcady, V.; Sharifzadeh, M.; Sarmadivaleh, M. Strength and Damage Response of Sandstone and Granodiorite under Different Loading Conditions of Multistage Uniaxial Cyclic Compression. *Int. J. Geomech.* **2020**, *20*, 04020159. [CrossRef]
2. Liu, Y.; Dai, F.; Dong, L.; Xu, N.; Feng, P. Experimental Investigation on the Fatigue Mechanical Properties of Intermittently Jointed Rock Models Under Cyclic Uniaxial Compression with Different Loading Parameters. *Rock Mech. Rock Eng.* **2018**, *51*, 47–68. [CrossRef]
3. Si, X.; Gong, F. Strength-weakening effect and shear-tension failure mode transformation mechanism of rockburst for fine-grained granite under triaxial unloading compression. *Int. J. Rock Mech. Min. Sci.* **2020**, *131*, 104347. [CrossRef]
4. Wang, Y.; Zhao, L.; Han, D.-h.; Qin, X.; Ren, J.; Wei, Q. Micro-mechanical analysis of the effects of stress cycles on the dynamic and static mechanical properties of sandstone. *Int. J. Rock Mech. Min. Sci.* **2020**, *134*, 104431. [CrossRef]
5. Gong, F.; Yan, J.; Li, X.; Luo, S. A peak-strength strain energy storage index for rock burst proneness of rock materials. *Int. J. Rock Mech. Min. Sci.* **2019**, *117*, 76–89. [CrossRef]
6. Wu, Q.-h.; Weng, L.; Zhao, Y.-l.; Feng, F. Influence of infilling stiffness on mechanical and fracturing responses of hollow cylindrical sandstone under uniaxial compression tests. *J. Cent. South Univ.* **2021**, *28*, 2485–2498. [CrossRef]
7. Zhang, Q.B.; Zhao, J. A Review of Dynamic Experimental Techniques and Mechanical Behaviour of Rock Materials. *Rock Mech. Rock Eng.* **2014**, *47*, 1411–1478. [CrossRef]
8. Bello, I.; Martínez-Abella, F.; Wardeh, G.; González-Fonteboa, B. Complete stress-strain analysis of masonry prisms under compressive loading-unloading cycles through digital image correlation. *Eng. Struct.* **2024**, *298*, 117088. [CrossRef]
9. Liu, Z.; Cao, P.; Zhao, Q.; Cao, R.; Wang, F. Deformation and damage properties of rock-like materials subjected to multi-level loading-unloading cycles. *J. Rock Mech. Geotech. Eng.* **2023**, *15*, 1768–1776. [CrossRef]
10. Meng, Q.-b.; Liu, J.-f.; Huang, B.-x.; Zhang, Z.-z.; Long, J.-k.; Wu, J.-y. Experimental analysis of the height–diameter ratio effect of rock energy under uniaxial cyclic loading–unloading conditions. *Bull. Eng. Geol. Environ.* **2023**, *82*, 283. [CrossRef]
11. Zhang, C.; Wang, Y.; Ruan, H.; Ke, B.; Lin, H. The strain characteristics and corresponding model of rock materials under uniaxial cyclic load/unload compression and their deformation and fatigue damage analysis. *Arch. Appl. Mech.* **2021**, *91*, 2481–2496. [CrossRef]
12. Gong, F.-Q.; Wu, C.; Luo, S.; Yan, J.-Y. Load–unload response ratio characteristics of rock materials and their application in prediction of rockburst proneness. *Bull. Eng. Geol. Environ.* **2019**, *78*, 5445–5466. [CrossRef]
13. Zhou, X.-P.; Jiang, D.-C.; Zhao, Z. Digital Evaluation of Micro-Pore Water Effects on Mechanical and Damage Characteristics of Sandstone Subjected to Uniaxial, Cyclic Loading–Unloading Compression by 3D Reconstruction Technique. *Rock Mech. Rock Eng.* **2022**, *55*, 147–167. [CrossRef]
14. Huang, Y.; Xie, T.; Xu, X.; Ding, Q. Experimental study of creep Poisson ratio of hydraulic concrete under multi-age loading-unloading conditions. *Constr. Build. Mater.* **2021**, *310*, 125269. [CrossRef]
15. Tian, W.-L.; Yang, S.-Q.; Wang, J.-G.; Dong, J.-P. Failure behavior of the thermal treated granite under triaxial cyclic loading–unloading compression. *Geomech. Geophys. Geo-Energy Geo-Resour.* **2021**, *7*, 19. [CrossRef]
16. Yuan, C.; Guo, Y.; Wang, W.; Cao, L.; Fan, L.; Huang, C. Study on "Triaxial Loading-Unloading-Uniaxial Loading" and Microscopic Damage Test of Sandstone. *Front. Earth Sci.* **2020**, *8*, 78. [CrossRef]
17. Wang, C.; Zhang, D.; Yu, B.; Li, S. Deformation and Seepage Characteristics of Coal Under True Triaxial Loading–Unloading. *Rock Mech. Rock Eng.* **2023**, *56*, 2673–2695. [CrossRef]
18. Tan, H.; Liu, H.; Shi, X.; Ma, H.; Qiu, X.; Guo, Y.; Ban, S. Mechanical and Acoustic Response of Low-Permeability Sandstone under Multilevel Cyclic Loading-Unloading Stress Paths. *Energies* **2023**, *16*, 6821. [CrossRef]
19. Ning, Z.; Xue, Y.; Li, Z.; Su, M.; Kong, F.; Bai, C. Damage Characteristics of Granite Under Hydraulic and Cyclic Loading–Unloading Coupling Condition. *Rock Mech. Rock Eng.* **2022**, *55*, 1393–1410. [CrossRef]
20. Isakov, M. Cyclic Impact Test with the Hopkinson Bar: Application to Granite. *J. Dyn. Behav. Mater.* **2024**, *10*, 75–91. [CrossRef]
21. Sun, B.; Yang, P.; Liu, S.; Zeng, S. Impact dynamic characteristics and constitutive model of granite damaged by cyclic loading. *J. Mater. Res. Technol.* **2023**, *24*, 333–345. [CrossRef]
22. Zha, X.; Jiang, F.; Xu, X. Investigating the high frequency fatigue failure mechanisms of mono and multilayer PVD coatings by the cyclic impact tests. *Surf. Coat. Technol.* **2018**, *344*, 689–701. [CrossRef]
23. Dai, B.; Shan, Q.-w.; Chen, Y.; Luo, X.-y. Mechanical and energy dissipation characteristics of granite under cyclic impact loading. *J. Cent. South Univ.* **2022**, *29*, 116–128. [CrossRef]
24. Luo, N.; Suo, Y.; Zhang, H.; Chai, Y.; Zhai, C.; Qu, Z.; Bai, G. On dynamic behaviors and failure of bedding coal rock subjected to cyclic impact. *Explos. Shock. Waves* **2023**, *43*, 14. (In Chinese). [CrossRef]
25. Fan, L.F.; Li, H.; Xi, Y.; Wang, M. Effect of Cyclic Impact on the Dynamic Behavior of Thermally Shocked Granite. *Rock Mech. Rock Eng.* **2024**, *57*, 4473–4491. [CrossRef]
26. Du, K.; Yang, C.; Su, R.; Tao, M.; Wang, S. Failure properties of cubic granite, marble, and sandstone specimens under true triaxial stress. *Int. J. Rock Mech. Min. Sci.* **2020**, *130*, 104309. [CrossRef]
27. Jiang, Q.; Su, G.; Feng, X.-t.; Chen, G.; Zhang, M.-z.; Liu, C. Excavation Optimization and Stability Analysis for Large Underground Caverns Under High Geostress: A Case Study of the Chinese Laxiwa Project. *Rock Mech. Rock Eng.* **2019**, *52*, 895–915. [CrossRef]

28. Wang, S.; Li, X.; Yao, J.; Gong, F.; Li, X.; Du, K.; Tao, M.; Huang, L.; Du, S. Experimental investigation of rock breakage by a conical pick and its application to non-explosive mechanized mining in deep hard rock. *Int. J. Rock Mech. Min. Sci.* **2019**, *122*, 104063. [CrossRef]
29. Bobet, A.; Einstein, H.H. Fracture coalescence in rock-type materials under uniaxial and biaxial compression. *Int. J. Rock Mech. Min. Sci.* **1998**, *35*, 863–888. [CrossRef]
30. Choo, J.; Sun, Y.; Fei, F. Size effects on the strength and cracking behavior of flawed rocks under uniaxial compression: From laboratory scale to field scale. *Acta Geotech.* **2023**, *18*, 3451–3468. [CrossRef]
31. Ge, Y.; Ren, G.; Zhang, C.; Shi, Y.; Zhang, L. Study on Damage Characteristics and Failure Modes of Gypsum Rock under Dynamic Impact Load. *Materials* **2023**, *16*, 3711. [CrossRef] [PubMed]
32. Chen, X.; Tang, M.; Tang, C. Effect of Confining Pressure on the Damage Evolution and Failure Behaviors of Intact Sandstone Samples During Cyclic Disturbance. *Rock Mech. Rock Eng.* **2022**, *55*, 19–33. [CrossRef]
33. Liu, E.; He, S.; Xue, X.; Xu, J. Dynamic Properties of Intact Rock Samples Subjected to Cyclic Loading under Confining Pressure Conditions. *Rock Mech. Rock Eng.* **2011**, *44*, 629–634. [CrossRef]
34. Song, H.; Duan, C.; Zhao, Y.; Teng, T.; Hu, S.; Wu, Y.; Gong, Z.; Jiang, Y.; Wu, J. Effect of cyclic loading-unloading on the mechanical anisotropy of coal under uniaxial compressive condition. *Bull. Eng. Geol. Environ.* **2024**, *83*, 131. [CrossRef]
35. Cao, Y.; Yan, Q.; Zhang, S.; Cai, F. Experimental Research on Anisotropy Characteristics of Shale under Triaxial Incremental Cyclic Loading and Unloading. *Appl. Sci.* **2024**, *14*, 2602. [CrossRef]
36. Xiao, J. Theoretical and Experimental Investigation on Fatigue Properties of Rock Under Cyclic Loading. Ph.D. Thesis, Central South University, Changsha, China, 2010.
37. Wang, R.; Wei, C.; Liu, J. Macro and micro characteristics of jointed sandstone under cyclic loading and unloading. *Chin. J. Rock Mech. Eng.* **2023**, *42*, 810–820. [CrossRef]
38. Zhang, W.; Guo, W.; Wang, Z. Influence of lateral pressure on mechanical behavior of different rock types under biaxial compression. *J. Cent. South Univ.* **2022**, *29*, 3695–3705. [CrossRef]

materials

MDPI

Article

Mechanical Properties and Damage Constitutive Model of Saturated Sandstone Under Freeze–Thaw Action

Meimei Feng [1,2], Xiaoxiao Cao [3,*], Taifeng Wu [2] and Kangsheng Yuan [2]

1 State Key Laboratory of Intelligent Construction and Healthy Operation and Maintenance of Deep Underground Engineering, China University of Mining and Technology, Xuzhou 221116, China; fengmeimei@cumt.edu.cn

2 School of Mechanics & Civil Engineering, China University of Mining & Technology, Xuzhou 221116, China; ts24030028a31@cumt.edu.cn (T.W.); ts20030220p31@cumt.edu.cn (K.Y.)

3 Laboratory of Rock Engineering & Mining Machinery, Department of Earth Resources Engineering, Kyushu University, Fukuoka 8190395, Japan

* Correspondence: cao.xiaoxiao.037@s.kyushu-u.ac.jp

Abstract: In order to investigate the impact of freeze–thaw damage on sandstone under the coupling of ground stress and pore water pressure, three types of porous sandstone were subjected to freezing at different negative temperatures (−5 °C, −10 °C, −15 °C, and −20 °C). Subsequently, hydraulic coupling triaxial compression tests were conducted on the frozen and thawed sandstone. We analyzed the effects of porosity and freezing temperature on the mechanical properties of sandstone under hydraulic coupling and performed nuclear magnetic resonance tests on sandstone samples before and after freezing and thawing. The evolution of the pore structure in sandstone at various freezing and thawing stages was studied, and a statistical damage constitutive model was established to validate the test results. The results indicate that the stress–strain curves of sandstone samples under triaxial compression after a freeze–thaw cycle exhibit minimal changes compared to those without freezing at normal temperature. The peak deviator stress shows a decreasing trend with decreasing freezing temperature, particularly between −5 °C and −10 °C, and then gradually stabilizes. The elastic modulus of sandstone with different porosity decreases with the decrease in freezing temperature, and the decrease is more obvious in the range of −5 °C~−10 °C, decreasing by 2.33%, 6.11%, and 10.5%, respectively. Below −10 °C, the elastic modulus becomes similar to that at −10 °C, and the change tends to stabilize. The nuclear magnetic porosity of sandstone samples significantly increases after freezing and thawing. The smaller the initial porosity, the greater the rate of change in nuclear magnetic porosity after a freeze–thaw cycle. The effects of freeze–thaw damage on the T2 distribution of sandstone with different porosity levels vary. We established a statistical damage constitutive model considering the combined effects of freeze–thaw damage, ground stress, and pore water pressure. The compaction coefficient K was introduced into the constitutive model for optimization. The change trend of the theoretical curve closely aligns with that of the test curve, better characterizing the stress–strain relationship of sandstone under complex pressure environments. The research findings can provide a scientific basis for wellbore wall design and subsequent maintenance in complex environments.

Keywords: freeze–thaw; water-force coupling; mechanical properties; nuclear magnetic resonance; damage model

1. Introduction

Geotechnical engineering construction inevitably faces challenges in complex geological environments, including high groundwater levels and low surrounding rock strength. Sudden events such as water inrushes, mud outbursts, and significant deformations of surrounding rock pose threats to construction safety [1,2]. The freezing method, commonly employed in China, is an effective means to control groundwater [3]. Concurrently, the mechanical properties of rocks are significantly enhanced after freezing compared to their pre-frozen state [4,5]. The deformation and strength characteristics of rocks after thawing, influenced by the combined effects of water pressure and ground stress [6,7], directly impact the design and subsequent maintenance of wellbore walls. Therefore, it is necessary to conduct hydraulic coupled triaxial compression tests on rocks after freeze–thaw cycles, and explore and analyze the mechanical characteristics and influencing factors of sandstone under the coupled effects of water pressure and confining pressure.

Several researchers, both nationally and internationally, have investigated the influence of freeze–thaw cycles on the physical and mechanical properties of rocks. For instance, Khanlari G et al. [8] conducted P-wave velocity, porosity, and uniaxial compressive strength tests on five different types of sandstone after freeze–thaw cycles, studying the impact on the physical and mechanical properties of red sandstone in central Iran. Cao et al. [9] discussed the effects of hydrochemical corrosion and freeze–thaw cycles on the dynamic mechanical properties and microstructure of rocks. Similarly, Gao et al. [10] investigated changes in density, ultrasonic velocity, and both static and dynamic mechanical behavior of sandstone subjected to 10 freeze–thaw cycles at various low temperatures, revealing the effects on sandstone density, P-wave velocity, stress–strain relationships, static and dynamic uniaxial compressive strength, static elastic modulus, and dynamic energy absorption. In related work, Jamshidi et al. [11] examined multiple mechanical indicators of granite post-freeze–thaw cycles, finding the greatest impact on tensile strength and the smallest effect on uniaxial compressive strength. Additional studies focused on the effects of freeze–thaw cycles on soil and rock under varying conditions. Shi et al. [12] quantitatively analyzed the damage degree of red sandstone subjected to freeze–thaw cycles using triaxial compression tests, incorporating mechanical parameters like peak stress, elastic modulus, and Poisson's ratio, and established a damage constitutive model for freeze–thaw loads. In cold regions, freeze–thaw cycles and water environments usually have a long-term impact on the mechanical effects of rocks. Therefore, many scholars have conducted targeted research on saturated rocks under freeze–thaw cycles. A large number of scholars' research on freeze–thaw saturated rocks mainly focuses on the changes in physical properties such as rock porosity, density, and wave velocity [13–15]. Studying the impact of freeze–thaw cycles on the physical properties of saturated rocks is of great significance for practical engineering. During the freeze–thaw process, the pore water inside the rock expands after freezing, and the frost heave force generated by it causes the cracks inside the rock to gradually expand. In view of this feature, Hou et al. [16] improved the physical properties of sandstone by the liquid nitrogen freeze–thaw method, thereby improving the production capacity of tight sandstone gas. In order to evaluate the durability of actual rock mass engineering such as wellbore walls and tunnels, Huang et al. [17] proposed a discrete element method that can incorporate smaller and different sized particles to evaluate the durability impact of freeze–thaw cycle rock mass engineering.

These studies provide valuable insights into how freeze–thaw cycling affects physical and mechanical properties, but further investigation is required to understand the microstructural damage induced by freeze–thaw processes. In an effort to understand the underlying mechanisms of microstructural damage caused by freeze–thaw action, numerous researchers have employed advanced testing techniques. Liu et al. [18,19] used three-dimensional digital rock technology to analyze microscale pore structure changes in rock samples during freeze–thaw processes under varying confining pressures and temperatures. Their triaxial seepage tests on sandstone, along with acoustic emission monitoring, provided insights into the mechanical and hydraulic behavior of mudstone post-freeze–thaw cycling. In related work, Chen et al. [20] employed triaxial compression tests, acoustic emission monitoring, and mercury intrusion porosimetry to investigate the impact of freeze–thaw cycles on the compressive strength and internal microstructure of sandstone. At the macroscopic level, the failure mode of sandstone shifted from splitting failure to shear failure, while microscopic analysis showed an increase in the proportion of large pores. Song et al. [21] expanded this research by studying the damage mechanical properties of red sandstone with varying saturation levels under freeze–thaw cycles. Using scanning electron microscopy and uniaxial compression tests on red sandstone samples with five different saturation levels, they developed macroscopic statistical damage equations for freeze–thaw failure in rocks with different saturation levels. These studies collectively highlight the need for a deeper understanding of freeze–thaw-induced microstructural damage, as it plays a critical role in the mechanical degradation of rocks. While significant progress has been made in freeze–thaw research, much work remains to be done, particularly in developing constitutive models to quantify freeze–thaw damage. Numerous scholars have sought to characterize freeze–thaw damage using various parameters and have developed constitutive models for rocks subjected to freeze–thaw cycles [22]. For instance, Feng et al. [23] investigated changes in chemical composition, microstructure, and damage mechanical properties of sandstone subjected to freeze–thaw cycles in acidic environments, utilizing X-ray diffraction, scanning electron microscopy, and uniaxial compression tests. They quantified the strength damage of sandstone using longitudinal wave velocity and optimized a statistical damage model. Zheng et al. [24] proposed a novel freeze–thaw damage model for rocks, validated through experiments on Sichuan sandstone and granite under triaxial compression, effectively capturing nonlinear mechanical behavior and incorporating void compaction and confining pressure influences, providing insights into deformation and failure mechanisms under freeze–thaw cycles. Jia et al. [25] developed a mesoscopic damage constitutive model for granite under freeze–thaw cycles (FTCs), based on tests of granite from Northeast China. The study found that the first 20 FTCs caused significant damage, with slower degradation afterward. The model, incorporating mesoscopic parameters and correction coefficients, performed well in uniaxial compression tests.

Research on the impact of freeze–thaw cycles on the fundamental physical and mechanical properties of rocks has been relatively extensive. However, there is a scarcity of studies addressing the mechanical properties of rocks and the evolution of damage under the coupled effects of water and mechanics following rock freeze–thaw cycles. For instance, during the construction process involving the freezing method, engineering rock continues to experience the combined influences of geostress and water pressure. In this study, freeze–thaw tests were conducted on three sandstone samples with different porosities at different negative temperatures, and then hydraulic coupled triaxial compression tests and nuclear magnetic resonance tests were conducted on the same type of sandstone. The study focused on analyzing the effects of freeze–thaw on the mechanical properties of sandstone, and studied the microevolution mechanism of the sandstone pore structure during freeze–thaw cycles. In addition, a statistical damage constitutive model that comprehensively considers the effects of freeze–thaw damage, ground stress, and pore water pressure was established. The research results provide a scientific basis for shaft wall design and subsequent maintenance, especially in special geological environments such as alpine areas, deep mines, and frozen soil areas, where the effects of freeze–thaw cycles on rock mechanical behavior are particularly significant. The constitutive model established through this experiment can provide theoretical support for underground engineering, tunnel construction, shaft wall reinforcement, etc., in these areas, and help predict the mechanical degradation characteristics of rocks under freeze–thaw cycles, providing a basis for the optimization of construction and maintenance plans.

2. Materials and Methods

2.1. Sample Preparation and Handling

The experiment utilized three different porosity sandstone samples, sourced from a coal mine in western China. According to the specifications set by the International Society for Rock Mechanics (ISRM), rocks with consistent texture and no surface cracks were selected [26]. A cylindrical standard rock sample of diameter 50 mm and height 100 mm was employed, with both ends polished to meet the criteria for parallel accuracy ($\leq$0.05 mm) and vertical accuracy ($\leq$0.25°). To reduce the discreteness of the experimental results, the processed samples underwent acoustic testing, and those exhibiting similar and more stable wave velocities were selected for the experiment.

Preceding the experiment, the sandstone sample was heated to 110 °C in a drying oven and dried for 24 h. Upon removal and cooling, the sample was weighed to determine its dry density. Subsequently, the cooled sandstone sample underwent vacuum saturation in a saturation barrel. The water surface in the barrel exceeded 5 cm above the upper surface of the sample, and the saturation pressure was set at 0.1 MPa. Once no bubbles emerged from the rock sample's surface and the saturation time exceeded 12 h, the sample was considered saturated. It was then removed and weighed to calculate the saturation density. The initial porosity of sandstone is calculated by the difference between saturated density and dry density. Three sandstone samples with varying porosities were selected to measure basic mechanical properties, such as uniaxial compressive strength (UCS). The fundamental physical parameters are detailed in Table 1.

Table 1. Basic physical parameters of sandstone samples.

Sample	Dry Density (g/cm^3)	Saturation Density (g/cm^3)	Porosity n (%)	UCS (MPa)
A	2.37	2.41	3.79	76.59
B	2.23	2.33	11.61	29.69
C	2.16	2.32	16.21	18.99

The freeze–thaw process was conducted on a fully automatic low-temperature freeze–thaw cycle testing machine. Referring to the temperature data of the freezing method construction, we set the freezing temperatures to −5 °C, −10 °C, −15 °C, and −20 °C. To ensure sufficient freezing of the rock samples, the low-temperature constant temperature time was set to 12 h. After the freezing was completed, the frozen sandstone samples were immersed in room-temperature water for melting and backup, and the thawing time was also set to 12 h. The detailed sample processing process is shown in Figure 1. According to the distribution pattern of geostress in China and the feasibility of the experimental system, the pore water pressure [27], horizontal geostress, and vertical geostress of the sample were taken as 3 MPa, 4.5 MPa, and 7.5 MPa, respectively. The cylindrical rock samples of diameter 25 mm and height 50 mm were selected for nuclear magnetic resonance testing under the same conditions as the triaxial compression sample.

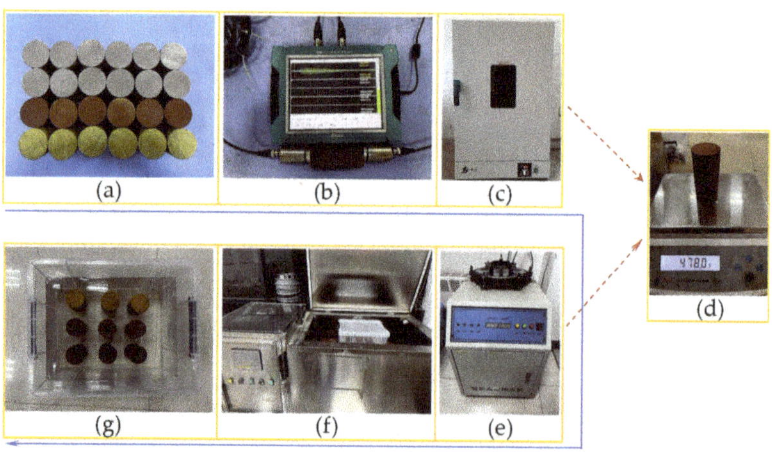

Figure 1. Sample preparation: (**a**) rock samples; (**b**) ultrasonic testing; (**c**) drying oven; (**d**) sample weighing; (**e**) saturation; (**f**) freezing–thawing box; (**g**) thawing of samples.

2.2. Test Equipment and Process

This experiment primarily utilized the high-pressure environment rock and soil triaxial test system (Figure 2), comprising a high-pressure triaxial pressure chamber, a confining water pressure control system, and a control monitoring system. The system is capable of delivering an axial load of up to 400 kN. The confining pressure and water pressure control system can generate a maximum pressure of 64 MPa, employing volume control with an accuracy of 0.1%. Each component can be operated and controlled independently, and they can also be cross-controlled through the control monitoring system. Real-time monitoring of data indicators from each system ensures the stability of various pressures throughout the test process.

Figure 2. Triaxial test system: (**a**) rock sample installation; (**b**) control and monitoring system; (**c**) pressure control system.

The rock sample, having undergone a single freeze–thaw cycle, was placed into the triaxial pressure chamber and connected to the inlet pipeline and displacement sensor. First, the sample was loaded with confining pressure and initial axial pressure, and then observed the stability of the pressure through the control detection system before loading the water pressure. The process of hydraulic loading was quite important and the program was complex. The outlet pressure and inlet pressure were both initially set at 0 MPa, with the target value subsequently adjusted. Once the inlet and outlet volumes were equal, the outlet pressure was set to the target value and maintained for a specific duration to ensure the stability of the internal water pressure within the sample. This allowed the commencement of the test, where axial loading was carried out using the displacement loading method at a loading speed of 0.01 mm/s.

3. Experimental Results and Analysis

3.1. Deformation Characteristics of Sandstone and Changes in Triaxial Compressive Strength

Figure 3 depicts the stress–strain curves of three types of porous sandstone undergoing hydraulic coupling at various negative temperatures following a single freeze–thaw test under triaxial compression. Notably, the peak deviatoric stress for all three types of sandstone decreased, with the attenuation amplitude exhibiting a positive correlation with the porosity of the sandstones under different negative temperature freeze–thaw conditions.

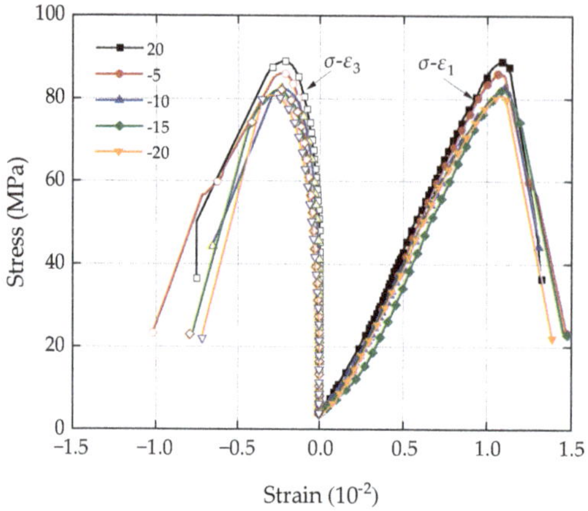

(**a**)

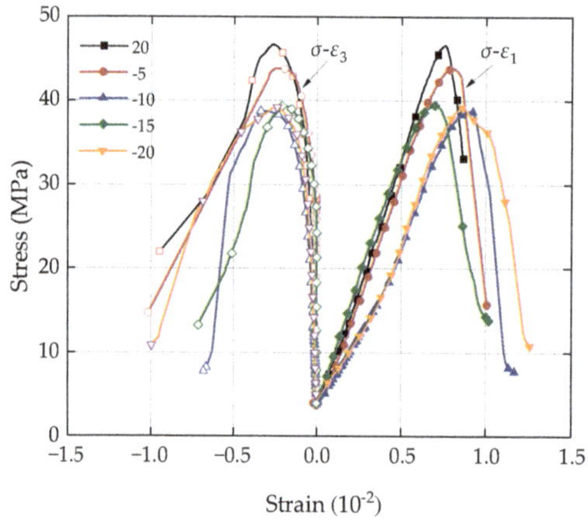

(**b**)

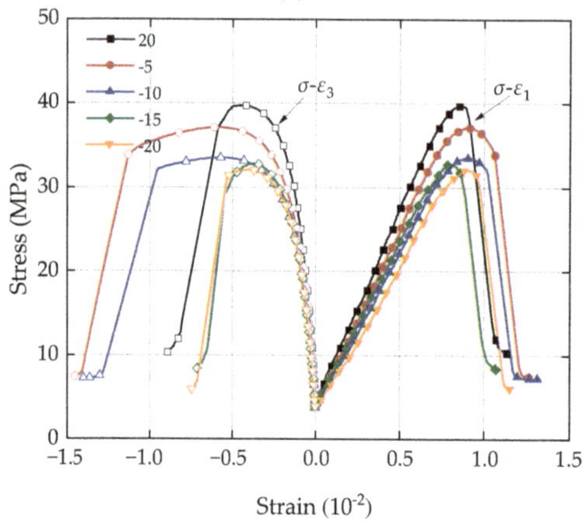

(**c**)

Figure 3. Triaxial compression stress–strain curves of sandstone with different porosity: (**a**) n = 3.79%; (**b**) n = 11.61%; (**c**) n = 16.21%.

For instance, in the sandstone sample group with smaller porosity (n = 3.79%), the triaxial compression stress–strain curve after one freeze–thaw cycle follows a similar trend to the stress–strain curve of the sandstone sample at room temperature. The change in peak deviatoric stress is relatively minor in this scenario. However, in the sandstone sample group with porosity n = 11.61% and n = 16.21%, the peak deviator stress of the sample attenuates significantly in the range of freezing and thawing temperature decreasing from room temperature to −10 °C, while the changes in other situations are not significant.

Figure 4 illustrates the relationship between the peak deviatoric stress of sandstone with different porosities and the variation in freezing temperature. The chart reveals the following observations: (1) The peak deviator stress of sandstones with different porosities decreased from 88.94 MPa, 46.56 Mpa, and 39.72 Mpa at room temperature to 80.88 Mpa, 39.16 Mpa, and 32.08 Mpa at −20 °C, indicating a consistent decreasing trend after undergoing a freeze–thaw cycle. (2) The peak deviatoric stress of the sandstones experiences a significant decrease from room temperature to −10 °C, with the attenuation becoming less pronounced within the range of −10 °C to −20 °C. This suggests that the freezing temperature has a considerable impact on the triaxial compressive strength of sandstone after thawing, especially at temperatures of −5 °C and −10 °C. When the freezing temperature drops below −10 °C, the decrease in the peak strength of sandstone slows down significantly. The primary reason for the strength attenuation post-freeze–thaw lies in the frost heave effect during freezing, which weakens the bonding state between sandstone particles. As the freezing process unfolds, free water in the external pores freezes first, isolating the connection between the internal and external water of the sample. As the freezing process progresses, unfrozen water migrates to the interior of the sample, causing the central water pressure of the sample to increase, further expanding the small-pore cracks, and ultimately completely freezing the high-pressure area in the center of the sample and the free water in the smaller pores. Upon melting, the frost heave effect subsides, leading to a decrease in internal pressure within pores and cracks. The effect of geostress and pore water pressure weakens the rock cementation ability, resulting in significant peak deviatoric stress attenuation in sandstone samples, particularly at higher freezing temperatures.

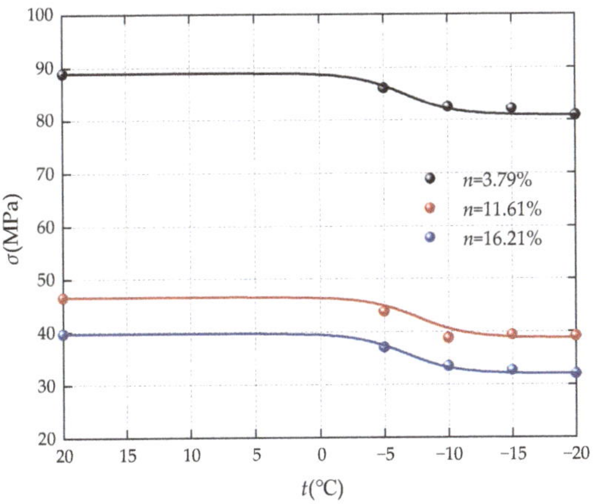

Figure 4. Relationship curve between peak deviatoric stress and freezing temperature of sandstone.

From the above analysis, it can be seen that the peak deviatoric stress of sandstone samples decreases to varying degrees before and after freeze–thaw cycles. Therefore, the reduction in compressive strength after a freeze–thaw cycle is used to describe the freeze–thaw damage of saturated sandstone samples. The relationship between the compressive strength loss rate and the freezing temperature is shown in Figure 5.

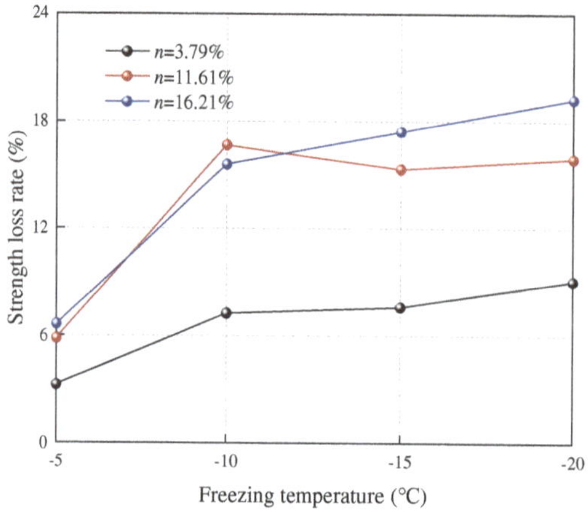

Figure 5. Relationship curve between sandstone strength loss rate and freezing temperature.

Figure 5 reveal a noticeable increase in the triaxial compressive strength loss rate of sandstone samples as the freezing temperature decreases. In the experimental group with a porosity of n = 3.79%, the triaxial compressive strength loss rate of sandstone rises from 3.23% at −5 °C to 9.06% at −20 °C, reflecting a 5.83% increase. Meanwhile, in the experimental group with a porosity of n = 16.21%, the triaxial compressive strength loss rate of sandstone elevates from 6.63% at −5 °C to 19.24% at −20 °C, indicating a 12.61% increase. These results underscore the significant impact of freezing temperature on the triaxial compressive strength of sandstone, with higher initial porosity leading to a greater strength loss rate. When n = 3.79%, the strength loss rate at −10 °C reaches 79.9% of the maximum loss rate; when n = 11.61%, the strength loss rate at −10 °C reaches 100% of the maximum loss rate; and when n = 16.21%, the strength loss rate at −10 °C reaches 80.9% of the maximum loss rate. Combined with Figure 5, it can be found that the strength loss rate of the sandstone sample changes greatly at −5 °C and −10 °C, and the strength loss rate at −15 °C and −20 °C changes less than that at −10 °C, and the change range of the loss rate is also smaller.

3.2. *Elastic Modulus and Poisson's Ratio of Sandstone with Different Porosity*

Figure 6 illustrates the relationship between the elastic modulus and Poisson's ratio of the three sandstones and the temperature change during freezing. Key observations from the figure and charts include the following: (1) The elastic modulus of the three types of sandstone (n = 3.79%, n = 11.61%, and n = 16.21%) with different porosities decreases from 4.71 Gpa, 6.42 Gpa, and 9.46 Gpa without freezing to 3.93 Gpa, 5.79 Gpa, and 8.9 Gpa at −20 °C. This indicates a decrease in the elastic modulus of sandstone with varying porosities as the freezing temperature decreases, highlighting the significant impact of freezing temperature on the elastic modulus of sandstone. (2) The Poisson ratio of the three porosity sandstones is minimally affected by freezing temperature, fluctuating within a certain range; notably, the Poisson ratio of high-porosity sandstones is larger than that of low-porosity sandstones. (3) The elastic modulus of the three sandstones decreased significantly from room temperature to −5 °C, decreasing by 2.33%, 6.11%, and 10.5%, respectively, and then further decreased from −10 °C to −20 °C. The results suggest that the temperature during freezing has a pronounced effect on the elastic modulus of sandstone after melting, particularly at −5 °C. Beyond −10 °C, the freezing temperature has a diminished impact on the elastic modulus of sandstone. The attenuation pattern of the elastic modulus in sandstone samples mirrors that of peak deviatoric stress.

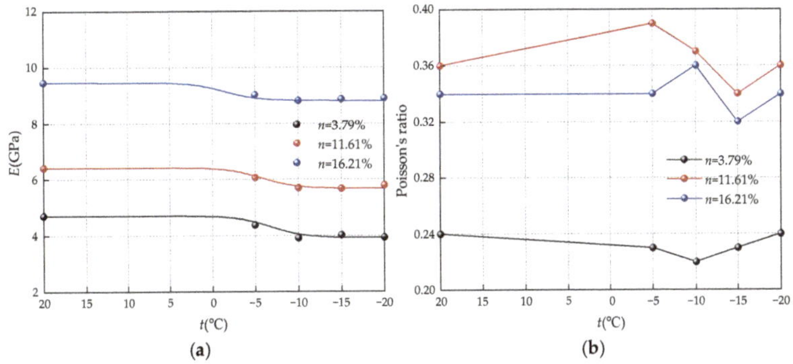

Figure 6. Variation of elastic modulus and Poisson's ratio with freezing temperature: (**a**) *E*; (**b**) Poisson's ratio.

3.3. Evolution Characteristics of Sandstone Pore Structure Based on Nuclear Magnetic Resonance (NMR) Technology

To examine the alterations in the pore structure of sandstone samples before, during, and after freezing, a low-field nuclear magnetic resonance analysis system was employed. Pore structure scanning tests were conducted on sandstone samples with varying porosities (n = 3.79%, 11.61%, and 16.21%) in the pre-freezing, freezing, and post-thawing stages. The changes in the transverse relaxation time (T2) spectrum distribution of sandstone samples with different porosities during various freezing stages were compared and analyzed, shedding light on the damage evolution mechanism of the sandstone pore structure.

3.3.1. Changing Law of Nuclear Magnetic Porosity of Sandstone

The primary principle of nuclear magnetic resonance measurement of rock porosity relies on determining the total number of hydrogen nuclei in the sample. Since nuclear magnetic resonance measures the porosity of the same sample at different stages, nuclear magnetic porosity can directly reflect the evolution process of porosity during the freezing process of sandstone. Table 2 presents the nuclear magnetic porosity of three types of sandstone samples with different porosity before freezing, after freezing, and after thawing, along with the change rate of nuclear magnetic porosity after freezing and thawing compared to that before freezing.

Table 2. Nuclear magnetic porosity and change rate of sandstone under different conditions.

Porosity (%)	Rock Specimen	Nuclear Magnetic Porosity (%)			Nuclear Magnetic Porosity Change Rate (%)	
		Before Freezing	In Freezing	After Melting	In Freezing	After Melting
3.79	N-1	9.58	10.83	10.21	13.09	6.54
	N-2	8.50	9.28	9.01	9.25	6.10
11.61	N-3	13.15	14.28	13.78	8.64	4.77
	N-4	19.57	20.48	20.21	4.63	3.26
16.21	N-5	29.00	30.36	29.50	4.66	1.70
	N-6	30.89	32.44	31.70	5.01	2.61

Observations from the table include the following: (1) Among the tested rock samples with different porosities, the nuclear magnetic porosity of the samples increased after freezing and thawing compared to before freezing, but the increase varied. For instance, the nuclear magnetic porosity of N-1 samples increased by 1.25% and 0.63% after freezing and thawing, with change rates of 13.09% and 6.54%, respectively. N-5 rock samples exhibited the lowest change rates, with increases of 4.66% and 1.70% after freezing and thawing. (2) The change rate of nuclear magnetic porosity after freezing and thawing for sandstone samples with small porosity is generally higher than that for sandstone with large porosity, indicating a negative correlation between the change rate of nuclear magnetic porosity after freezing and thawing and porosity. (3) The nuclear magnetic porosity in frozen sandstone is greater than that after thawing, suggesting that the volume of pores in frozen sandstone is larger than that after thawing.

The NMR-measured porosity primarily results from the summation of water signals within cracks of different sizes. Negative-temperature freezing causes the originally present free water in the pores to freeze and solidify, leading to volume expansion. This expansion exerts pressure on the particle surface in the pore wall, enlarging the volume occupied by the pores and even creating new tiny cracks. This process is the fundamental reason for the increase in rock porosity after freezing. Upon thawing, the frozen ice in the pores undergoes liquefaction, resulting in volume reduction and the withdrawal of pressure on the pore wall. The tiny cracks caused by frost heave close again, and the volume of larger pores diminishes with the disappearance of frost heave. Consequently, the nuclear magnetic porosity of the sample after melting is smaller than during the freezing process. Additionally, during vacuum saturation of the melted sandstone, free water re-enters the nascent cracks and tiny pores due to pressure. This action results in an increase in nuclear magnetic porosity during detection.

3.3.2. NMR T2 Spectrum Analysis of Sandstone

The T2 value corresponds to pore size, and the T2 value distribution reflects information about the pore structure inside the rock sample. Changes in the T2 spectrum of sandstone samples after freezing and thawing indicate the impact of freezing on the evolution of the pore structure. Figure 7a–c depict the T2 spectral distribution curves and cumulative nuclear magnetic porosity of rock samples with three porosity levels before and after freezing, respectively. Observations from the figures include the following:

(1) For the small initial porosity ($n = 3.79\%$), as shown in Figure 7a, after the sandstone sample undergoes freeze–thaw cycles, the first peak dominates the T2 spectrum, with the second peak contributing minimally. This suggests a scarcity of large and medium-sized pores in sandstone with small porosity, where small and medium-sized pores constitute the main components of the pore structure. Compared to the T2 curve of sandstone before freezing, the post-thaw curve notably changes on the right side of the first spectral peak, generally surpassing the pre-freezing curve. This indicates an increase in the number of corresponding-size pores on the right side of the spectral peak after freeze–thaw cycles. On the left side of the first spectral peak, T2 curves of rock samples before and after freezing largely coincide, indicating minimal impact on rocks with small pores due to freeze–thaw action.

(2) For an initial porosity of $n = 11.61\%$, as shown in Figure 7b, the first two spectral peaks of T2 distribution increase after freeze–thaw cycles, and the curves on both sides of the peaks coincide. This suggests an increase in the number of small and medium-sized pores in the rock sample due to freeze–thaw action, causing damage to the pore structure.

(3) When the initial porosity is 16.21%, the second peak value of T2 distribution significantly surpasses the pre-freezing peak value after freeze–thaw cycles. As T2 is proportional to the pore size (r), this indicates a substantial increase in the number of large-size pores in the rock sample with large porosity due to freeze–thaw action. With the increase in T2 value, the signal amplitude value of the rock sample after thawing appears to shift to the right. This phenomenon occurs because smaller-sized pores develop into larger ones under frost heave, causing the value at the front end of the T2 curve to decrease after thawing while the value at the end increases.

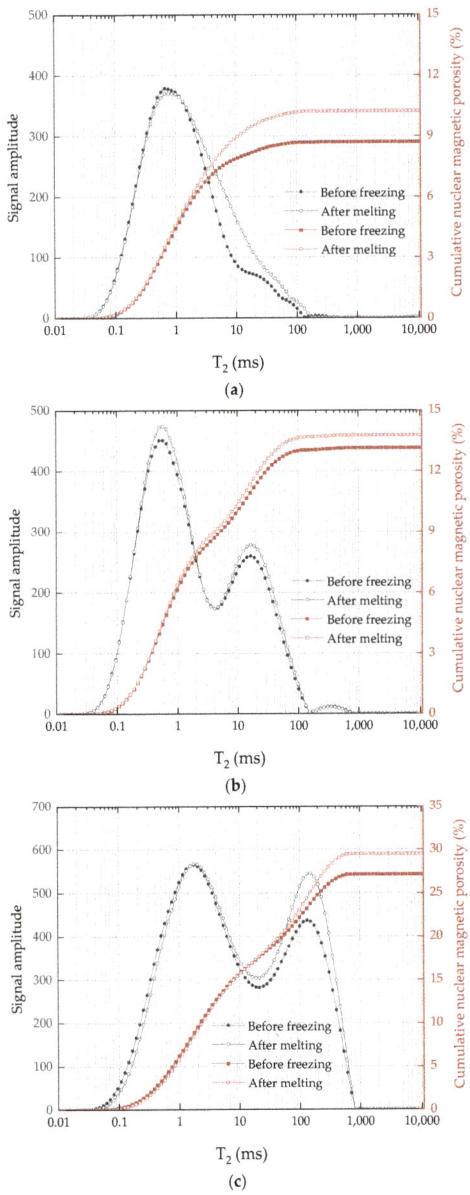

Figure 7. T2 spectrum distribution curve and cumulative nuclear magnetic porosity of sandstone before and after freezing: (**a**) n = 3.79%; (**b**) n = 11.61%; (**c**) n = 16.21%.

3.4. Statistical Damage Constitutive Model

3.4.1. Construction of Statistical Damage Constitutive Model

During the formation of rocks over a long period of time, there are various defects in the interior, including various unevenly distributed through-cracks and pores, which provide a place for the flow of groundwater in the underground aquifer. In the complex underground pressure environment, ground stress and pore water pressure have different degrees of complex effects on the failure of porous rocks. In order to explore the damage evolution mechanism of sandstone before and after freeze–thaw cycling under complex stress, based on the existing research results, this section established a statistical damage constitutive model that considers the comprehensive effects of freeze–thaw damage, ground stress, and pore water pressure.

First, the influence of pore water pressure is considered. This paper refers to the statistical damage model of rock considering pore water pressure obtained by Wang et al. [28] based on the effective stress principle and statistical damage theory:

$$\sigma_{1t} = [E\varepsilon_{1t} + (1-2\mu)(\sigma_3 - P_w)]\cdot \exp\left[-\left(\frac{F}{F_0}\right)^m\right] + (2\mu - 1)(\sigma_3 - P_w) \quad (1)$$

$$F = \frac{[E\varepsilon_{1t} + (1-2\mu)(\sigma_3 - P_w)]}{\sigma_{1t} + (1-2\mu)(\sigma_3 - P_w)}\cdot\left[\frac{\sin\varphi(\sigma_{1t} + 3\sigma_3 - 3P_w)}{\sqrt{9+3\sin^2\varphi}} + \frac{\sigma_{1t}}{\sqrt{3}}\right] \quad (2)$$

where σ_{1t} is the axial deviator stress, E and μ are elastic modulus and Poisson's ratio respectively, ε_{1t} is the actual strain measured in the experiment, P_w is the pore water pressure, F represents the random distribution variable of rock element strength, F_0, m are Weibull parameters, and φ is the internal friction angle of the rock.

In the coupling triaxial compression test of in situ stress and pore water pressure, the sample is first loaded with simulated in situ stress, and the values of axial pressure and confining pressure are different. Then, pore water pressure is loaded. The processing of rock stress–strain relationship starts after simulating the environmental stress on the sample, but in fact, the initial axial strain of the sample has already occurred during the loading process of in situ stress. According to the effective stress principle and generalized Hooke's law, the following can be concluded:

$$\varepsilon_{10} = \frac{1}{E}[\sigma_{10} - 2\mu\sigma_3 - (1-2\mu)P_w] \quad (3)$$

where σ_{10} is the simulated vertical geostress on the rock, and the actual strain ε_1 that occurs in the axial direction of the rock is the sum of the experimental measurement values ε_{1t} and the axial strain ε_{10} that occurs during the simulated geostress process, namely:

$$\varepsilon_1 = \varepsilon_{10} + \varepsilon_{1t} \quad (4)$$

By bringing Equations (3) and (4) into Equations (1) and (2), the statistical damage constitutive model of sandstone considering pore water pressure and in situ stress conditions can be obtained:

$$\sigma_{1t} = [E\varepsilon_{1t} + \sigma_{10} - 2\mu\sigma_3 - (1-2\mu)P_w]\cdot \exp\left[-\left(\frac{F}{F_0}\right)^m\right] + (2\mu - 1)(\sigma_3 - P_w) \quad (5)$$

$$F = \frac{E\varepsilon_{1t} + \sigma_{10} - 2\mu\sigma_3 - (1-2\mu)P_w}{\sigma_{1t} + (1-2\mu)(\sigma_3 - P_w)}\cdot\left[\frac{\sin\varphi(\sigma_{1t} + 3\sigma_3 - 3P_w)}{\sqrt{9+3\sin^2\varphi}} + \frac{\sigma_{1t}}{\sqrt{3}}\right] \quad (6)$$

Assuming that rock mass failure follows a Weibull distribution, the damage variable D of rock under load can be expressed as follows:

$$D = 1 - \exp\left[-\left(\frac{F}{F_0}\right)^m\right] \quad (7)$$

Due to freeze–thaw action, damage has occurred inside the rock, so the two damage factors of freeze–thaw cycling and loading should be considered comprehensively to establish the damage constitutive model.

The freeze–thaw damage variable D_r is defined by the elastic modulus of rock before and after freeze–thaw cycling, and can be expressed as follows:

$$D_r = 1 - \frac{E_r}{E_0} \tag{8}$$

where E_r is the elastic modulus of the rock after one freeze–thaw cycle, and E_0 is the elastic modulus of the rock that has not undergone freeze–thaw cycles.

Considering the above factors, the total damage variable D_t of rock subjected to freeze–thaw cycling and loading damage can be expressed as follows:

$$D_t = 1 - (1 - D_r)(1 - D) \tag{9}$$

$$D_t = D_r + D - D_r D \tag{10}$$

Combining Equations (7) and (8), the damage variable of sandstone after freeze–thaw cycling can be expressed as follows:

$$D_t = 1 - \frac{E_r}{E_0} \exp\left[-\left(\frac{F}{F_0}\right)^m\right] \tag{11}$$

According to Equation (5), the statistical damage constitutive model of sandstone after a freeze–thaw process can be obtained considering pore water pressure and geostress conditions:

$$\sigma_{1t} = [E\varepsilon_{1t} + \sigma_{10} - 2\mu\sigma_3 - (1-2\mu)P_w]\frac{E_r}{E_0} \cdot \exp\left[-\left(\frac{F}{F_0}\right)^m\right] + (2\mu - 1)(\sigma_3 - P_w) \tag{12}$$

In the early stage of stress loading on rocks, various internal defects such as cracks and pores will compress and close, appearing as concave shapes in the front of the stress–strain curve. Considering the compaction process, the compaction coefficient K [29] is introduced for correction:

$$K = \begin{cases} \log_n\left[\frac{(n-1)\varepsilon}{\varepsilon_y} + 1\right] & (\varepsilon \leq \varepsilon_y) \\ 1 & (\varepsilon \geq \varepsilon_y) \end{cases} \tag{13}$$

where ε_y is the axial strain corresponding to the peak stress of the sandstone sample, and n is the constant parameter obtained from the experimental data. After adding the compaction factor, Equation (13) can be expressed as follows:

$$\sigma_{1t} = [KE\varepsilon_{1t} + \sigma_{10} - 2\mu\sigma_3 - (1-2\mu)P_w]\frac{E_r}{E_0} \cdot \exp\left[-\left(\frac{F}{F_0}\right)^m\right] + (2\mu - 1)(\sigma_3 - P_w) \tag{14}$$

3.4.2. Determination of Model Parameters

In the established statistical damage constitutive model for sandstone after a single freeze–thaw cycle, the parameters to be determined are m and F_0, while the remaining parameters can be obtained through experimental data processing. Equation (14) can be transformed into the following:

$$\left[\frac{\sigma_{1t} - (2\mu - 1)(\sigma_3 - P_w)}{E\varepsilon_{1t} + \sigma_{10} - 2\mu\sigma_3 - (1-2\mu)P_w}\right]\frac{E_0}{E_r} = \exp\left[-\left(\frac{F}{F_0}\right)^m\right] \tag{15}$$

Taking two logarithms on both sides of the Equation (16) above gives:

$$\ln\left\{\ln\left[\frac{E\varepsilon_{1t} + \sigma_{10} - 2\mu\sigma_3 - (1-2\mu)P_w}{\sigma_{1t} - (2\mu - 1)(\sigma_3 - P_w)}\frac{E_r}{E_0}\right]\right\} = \text{mln}F - \text{mln}F_0 \tag{16}$$

Let

$$Y = \ln\left\{\ln\left[\frac{E\varepsilon_{1t} + \sigma_{10} - 2\mu\sigma_3 - (1-2\mu)P_w}{\sigma_{1t} - (2\mu-1)(\sigma_3 - P_w)}\frac{E_r}{E_0}\right]\right\} \quad (17)$$

$$X = \ln F \quad (18)$$

$$B = m \ln F_0 \quad (19)$$

Equation (16) can be transformed into the following:

$$Y = m \cdot X - B \quad (20)$$

Based on experimental data, obtain Y and X values and perform linear fitting to obtain m and B values. Then, according to the equation

$$F_0 = \exp(B/m) \quad (21)$$

the value of parameter F_0 can be obtained, which can then determine the statistical damage constitutive model of rocks under the action of in situ stress and pore water pressure after a single freeze–thaw cycle. The parameters of the statistical damage constitutive model for sandstone with different porosity values after four kinds of cryogenic freezing and thawing are shown in Table 3.

Table 3. Parameters of statistical damage constitutive model of sandstone under in situ stress and pore water pressure.

Porosity (%)	Freezing Temperature (°C)	Elastic Modulus (GPa)	Poisson's Ratio	Peak Strain (%)	F_0	m
3.79	−5	9.021	0.233	1.0743	99.564	9.049
	−10	8.813	0.220	1.1061	97.211	9.175
	−15	8.867	0.225	1.097	96.438	9.832
	−20	8.903	0.215	1.0925	95.645	9.257
11.61	−5	6.059	0.392	0.8041	54.655	6.646
	−10	5.693	0.373	0.9164	53.306	7.452
	−15	5.667	0.343	0.6964	50.886	6.672
	−20	5.790	0.352	0.8664	50.642	5.283
16.21	−5	4.375	0.340	0.9028	47.580	6.629
	−10	3.919	0.363	0.9043	45.323	7.347
	−15	4.016	0.316	0.8057	41.711	7.360
	−20	3.926	0.301	0.9043	41.358	7.931

3.4.3. Verification of the Statistical Damage Composition Model

Figures 8–10 present a comparison between the theoretical and experimental curves of three constitutive models applied to sandstone with varying porosities. Observing the figures, it becomes evident that the experimental curves for porosities of 3.79% and 11.67% align more closely with the theoretical model curve that includes an additional compaction coefficient K. Conversely, the experimental curve for sandstone with a porosity of 16.21% more closely resembles the theoretical model curve without the added compaction coefficient K. This observation may stem from the fact that high-porosity sandstone samples exhibit weak inter-particle cementation and lower rock strength. Following freeze–thaw damage, internal damage intensifies, leading to a higher compression degree in the early stage of loading failure for high-porosity sandstone samples due to simulated in situ stress and pore water pressure loading. Consequently, the stress–strain curve obtained from the experiment directly transitions through the compaction stage into the elastic deformation stage. However, the simulated in situ stress loading did not fully compact the other two samples with smaller porosities, and these samples underwent a compaction process during the loading failure stage, bringing them closer to the theoretical model with a compaction coefficient K.

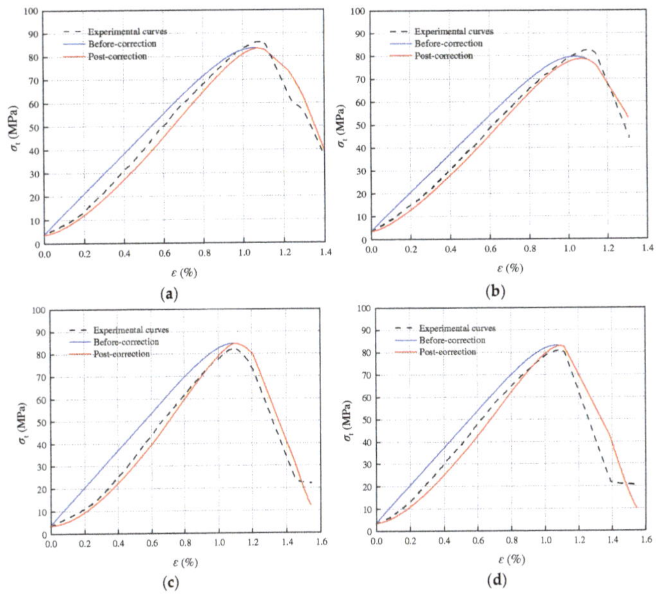

Figure 8. Porosity 3.79% sandstone statistical damage constitutive model validation: (**a**) T = −5 °C; (**b**) T = −10 °C; (**c**) T = −15 °C; (**d**) T = −20 °C.

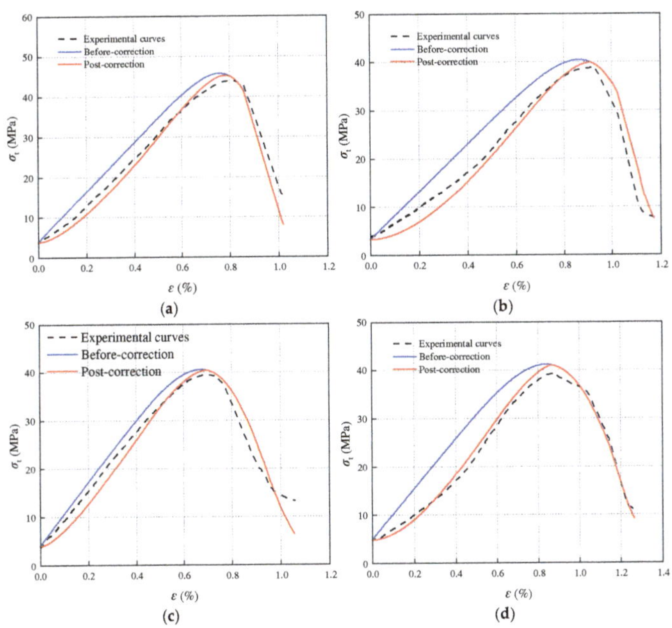

Figure 9. Porosity 11.61% sandstone statistical damage constitutive model validation: (**a**) T = −5 °C; (**b**) T = −10 °C; (**c**) T = −15 °C; (**d**) T = −20 °C.

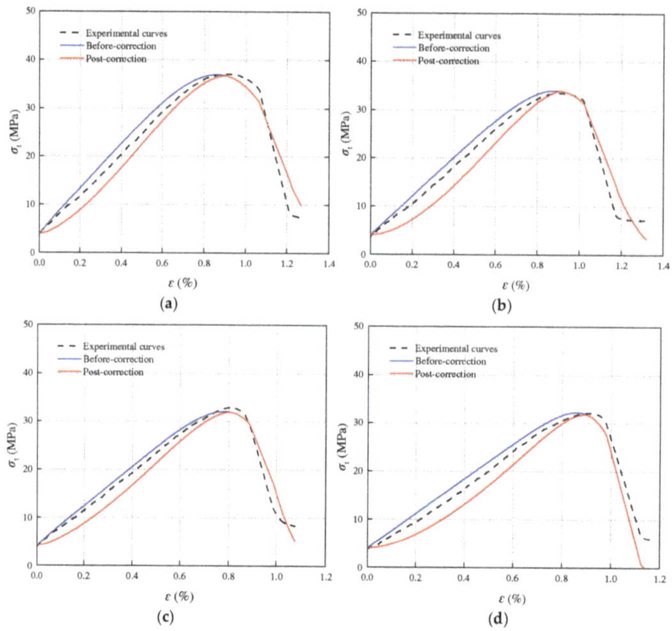

Figure 10. Porosity 16.21% sandstone statistical damage constitutive model validation: (**a**) T = −5 °C; (**b**) T = −10 °C; (**c**) T = −15 °C; (**d**) T = −20 °C.

In summary, the statistical damage constitutive model, incorporating the freeze–thaw damage variable D_r, proves to be more effective in characterizing the stress–strain relationship of sandstone under the complex conditions of freeze–thaw action, in situ stress, and pore water pressure.

3.5. Discussion

This study focuses on the effects of freeze–thaw cycles on the mechanical properties and pore structure evolution of sandstone, and proposes some new insights by combining experimental results and statistical damage constitutive models. Although the results have high scientific and application value, further analysis and discussion will help deepen the research significance and reveal future research directions.

Through freeze–thaw experiments, it was found that the peak deviatoric stress and elastic modulus both decreased with the decrease in freezing temperature, especially in the range of −5 °C to −10 °C, and tended to be stable below −10 °C. This phenomenon may be closely related to the frost heave effect. When the temperature is above −10 °C, the frost heave of water in the pores has a strong destructive effect on the binding force between rock particles; when it is below −10 °C, further temperature reduction mainly causes the already frozen pore water to solidify, and the generation of new cracks is limited. Nuclear magnetic resonance analysis shows that the porosity changes in sandstone with lower porosity after freezing and thawing are more significant. The reason for this may be that the small-pore samples are more significantly affected by the frost heave pressure, and the internal cracks expand more efficiently, resulting in a rapid increase in porosity. However, the samples with higher porosity have a relatively small change range due to the larger original pore space and limited pore expansion. This discovery provides a new perspective for understanding the evolution mechanism of sandstone damage under different initial porosity conditions. The damage constitutive model proposed in this study combines the coupling effects of freeze–thaw cycles, geostress, and pore water pressure well, and systematically describes the complexity of sandstone mechanical behavior. The good agreement between the experimental results and the theoretical model curve proves the effectiveness of the model, especially in small-porosity and medium-porosity samples. However, for high-porosity sandstone, the fitting accuracy between the theoretical curve and the experimental curve is low in the initial compaction stage, which may be related to the failure to completely compact the high-porosity sample when simulating geostress loading. In the future, the model can be further optimized by introducing correction parameters such as pore distribution uniformity or freeze–thaw cycle number.

The research results have important guiding significance for engineering practices in complex frozen soil environments such as shaft wall design and tunnel construction. However, the samples used in the study only include sandstones with three porosities, and the experimental conditions are a single freeze–thaw cycle, which limits the wide applicability of the model. For example, the cumulative effect of freeze–thaw cycles and the influence of acidic or saline water environments on the pores and mechanical properties of sandstone have not been involved. In addition, although low-field nuclear magnetic resonance technology is more sensitive to the detection of pore structure changes, its combined analysis with other microscopic detection techniques (such as CT scanning) can help further verify the reliability of the experimental results. This study provides a new theoretical basis for the evolution of mechanical damage of sandstone under freeze–thaw conditions, but further exploration is needed in the future. On the one hand, the experimental conditions can be extended to multiple freeze–thaw cycles, different water chemical environments, or more complex temperature change conditions; on the other hand, the freeze–thaw damage behavior of other rock types can be considered to improve the universality of the model. At the same time, combining macroscopic mechanical experiments with microscopic detection methods, such as scanning electron microscopy (SEM) or X-ray diffraction (XRD), can further reveal the intrinsic mechanism of freeze–thaw damage.

In summary, the experimental results and model verification of this study provide important data support for the study of the mechanical behavior of sandstone in freeze–thaw environments. Through the deepening of subsequent research and optimization of models, this field will better serve actual engineering needs.

4. Conclusions

To investigate the mechanical properties of sandstone under the coupling of water pressure and confining pressure after a single freeze–thaw cycle, water-force coupling triaxial compression tests were conducted on sandstone having three types of porosity at various negative temperatures. The effects of porosity and freezing temperature on the compressive strength and elastic modulus of sandstone were studied. Microscopic damage characteristics and the damage evolution mechanism of sandstone were analyzed through nuclear magnetic resonance testing and a statistical damage constitutive model. The main conclusions are outlined below:

(1) The stress–strain curve of sandstone samples under triaxial compression after freeze–thaw shows little change compared to that without freezing at normal temperatures. The peak deviator stress exhibits a decreasing trend with the reduction in freezing temperature, particularly pronounced between −5 °C and −10 °C, with minor influence when the temperature drops below −10 °C.

(2) The overall elastic modulus of sandstone with different porosities decreases with decreasing freezing temperature. The elastic modulus experiences a significant decrease from −5 °C to −10 °C. Below −10 °C, the elastic modulus remains similar to that at −10 °C, showing little change. The relationship between Poisson's ratio and the freezing temperature for three porosities of sandstone is irregular.

(3) Nuclear magnetic porosity of sandstone samples significantly increases after both freezing and thawing processes. The smaller the initial porosity of the rock sample, the greater the rate of change in nuclear magnetic porosity after freeze–thaw cycling. Differences exist in the influence of freeze–thaw damage on the T2 distribution of sandstone with different porosities.

(4) A statistical damage constitutive model, considering the combined effects of freeze–thaw damage, geostress, and pore water pressure, was established. The compaction coefficient K was introduced into the model for optimization. The theoretical curve and experimental curve trends are relatively close, providing a better characterization of the stress–strain relationship of sandstone in complex pressure environments.

(5) Although this study systematically analyzed the mechanical properties and damage evolution mechanism of sandstone under freeze–thaw action and established a statistical damage constitutive model, there are still some limitations. First, the experiment only targeted sandstone samples with three porosities, and the number of freeze–thaw cycles was limited, which failed to fully reflect the cumulative effect of multiple freeze–thaw cycles on the mechanical properties of sandstone. In addition, the research environment was mainly based on a single hydraulic coupling condition, and the potential influence of water chemistry (such as acidity or salinity) on sandstone damage was not considered. Although low-field nuclear magnetic resonance technology reveals the evolution law of the pore structure, the joint verification with other microscopic detection methods is still insufficient, which may limit the comprehensiveness of the results. Future research can be expanded in terms of the following aspects: first, increase the number of freeze–thaw cycles and consider the long-term effects of different environmental conditions (such as acidic water or salt freezing) on sandstone damage; second, conduct comparative studies on multiple rock types to improve the applicability and universality of the model; and third, combine more accurate microscopic detection techniques (such as scanning electron microscopy and CT scanning) to deeply reveal the freeze–thaw damage mechanism from a multi-scale level. Through the above improvements, the research results will better serve the engineering design and maintenance in complex geological environments.

Author Contributions: Conceptualization, X.C.; Methodology, M.F.; Validation, X.C.; Formal analysis, K.Y.; Investigation, T.W. and K.Y.; Data curation, T.W.; Writing—original draft, M.F. and X.C.; Writing—review & editing, X.C.; Funding acquisition, M.F. All authors have read and agreed to the published version of the manuscript.

Funding: Financial support for this work was provided by Fundamental Research Funds for the Central Universities (No. 2020ZDPYMS35).

Institutional Review Board Statement: Not applicable.

Informed Consent Statement: Not applicable.

Data Availability Statement: The original contributions presented in the study are included in the article, further inquiries can be directed to the corresponding author.

Acknowledgments: The authors would like to thank the editor for handling this submission and the anonymous referees for reading the manuscript.

Conflicts of Interest: The authors declare no conflicts of interest.

References

1. Ma, D.; Duan, H.; Cai, X.; Li, Z.; Li, Q.; Zhang, Q. A Global Optimization-Based Method for the Prediction of Water Inrush Hazard from Mining Floor. *Water* **2018**, *10*, 1618. [CrossRef]
2. Wang, X.; Zhang, Y.; Zhang, Q.; Wei, Y.; Liu, W.; Jiang, T. Space-Time Evolution Characteristics of Deformation and Failure of Surrounding Rock in Deep Soft Rock Roadway. *Sustainability* **2022**, *14*, 12587. [CrossRef]
3. Wang, H.; Wang, F.; Sun, J.; Cheng, Z.; Wang, Y.; Cao, Y. New strategy for evaluating the spatiotemporal distribution of groundwater resource quantity under seasonal freeze/thaw in mountainous areas. *J. Hydrol.* **2023**, *616*, 128850. [CrossRef]
4. Davarpanah, S.M.; Török, Á.; Vásárhelyi, B. Review on the Mechanical Properties of Frozen Rocks. *Rud. Geološko-Naft. Zb.* **2022**, *37*, 83–96. [CrossRef]
5. Wang, T.; Jia, H.; Sun, Q.; Lu, T.; Tang, L.; Shen, Y. Pressure melting of pore ice in frozen rock under compression. *Cold Reg. Sci. Technol.* **2023**, *210*, 103856. [CrossRef]
6. Liu, T.; Zhang, C.; Cao, P.; Zhou, K. Freeze-thaw damage evolution of fractured rock mass using nuclear magnetic resonance technology. *Cold Reg. Sci. Technol.* **2020**, *170*, 102951. [CrossRef]

7. Liu, T.; Zhang, C.; Li, J.; Zhou, K.; Ping, C. Detecting freeze–thaw damage degradation of sandstone with initial damage using NMR technology. *Bull. Eng. Geol. Environ.* **2021**, *80*, 4529–4545. [CrossRef]
8. Khanlari, G.; Sahamieh, R.Z.; Abdilor, Y. The effect of freeze–thaw cycles on physical and mechanical properties of Upper Red Formation sandstones, central part of Iran. *Arab. J. Geosci.* **2014**, *8*, 5991–6001. [CrossRef]
9. Cao, X.; Feng, M.; Yuan, K.; Davarpanah, A. Dynamic Mechanical Properties and Damage Mechanism of Freeze-Thaw Sandstone under Acid Corrosion. *Geofluids* **2021**, *2021*, 7335284. [CrossRef]
10. Gao, J.; Xu, C.; Xi, Y.; Fan, L. Degradation of Mechanical Behavior of Sandstone under Freeze-Thaw Conditions with Different Low Temperatures. *Appl. Sci.* **2021**, *11*, 10653. [CrossRef]
11. Jamshidi, A. Predicting the Strength of Granitic Stones after Freeze–Thaw Cycles: Considering the Petrographic Characteristics and a New Approach Using Petro-Mechanical Parameter. *Rock Mech. Rock Eng.* **2021**, *54*, 2829–2841. [CrossRef]
12. Shi, L.; Liu, Y.; Meng, X.; Zhang, H.; Dede, T. Study on Mechanical Properties and Damage Characteristics of Red Sandstone under Freeze—Thaw and Load. *Adv. Civ. Eng.* **2021**, *2021*, 8867489. [CrossRef]
13. Lan, Y.; Gao, H.; Zhao, Y. Pore Structure Characteristics and Strength Variation of Red Sandstone under Freeze-Thaw Cycles. *Materials* **2022**, *15*, 3856. [CrossRef] [PubMed]
14. Liu, Y.; Cai, Y.; Huang, S.; Guo, Y.; Liu, G. Effect of water saturation on uniaxial compressive strength and damage degree of clay-bearing sandstone under freeze-thaw. *Bull. Eng. Geol. Environ.* **2019**, *79*, 2021–2036. [CrossRef]
15. Xiong, X.; Gao, F.; Zhou, K.; Li, C.; Zhang, Y. Dynamic tensile mechanical properties of water-saturated and frozen sandstone after freeze-thaw fatigue damage treatment. *J. Mater. Res. Technol.* **2023**, *24*, 9323–9338. [CrossRef]
16. Hou, P.; Su, S.; Liang, X.; Gao, F.; Cai, C.; Yang, Y.; Zhang, Z. Effect of liquid nitrogen freeze–thaw cycle on fracture toughness and energy release rate of saturated sandstone. *Eng. Fract. Mech.* **2021**, *258*, 108066. [CrossRef]
17. Huang, C.; Zhu, C.; Ma, Y.; Aluthgun Hewage, S. Investigating Mechanical Behaviors of Rocks Under Freeze–Thaw Cycles Using Discrete Element Method. *Rock Mech. Rock Eng.* **2022**, *55*, 7517–7534. [CrossRef]
18. Liu, B.; Ma, Y.; Zhang, G.; Xu, W. Acoustic emission investigation of hydraulic and mechanical characteristics of muddy sandstone experienced one freeze-thaw cycle. *Cold Reg. Sci. Technol.* **2018**, *151*, 335–344. [CrossRef]
19. Liu, B.; Ma, Y.; Liu, N.; Han, Y.; Li, D.; Deng, H. Investigation of pore structure changes in Mesozoic water-rich sandstone induced by freeze-thaw process under different confining pressures using digital rock technology. *Cold Reg. Sci. Technol.* **2019**, *161*, 137–149. [CrossRef]
20. Chen, L.; Li, K.; Song, G.; Zhang, D.; Liu, C. Effect of freeze-thaw cycle on physical and mechanical properties and damage characteristics of sandstone. *Sci. Rep.* **2021**, *11*, 12315. [CrossRef]
21. Song, Y.; Tan, H.; Yang, H.; Chen, S.; Che, Y.; Chen, J. Fracture evolution and failure characteristics of sandstone under freeze-thaw cycling by computed tomography. *Eng. Geol.* **2021**, *294*, 106370. [CrossRef]
22. Qu, D.; Li, D.; Li, X.; Luo, Y.; Xu, K. Damage evolution mechanism and constitutive model of freeze- thaw yellow sandstone in acidic environment. *Cold Reg. Sci. Technol.* **2018**, *155*, 174–183. [CrossRef]
23. Feng, M.; Cao, X.; Yuan, K.; Zhang, M.; Li, Z. Influence of acidic environment on damage mechanism of sandstone under different climatic temperatures. *Bull. Eng. Geol. Environ.* **2022**, *81*, 452. [CrossRef]
24. Zheng, Y.; Jia, C.; Lei, M.; Huang, J.; Shi, C. Investigation of the Constitutive Damage Model of Rock Under the Coupled Effect of Freeze–Thaw Cycles and Loading. *Rock Mech. Rock Eng.* **2023**, *57*, 1861–1879. [CrossRef]
25. Jia, S.; Yu, Q.; Yin, H.; Dai, Z.; Yin, S.; Kong, Y.; Thanh, H.V.; Soltanian, M.R. Analysis of damage evolution and study on mesoscopic damage constitutive model of granite under freeze–thaw cycling. *Bull. Eng. Geol. Environ.* **2024**, *83*, 236. [CrossRef]
26. Matsuki, K.; Hasibuan, S.; Takahashi, H. Specimen size requirements for determining the inherent fracture toughness of rocks according to the ISRM suggested methods. *Int. J. Rock Mech. Min. Sci. Geomech. Abstr.* **1991**, *28*, 365–374. [CrossRef]
27. Zhao, J.; Pan, R.; Zhang, W. Influence law of ground stress distribution law on dynamic disaster. *Coal Eng.* **2023**, *55*, 69–74.
28. Wang, W.; Tian, Z.; Zhu, Q.; Li, X.; Xu, W. Study of statistical damage constitutive model for rock considering pore water pressure. *Chin. J. Rock Mech. Eng.* **2015**, *34*, 3676–3682.
29. Wang, S.; Liao, H.; Chen, Y.; Fernandez-Steeger, T.M.; Du, X.; Xiong, M.; Liao, S. Damage Evolution Constitutive Behavior of Rock in Thermo-Mechanical Coupling Processes. *Materials* **2021**, *14*, 7840. [CrossRef]

applied sciences

MDPI

Article

Automatic Identification of Rock Discontinuity Sets by a Fuzzy C-Means Clustering Method Based on Artificial Bee Colony Algorithm

Peng Li [1,2], Tianqi Chen [3], Yan Liu [1], Meifeng Cai [1], Liang Sun [3], Peitao Wang [1], Yu Wang [1] and Xuepeng Zhang [2,*]

[1] Key Laboratory of Ministry of Education for Efficient Mining and Safety of Metal Mines, University of Science and Technology Beijing, Beijing 100083, China; pengli@ustb.edu.cn (P.L.)

[2] State Key Laboratory of Strata Intelligent Control and Green Mining Co-Founded by Shandong Province and The Ministry of Science and Technology, Shandong University of Science and Technology, Qingdao 266590, China

[3] Key Laboratory of Intelligent Bionic Unmanned Systems of Ministry of Education, University of Science and Technology Beijing, Beijing 100083, China

* Correspondence: zhangxuepeng0722@126.com

Abstract: The identification and classification of rock discontinuities are crucial for studying rock mechanical properties and rock engineering optimization design and safety assessment. An improved artificial bee colony (ABC) algorithm is proposed and combined with the fuzzy C-means (FCM) clustering method to develop an FCM clustering method for automatically identifying rock discontinuity sets based on the ABC algorithm (FCM-ABC method). All the equations of the method are fully developed, and the methodology is presented in its entirety. Moreover, the rock structural planes are investigated in a gold mine in China using a ShapeMetriX 3D system. Based on the measured structural plane data, the specific calculation process, selection of parameters, effectiveness of grouping, and the dominant orientation of the proposed method for structural plane occurrence classification are analyzed and discussed, and satisfactory clustering results are achieved. This validates the validity and reliability of the method. Furthermore, multiple aspects of the excellent performance of this method for the identification of structural plane sets compared to traditional clustering methods are demonstrated. In addition, the significance of structural plane identification in the prevention and control of rock engineering disasters is discussed. This new method theoretically expands the technology of rock mass structural plane identification and has important application value in practical engineering.

Keywords: rock discontinuity sets; structural plane investigation; FCM-ABC method; automatic identification

1. Introduction

The grouping and orientation characterization of discontinuity (structural plane) sets is crucially important for rock mass characterization, geological disaster prevention and control, rock engineering applications (such as geothermal reservoirs, mines, tunnels, slopes, and water conservancy and hydropower), and earth science research [1,2]. Over a long period of geological evolution, the crust has undergone complex geological processes of construction and transformation, and various types and scales of structural planes, such as joints, fractures, and faults, are widely developed in the natural rock mass [3–6]. Because of the presence of discontinuity sets, the mechanical properties of rock masses exhibit significant anisotropic characteristics and are predominantly controlled by the structural plane. The occurrence, scale, density, morphology, and their combination of structural planes not only govern the stability of underground rock engineering but also dominate the shape, scale, and trend of the induced sliding body [7,8], because essentially, the structural plane changes the stress distribution mode in the rock mass. Also, the accurate identification of structural planes contributes significantly to the classification of the rock mass and the subsequent prevention and control of engineering geological disasters. As a consequence, on-site measurement and statistics of structural plane parameters in rock masses to detect structural plane characteristics and their combination distribution patterns are necessary prerequisites for quantitative analysis and the calculation of rock mass engineering stability.

In recent years, considerable efforts have been devoted to investigating effective methods for rock mass structural plane identification. Because of the strong randomness and uncertainty of the parameters of the structural plane, deterministic research methods are difficult to directly adopt. To analyze the regularity of developed structural planes, structural planes with some common characteristics are often classified, among which the most common is to group the structural planes and determine the dominant orientation according to their occurrence [9]. The traditional approaches for grouping structural planes generally adopt rose diagrams, pole plots, and isodensity plots. The representation of structural plane direction data and the identification of structural planes are usually conducted using techniques for the hemispherical projection of discontinuity poles (i.e., unit vectors with directions normal to structural planes) [10]. The advantage of this type of method lies in the ease of making an intuitive judgment on the distribution of the main structural planes, but the grouping results primarily depend on experience; in particular, the differences in results are even more pronounced when the boundaries between clusters are not clear [11,12]. Moreover, the visual clustering method of calculating density contour lines by calculating the number of poles falling within the reference circle is a commonly used method, but due to sampling bias, this method has some problems that need to be corrected [13,14]. In summary, these approaches may not be entirely satisfactory in some conditions, leading to the updating of alternative technologies for the automatic identification of structural planes.

Numerous attempts to address subjective issues have already been made using cluster analyses. For instance, Shanley and Mahtab [15] first proposed a clustering algorithm for structural plane orientation in 1976, which was then upgraded by Mahtab and Yegulalp [16] and Hammah and Curran [17] to develop an FCM clustering algorithm for the automatic identification of structural planes. Dershowitz et al. [18] described a random algorithm for clustering structural planes, which can handle the additional information provided; this algorithm is based on the probability distribution that defines each crack feature and involves the integration of these probability distribution functions. The above clustering algorithms are all dynamic clustering algorithms, which generally require a predetermined number of target sets to be divided and the determination of initial clustering centers for each type. Due to the significant impact of initial cluster centers on classification results, different initial cluster centers result in differences in the classification results. To this end, Cai et al. [7] proposed an FCM clustering method for rock mass discontinuity sets using a genetic algorithm, which avoided the subjectivity of artificially delineating classification boundaries. Zhou et al. [19] adopted a combination of fuzzy equivalence clustering and fuzzy soft clustering, using the optimal grouping results obtained from the fuzzy equivalence clustering method as the initial division of the fuzzy soft clustering approach, and using the clustering center obtained from the soft clustering method as the central orientation of each classification. Jimenez-Rodriguez and Sitar [20] used a spectral clustering algorithm to construct an affinity matrix for similarity measurement between discontinuities, used the eigenvalue of the affinity matrix to construct space vectors corresponding to the projection poles of discontinuities, and then employed the K-means method for classification. Behzad et al. [21] adopted the K-means clustering algorithm to group the multi-property data of discontinuity sets and applied the principal component analysis approach to calculate the eigenvalue of the covariance matrix of the measured data.

Additionally, statistical tools such as discriminant analysis, regressions, and decision analysis are used for exploratory analyses of structural plane data, and clustering methods that can combine other information besides structural plane orientation, such as planarity, weathering, spacing, or roughness, are proposed. Furthermore, methods of classifying structural plane sets using artificial intelligence technologies such as neural networks and machine learning have also been developed [22]. However, without prior information on the groups, data must be classified into homogeneous object groups, and cluster analysis is the most suitable tool for applications [17]. The clustering method is essentially a local search optimization approach, which is prone to getting stuck in local minima, thus dramatically affecting the clustering results of structural planes. Moreover, clustering analysis requires the number of clusters and fuzzy weighting index to be given in advance; how to scientifically and reasonably select these two parameters is currently an unresolved problem.

Against this background, new technical means need to be introduced to solve the above problems. In the present study, an improved ABC algorithm is proposed and combined with the FCM clustering method, and a new FCM clustering method for automatically identifying dominant rock mass structural planes based on the ABC algorithm (named the FCM-ABC method) is developed. All the equations of this approach are fully developed, and the methodology is presented in its entirety. Moreover, the structural planes of rock masses are investigated at multiple depth levels in a gold mine in China by using a ShapeMetriX 3D system (3GSM GmbH, Graz, Austria). Based on the measured data of structural planes, the specific calculation process, the selection of parameters, the effectiveness of grouping, and the determination of the dominant orientation of the proposed method for structural plane occurrence classification are analyzed and discussed, and multiple aspects including the higher recognition accuracy, faster convergence speed, and excellent performance of this method for the identification of structural plane sets compared to traditional clustering methods used in rock engineering applications are demonstrated. This method theoretically expands the technology of rock mass structural plane identification and has important application value in practical engineering.

2. FCM Clustering Algorithm

2.1. Measurement of Similarity

The FCM clustering algorithm was first proposed by Dunn [23] in 1974 and then extended by Bezdek [24]. In the FCM algorithm, the membership degree of each sample point to all class centers is obtained by optimizing the objective function, thereby determining the class membership of the sample points and achieving the automatic classification of sample data. Each sample is assigned a membership function belonging to each cluster, and the samples are classified according to their membership values. Among numerous fuzzy clustering algorithms, the FCM clustering algorithm is the most widely and successfully applied algorithm, which has been used in fields such as data mining and image processing.

A sample is a point in a feature space, and the distance between points reflects whether the corresponding samples belong to different types [11]. According to the distribution of main discontinuity sets in rock masses, the dip direction and dip angle parameters describing the occurrence of the structural planes are usually used as statistical indicators for cluster analysis. To facilitate analyses and comparisons, it is necessary to standardize the statistical indicators of each representative point. A good standardization method, while dimensionless, should also maintain the resolution of the original statistical indicators, that is, the magnitude of variability. For the clustering results and the proximity between the clustering center and the samples, the sine value of the included angle between normal vectors, similarity coefficient, or Euclidean distance are often used to represent them. Due to the characteristics of rotation invariance, sample features being unit vectors, and there being no need for the linear transformation of coordinate units, the Euclidean distance is adopted to measure the similarity between discontinuity sets in this study.

2.2. Characteristic Function

FCM clustering is to obtain the clustering center by minimizing the objective function, which is essentially the sum of the Euclidean distances (the sum of the squares of errors) from each point to each class. The objective function can be expressed as follows:

$$J_m = \sum_{i=1}^{c} \sum_{k=1}^{n} u_{ik}^m \|X_k - V_i\|^2 \tag{1}$$

where $X = \{X_1, X_2, \cdots, X_n\} \subset \mathbb{R}^2$ is a finite sample in the two-dimensional real number space $\mathbb{R}^2$, which contains n sample data subsets; $X_k = (x_{k1}, x_{k2}) \in \mathbb{R}^2$ is the sample feature vector, that is, the dip direction and dip angle parameter vector of the k-th sample; x_{k1} represents the dip direction; x_{k2} represents the dip angle; $V_i = (v_{i1}, v_{i2})$ is the i-th cluster center; c is the number of cluster centers; and u_{ik} represents the degree to which the k-th sample belongs to the i-th cluster center and satisfies $u_{ik} \in [0,1]$ and $\sum\limits_{i=1}^{c} u_{ik} = 1$.

When $X_k \neq V_i$, the following can be written:

$$u_{ik} = \left[\left(\frac{1}{\|X_k - V_i\|^2} \right)^{\frac{1}{m-1}} \right] \cdot \left[\sum_{j=1}^{c} \left(\frac{1}{\|X_k - V_j\|^2} \right)^{\frac{1}{m-1}} \right]^{-1} \tag{2}$$

$$V_i = \frac{\sum\limits_{k=1}^{n} u_{ik}^m X_k}{\sum\limits_{k=1}^{n} u_{ik}^m}, (1 \le i \le c, 1 \le k \le n) \tag{3}$$

where m is the fuzzy weighted index and $m \in [1, \infty)$, and j is a randomly selected subscript. The index m controls the allocation of membership degree and the fuzzy degree of clustering. The selection of m has been discussed in previous studies [17,25], and in general, taking $m = 2$ can meet the requirements.

By using the FCM clustering algorithm mentioned above, a fuzzy C-partition of X can be determined. This fuzzy processing method can precisely characterize the actual distribution of data, which is particularly suitable for situations in which various data points have overlapping distributions. Nevertheless, because the FCM clustering algorithm is realized through the iterative optimization of the objective function, the objective function J_m decreases in the iterative process, and the option of the initial clustering centers and the input order of samples have a remarkable impact on the final clustering results. If the initial clustering centers are imbalanced in the entire sample space, the FCM clustering algorithm will probably fall into local minima, which is particularly prominent when the data volume is large, especially in high-dimensional cases. Thus, it is crucially important to give the initial cluster center as accurately as possible. That is, using clustering methods to divide the occurrence data of structural planes requires a rough judgment of the grouping beforehand, including the number of clusters and the orientation of cluster centers. For this problem, the common countermeasure is to cluster with several different initial clustering centers and then choose the most satisfactory one as the final clustering result. When this method is used for data analyses of the structural plane orientation, it not only requires a large amount of work but also cannot guarantee the optimality of clustering results. In view of this, an attempt is made to use the ABC algorithm to solve the global optimization problem in the FCM clustering algorithm of rock mass structural planes.

3. FCM-ABC Method for Rock Mass Structural Plane Identification

3.1. Principle of the ABC Algorithm

The ABC algorithm, developed by Derviş Karaboğa [26] in 2005, is a meta-heuristic optimization algorithm that imitates the behavior of bee colonies. It seeks the optimal solution by simulating the search and food collection processes of bee colonies, aiming at solving complex optimization problems, especially those that are difficult to solve by traditional methods. The algorithm exhibits the characteristics of simple individual behavior, distributed control, and strong robustness and scalability and is not constrained by domain knowledge. Compared to other methods, such as genetic algorithms, differential evolution algorithms, and particle swarm optimization algorithms, the ABC algorithm has greatly improved its convergence speed and algorithm performance [27,28] and has been applied in related fields such as cluster analysis [29], large-scale problems and project design optimization [30], and constrained optimization problems [31]. Combining the ABC algorithm with the FCM clustering method has advantages in terms of its global search ability, diversity maintenance, flexible parameter adjustment, efficient convergence, and adaptability to large-scale problems.

According to the different divisions of labor, bee colonies are classified into hiring bees, following bees, and reconnaissance bees [32]. The main task of hiring bees is to explore potential solutions in the search space. They choose a honey source (candidate solution), collect information on the honey source, and choose the next honey source to explore according to the quality of the current solution. The following bees observe the actions of the hiring bees and decide which honey source is worthy of further study based on the quality of the solutions discovered by the hiring bees. They explore this honey source with the hiring bees to obtain more information. If a honey source contains a better solution, the following bees will inform other bees. Three strategies for finding new honey sources are provided, as illustrated in Figure 1. The search space of reconnaissance bees is not limited to certain areas. If reconnaissance bees discover a potential honey source, they will transmit the information to other bees so that other bees can further explore the area. Therefore, the kernel of the ABC algorithm involves the following three parts: the hiring bees search for honey sources; the following bees select honey sources with a certain probability of searching according to the honey source information shared by the hiring bees; and the reconnaissance bees randomly search in the search space. In the search and optimization process of the ABC algorithm, the entire process is repeated over and over again until the end condition is met (i.e., the optimal solution is obtained).

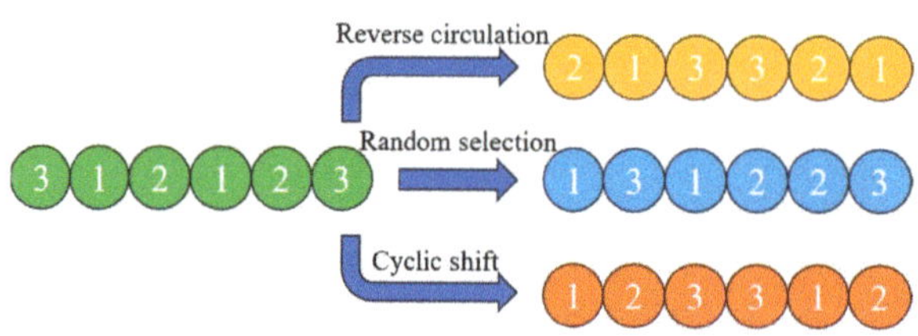

Figure 1. Illustration of the honey source update.

3.2. Improved ABC Algorithm

When using the ABC algorithm to solve optimization problems, the location of the honey source is abstracted as a point in the solution space, which represents the potential solution of the problem. The mass of the honey source g (g = 1, 2, . . ., SN) corresponds to the fitness value fit_g of the solution, where SN is the quantity of the honey sources. Let D be the dimension of the problem to be solved, and the position of the honey source g at t iterations can be expressed as follows:

$$L_g^t = \left[l_{g1}^t,\ l_{g2}^t,\ l_{g3}^t,\ \cdots,\ l_{gD}^t\right] \tag{4}$$

The initial location of the honey source g is stochastically produced in the search space based on Equation (5):

$$x_{gh} = Q_h + rand(0,\ 1)(G_h - Q_h) \tag{5}$$

where $x_{gh} \in (G_h, Q_h)$; G_h and Q_h are the lower and upper limits of the search space, respectively; h = 1, 2, . . ., D; and $rand$ stands for a random number between (0, 1).

In the ABC algorithm, the richness of the nectar contained in each honey source, that is, the fitness fit_g of the solution it represents, is as follows:

$$fit_g = \frac{1}{J_m + 1} \tag{6}$$

In which

$$J_m = \sum_{l=1}^{R} \sum_{g=1}^{S} m_{gl} d^2(P_g, w_l) \tag{7}$$

where J_m is the sum of the squares of the deviations; S is the number of structural planes; R is the number of groups divided by S structural planes; P_g is the g-th grouped structural plane; w_l is the center vector (i.e., group center) of the l-th grouped structural plane W_l; $d(P_g, w_l)$ is the distance between P_g and w_l; and m_{gl} represents the attribution relationship between P_g and W_l as follows:

$$m_{gl} = \begin{cases} 1\ (P_g \in W_l) \\ 0\ (P_g \notin W_l) \end{cases} \tag{8}$$

At the beginning of the search, the hiring bees search around the honey source g according to Equation (9) to generate a new honey source, that is, update the solution:

$$v_{gh} = x_{gh} + o_{gh}\left(x_{gh} - x_{bh}\right) \tag{9}$$

where v_{gh} is a new honey source (solution) obtained from the x_{gh} neighborhood search; b is a honey source different from g ($b \neq g$) randomly selected from SN honey sources; $b \in \{1, 2, \ldots, SN\}$; o_{gh} represents a random number in the range of [0, 1], which controls the generation range of the x_{gh} neighborhood; and h represents a random integer in [1, D], which implies that the hiring bees stochastically select one dimension to search.

The newly produced possible solution is compared with the original solution:

$$\text{New: } v_{gh} = \{v_{g1}, v_{g2}, \ldots, v_{gD}\} \tag{10}$$

$$\text{Old: } x_{gh} = \{x_{g1}, x_{g2}, \ldots, x_{gD}\} \tag{11}$$

The greedy selection strategy is adopted to preserve the better solution.

The choice of the solution by the following bees is to judge the fitness value of the solution by observing the swing dance of the leading bees and selecting which hiring bees to follow according to the selection probability. The fitness value fit_g is computed using the following equation:

$$fit_g = \begin{cases} \frac{1}{1+f_g}, f_g > 0 \\ 1 + |f_g|, f_g \leq 0 \end{cases} \tag{12}$$

where f_g represents the objective function value of the g-th solution.

The following bees follow the others according to the probability calculated by Equation (13) based on the honey source information shared by the hiring bees:

$$P_g = \frac{fit_g}{\sum\limits_{e=1}^{SN} fit_n} \tag{13}$$

where P_g represents the probability that the g-th honey source is chosen; and e denotes the maximum number of cycles.

In addition to the two control parameters of the number of honey sources and the maximum number of cycles, there is also another crucial control parameter in the ABC algorithm, i.e., the maximum number of times the same honey source is collected (*limit*), which is employed to record the number of times a solution is updated. If a solution has not been improved after *limit* consecutive cycles, it indicates that this solution is trapped into a local optimum and should be discarded, and then a new solution will be stochastically generated using Equation (5) to replace the original solution. The ABC algorithm defines two main behavior patterns that are essential for self-organization and swarm intelligence, namely, the positive feedback mechanism of recruiting more following bees to exploit richer honey sources and the negative feedback mechanism of abandoning honey source exhaustion [12]. The ABC algorithm combines global search and local search methods, achieving a good balance between honey source exploration and extraction for the bees, thereby greatly improving the performance of the algorithm.

In this study, the detailed calculation process of the FCM-ABC method is presented in Figure 2, which is automatically completed using a self-designed MATLAB program. The above approach also introduces the idea of a roulette wheel algorithm, which can provide better optimization direction and higher-quality candidate solutions in the iterative process of the ABC algorithm, strengthen the robustness and convergence speed of this algorithm, and balance its local search and global search capabilities.

3.3. Clustering Validity Verification

The clustering results obtained needed to be verified. At present, a variety of validity test indicators have been proposed. From the effectiveness of clustering, a good clustering result should be as clear of a division as possible [17,25]. Here, two test indicators, i.e., the classification entropy coefficient H_m and the fuzzy classification coefficient F_m, are employed to test the effectiveness of clustering, and the formulas for both are as follows [24]:

$$H_m = -\frac{1}{n}\sum_{k=1}^{n}\sum_{i=1}^{c} u_{ik}\log_a(u_{ik}) \tag{14}$$

$$F_m = \frac{1}{n}\sum_{k=1}^{n}\sum_{i=1}^{c} u_{ik}^2 \tag{15}$$

where the base number of logarithm a∈(1, ∞), and it is agreed that when $u_{ik} = 0$, $u_{ik}\log_a(u_{ik}) = 0$ (the natural logarithm is taken in this study). For the two test indicators H_m

and F_m, when $H_m \to 0$ and $F_m \to 1$, it indicates that the smaller the fuzziness value of the classification, the better the clustering effect.

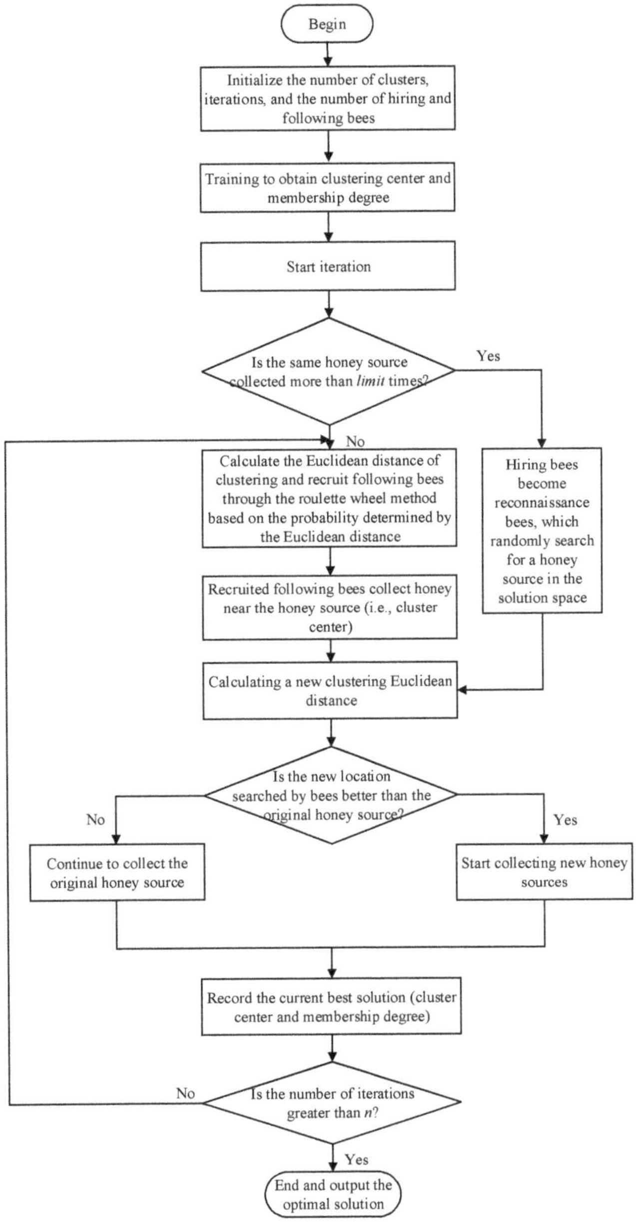

Figure 2. FCM clustering flowchart based on the ABC algorithm.

In addition, to make an objective judgment on the optimal clustering results, it is necessary to use multiple indicators to test the clustering effect because fewer evaluation indicators may lead to one-sided evaluation results. Thus, another evaluation indicator, i.e., the fuzzy hypervolume F_h, is also used to evaluate the clustering effect. A minimum F_h value indicates the optimal fuzzy partition. The calculation formula for F_h is as follows [33]:

$$F_h = \sum_{i=1}^{c} (\det(T_i))^{0.5} \tag{16}$$

In which, T_i is a fuzzy covariance matrix and the following can be written:

$$T_i = \frac{\sum_{j=1}^{n} (u_{ij})^m (X_j - V_i)(X_j - V_i)^{\mathrm{T}}}{\sum_{j=1}^{n} (u_{ij})^m} \tag{17}$$

According to the proposed FCM-ABC method, after the clustering effect test, the appropriate clustering number c obtained is the structural plane grouping number. Each group represents the dominant structural plane at the measurement point, and the clustering center stands for the mean normal vector of this group of the dominant structural planes in that group, which can be converted into the average occurrence according to Equation (18). Meanwhile, the sample size in each group can also reflect the different levels of importance of the dominant structural plane.

$$X_l = (\cos\alpha_l \sin\beta_l,\ \sin\alpha_l \sin\beta_l,\ \sin\alpha_l \cos\beta_l) \tag{18}$$

where α_l and β_l represent the dip direction and dip angle of the l-th structural plane, respectively, and $l\in[1, n]$.

4. Engineering Application

4.1. Discontinuity Set Survey in a Gold Mine

A gold mine is located in Shandong province, China, with a mining depth exceeding one kilometer underground. The structural planes such as joints and fissures in this mine area are quite developed (Figure 3), which seriously threatens the stability of roadways and stopes. Hence, investigating the scale, occurrence, and distribution of the structural planes in this mine is crucial for the failure mode analysis and stability evaluation of the surrounding rock. A 3D non-contact measuring system (ShapeMetriX 3D) for the geometric parameters of the rock mass is used to investigate the rock mass structural planes of the mine, which primarily contains a high-resolution standard DSLR zoom camera (Nikon Imaging Japan Inc., Tokyo, Japan), corresponding marker rod components, and a 3D visual analysis software system, as shown in Figure 4. This system can provide real 3D images of the rock mass surface and determine the geometric measurement data of the rock discontinuity sets by being combined with the matching image analysis software, thus recording the tunnel outline and the spatial position of the rock mass discontinuity. The system essentially relies on the digital image correlation technique, which organizes the digital information of the rock mass to obtain information on the rock mass structure plane occurrence.

Figure 5 shows the basic principle of the 3D non-contact photogrammetry. Firstly, the investigation area is determined, and the left and right views of the exposure surface are obtained through the calibrated high-pixel camera, such as the points $P_1(u, v)$ and $P_2(u, v)$. Afterwards, the two views in the picture are imported into a software analysis system, and the system automatically identifies the points $P_1(u, v)$ and $P_2(u, v)$ and reconstructs them in three dimensions to obtain 3D points $P(x, y, z)$ based on a series of techniques such as pixel point matching and image distortion deviation correction. Finally, the 3D model is synthesized and the direction and distance are realistic to obtain a true 3D digital model of the rock mass surface. According to the reconstructed 3D surface, the discontinuity sets of rock masses can be marked, such as the joint trace length, spacing, and fault distance, and thus, the geometric occurrence information of structural planes can be statistically analyzed.

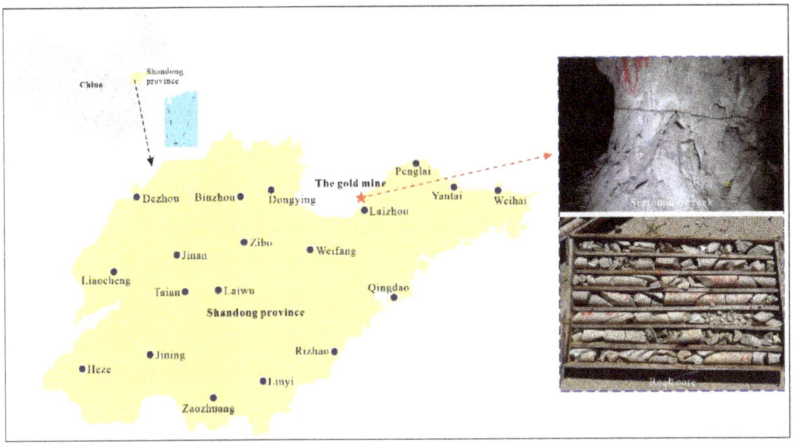

Figure 3. Illustrations of the geographical location and structural plane development of the gold mine.

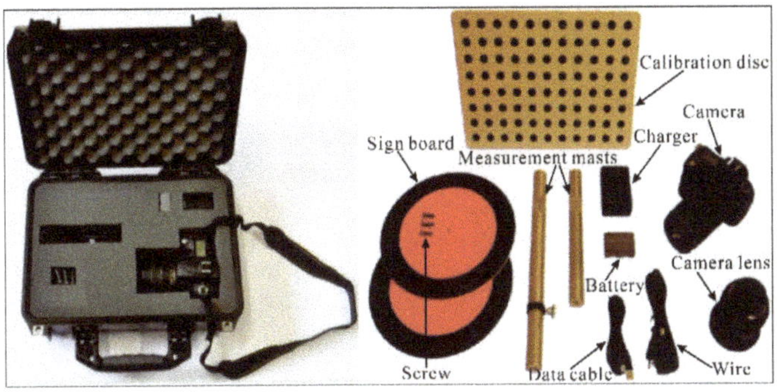

Figure 4. ShapeMetriX 3D measurement system.

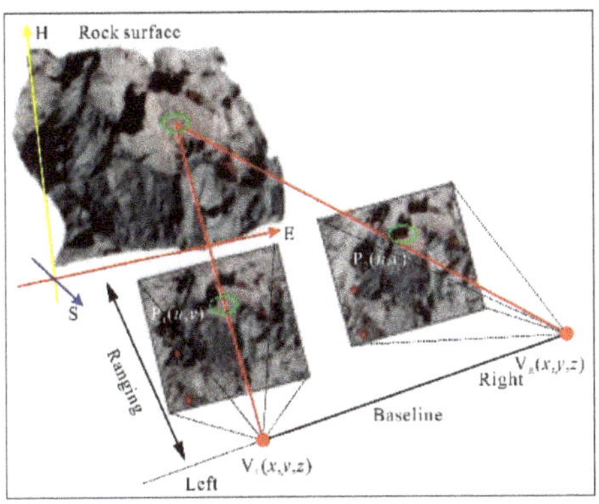

Figure 5. Imaging principle of the 3D surface measurement [9].

The structural planes of the rock masses are investigated at multiple depth levels in the gold mine by using the ShapeMetriX 3D system. The following is an example of measuring the structural planes at a measuring point at a depth of 765 m underground. The left and right views of the survey location are identified (Figure 6a). These two views are imported into the ShapeMetriX 3D software analysis system, the key measurement areas are delineated, the 3D models are synthesized, and the direction and distance are realistically transformed to obtain the 3D view of the rock mass surface (Figure 6b). On the synthesized 3D map, rock joints are grouped according to the distribution of the main joints and fractures, and different colors represent different groups. There are two groups of the main joint distribution, namely, the red group and the blue group in Figure 6c. Based on the spatial distribution and grouping of discontinuity sets, the stereographic projection map (Figure 6d) and the structural plane trace distribution map (Figure 6e), which can reflect the joint spacing, are drawn.

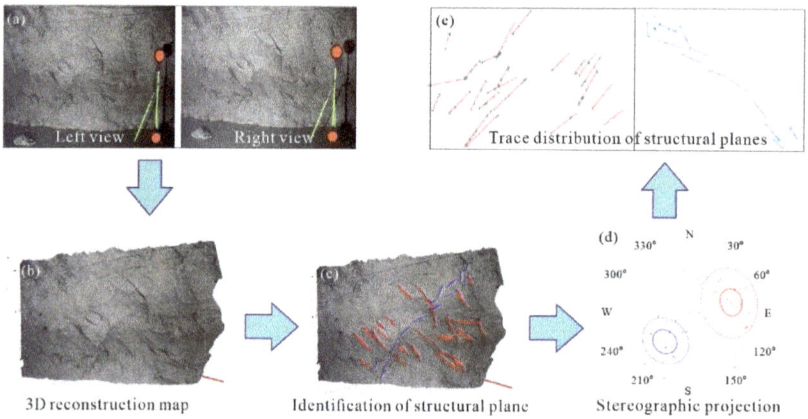

Figure 6. Illustration of information collection for structural planes of rock masses: (**a**) left and right views during photography; (**b**) 3D reconstruction map; (**c**) identification of structural plane; (**d**) stereographic projection; and (**e**) trace distribution of structural planes.

The spatial distribution of rock discontinuity sets can be described by spatial geometric parameters such as occurrence, shape, scale, and spacing, and these basic geometrical features can be used to determine the spatial shape characteristics of each group of rock mass discontinuities. The collected data are counted and calculated using the software of the ShapeMetriX 3D system, and the geometric parameters such as the trace length, dip direction, dip angle, and spacing of each group of joints are counted, respectively, and the statistical distribution law of each set of discontinuities is obtained, as listed in Table 1.

Table 1. Geometric parameter values of structural planes and their distribution models.

Group	Trace Length (m)			Dip Direction (°)			Dip Angle (°)			Spacing (m)		
	Distribution Model	Mean Value	Standard Deviation	Distribution Model	Mean Value	Standard Deviation	Distribution Model	Mean Value	Standard Deviation	Distribution Model	Mean Value	Standard Deviation
No. 1	Normal	0.44	0.103	Normal	251.88	39.4	Normal	56.65	16.06	Negative exponential	0.34	0.35
No. 2	Normal	0.25	0.06	Normal	49.21	19.75	Normal	42.70	16.92	Negative exponential	0.11	0.07

4.2. Verification of Discontinuity Sets' Clustering Effect

The measured data of the structural planes at a depth of 765 m in this gold mine are used to validate the effectiveness and reliability of the automatic identification method for the structural planes proposed in this study. The FCM-ABC method is employed to cluster the joint parameters of each group together. The measured group size is 81, and the relevant parameters are set as follows: $m = 2$, $c = 2$–7, $limit = 100$, and $e = 200$. The two-dimensional distribution of joints on the surface of the original rock mass is plotted in Figure 7, showing two significant groups of dominant structural planes, which are used for comparison. The clustering results of the field discontinuity data automatically computed by the self-designed MATLAB code using different c values are presented in Figure 8, and the obtained values of the three cluster validity test indicators under different c values are provided in Table 2. Apparently, for the tested sample data, from $c = 2$ to $c = 7$, when $c = 2$, the classification entropy coefficient H_m is the closest to 0, the fuzzy classification coefficient F_m is the closest to 1, and the fuzzy hypervolume F_h is the smallest, indicating that the clustering effect is the best in this case and the grouping result is very satisfactory. According to the calculation results of the cluster center when $c = 2$, the average occurrence of the two sets of predominant discontinuity sets is determined, i.e., the dip direction of the first set of discontinuities is 248.16°, and the dip angle is 58.17°; the second set of discontinuities has a dip direction of 52.33° and a dip angle of 44.86°. The clustering calculation results are quite comparable to the actual measured results (Table 1 and Figure 7), which implies that the FCM-ABC method has a high solving accuracy and also verifies the correctness of the new algorithm.

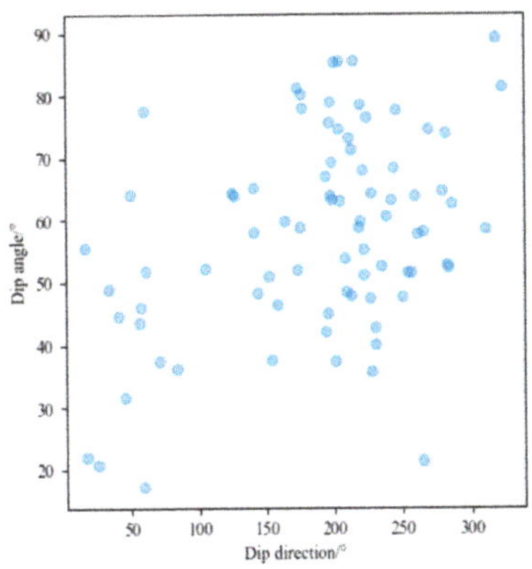

Figure 7. Distribution of joints on the surface of the original rock mass.

Table 2. Obtained values of the three cluster validity test indicators under different c values.

c	2	3	4	5	6	7
H_m	0.10	0.21	0.19	0.19	0.19	0.22
F_m	0.58	0.25	0.18	0.13	0.10	0.09
F_h	998.47	1288.00	1340.69	1297.49	1322.03	1459.50

Moreover, the correlation between the values of the cluster validity test indicators and c values is intuitively presented in Figure 9. Evidently, with the increase in the c value, H_m and F_h show an overall increasing trend, while F_m shows a decreasing trend. This indicates that the clustering effect is best when $c = 2$, which is identical to the real situation of the two dominant structural planes. The selection of the c value has a significant influence on the clustering results. Hence, choosing an appropriate c value is crucial. The proposed FCM-ABC method can accurately obtain the clustering center and membership degree through extensive training, which can effectively solve the problem of conventional clustering methods not being able to effectively select the appropriate initial clustering center. Accordingly, the new clustering approach can be applied to solve the problem of grouping the dominant structural planes of rock masses and reveal the geometric characteristics of the structural planes. In addition, the quality of clustering results will affect the convergence process to some extent. Figure 10 shows the process of searching for optimal solutions during the structural plane clustering calculation. The optimal solution is reached after around 125 cycles, and the actual time to find the optimum is approximately 0.02 s. The important feature of the ABC algorithm converging to the global optimal solution ensures that the proposed FCM-ABC method can quickly obtain the global optimal solution.

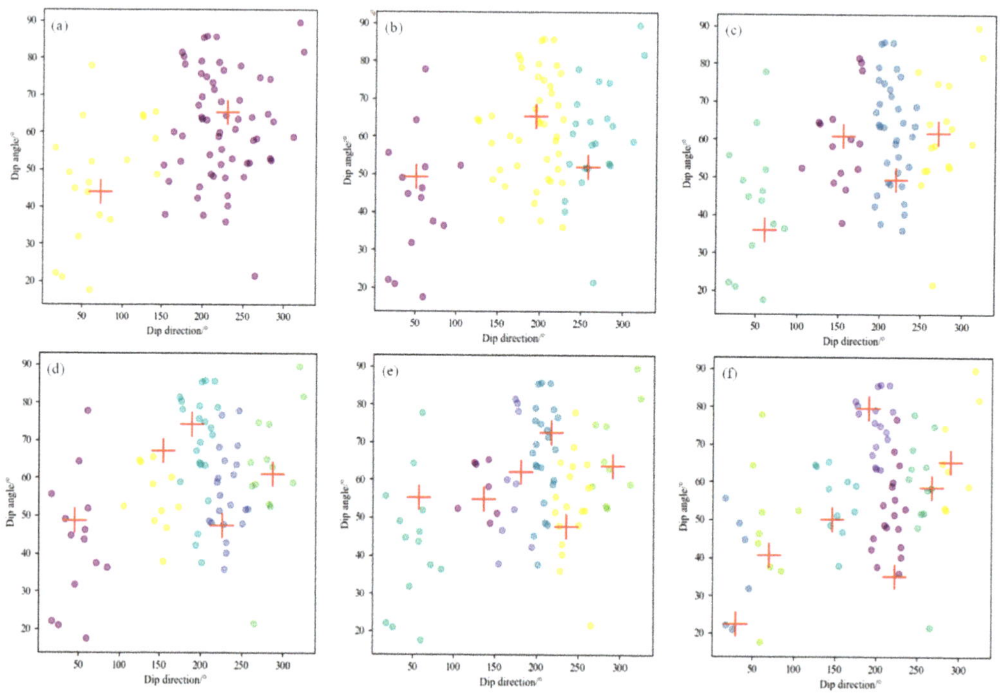

Figure 8. Clustering results of the field structural plane data based on different c values: **(a)** $c = 2$; **(b)** $c = 3$; **(c)** $c = 4$; **(d)** $c = 5$; **(e)** $c = 6$; and **(f)** $c = 7$.

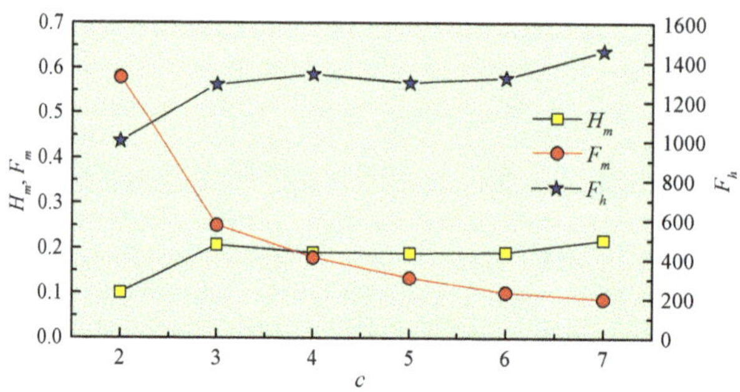

Figure 9. Relationship between the values of the cluster validity test indicators and c values.

To further validate the performance and advantages of the proposed FCM-ABC method, this method was compared with other clustering algorithms (the FCM algorithm, genetic algorithm, and neural network) based on the same data. The clustering results that achieve the optimal solutions obtained by different methods are listed in Table 3. Compared to the traditional clustering algorithms, the FCM-ABC method significantly reduces the convergence times, demonstrating a better algorithm performance and faster convergence speed. Consequently, the calculation time of the program can be greatly reduced in engineering applications, which is convenient for repeated calculations. Moreover, this new method recruits following bees through a roulette wheel algorithm according to the probability determined by the calculated Euclidean distance of the clustering and constantly updates the Euclidean distance, inducing the optimization process to approach the optimal solution more closely. While improving the convergence speed, it can also converge to better results, greatly improving the reliability and objectivity of the clustering results. Therefore, the approach that we developed has outstanding advantages in grouping analyses of structural planes, effectively avoids the shortcoming of conventional clustering algorithms easily falling into local optima, and has broad application prospects. This study provides an effective approach and method for the automatic identification and grouping of structural planes.

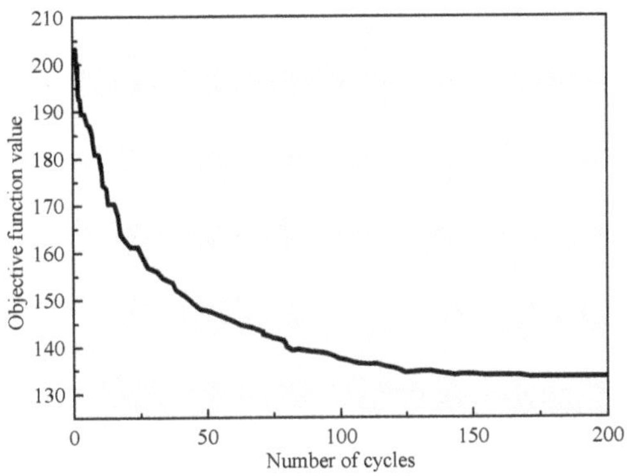

Figure 10. Process of searching for optimal solutions during the structural plane clustering calculation.

Table 3. Clustering results obtained by different clustering algorithms.

Algorithm	H_m	F_m	F_h	Iterations	Convergence Time/s
FCM-ABC	0.10	0.58	998.47	125	0.02
FCM	0.15	0.27	1195.91	308	8.99
Genetic algorithm	0.18	0.21	1453.83	562	306.69
Neural network	0.20	0.13	1501.55	40037	509.56

In the grouping calculation of structural planes, for different geological conditions, a clustering analysis can be performed as long as there are structural plane parameters of the dip direction and dip angle. Of course, other parameters can also be selected, but they have not been verified in this paper. For low-quality or incomplete structural plane data, they can be eliminated in advance manually or by compiling a program that meets certain rules to avoid affecting the accuracy of the clustering calculation results. Moreover, based on the clustering calculation via the FCM-ABC method using the dip direction and dip angle of the engineering structural plane in this study, the empirical thresholds of the three clustering effectiveness indicators H_m, F_m, and F_h could be recommended as $H_m < 0.15$, $F_m > 0.5$, and $F_h < 1000$, respectively, to evaluate whether the clustering calculation has reached the optimal solution. This may need to be adjusted according to different data types, specific engineering conditions, and actual requirements. In addition, the selection scheme for an appropriate number of clusters is generally predetermined with different numbers of partition groups (usually between two and ten) [34], and the corresponding grouping result is the best grouping result when the clustering effectiveness indicator values are optimal.

Notably, the above method is also used to cluster analyze the structural plane data (all of which are two groups of dominant structural planes) investigated at other depth levels in this gold mine, and good recognition results are also reached. Due to the space limitation, the clustering calculation results of the dominant grouping for these structural planes are not provided. Nevertheless, it should be emphasized that only the sample data containing two groups of dominant structural planes are used to analyze and verify the proposed FCM-ABC method in this study. To further verify the applicability of the proposed method, the sample data containing at least three groups of dominant structural planes will be used for the clustering analysis in the next step, which is also conducive to optimizing and perfecting the method. In addition, only two parameters, i.e., dip direction and dip angle, are employed to group the dominant structural planes of rock masses, and the proposed method should be used to identify the dominant structural planes based on more parameters in the future because more parameters can more accurately characterize the distribution and development laws of structural planes.

5. Discussion

Numerous studies [35–37] have indicated that the occurrence of rock engineering disasters (such as spalling, falling roofs, collapses, large deformations of surrounding rock, rockburst, and water inrush) is mostly due to insufficient attention given to the influence of structural planes in the rock mass. In practice, generally, rock engineering is not suitable for being built in rock masses with completely continuous structural planes but is usually built in rock masses with discontinuous structural planes. Discontinuous structural planes usually crisscross in rock masses, forming non-continuous bodies with complex structural characteristics. The influence of the structural plane on the mechanical response behavior of engineering rock masses is multifaceted. The morphology, size, spatial distribution, and combination characteristics of structural planes are key factors determining the mechanical and engineering properties of rock masses and also controlling the stability and failure mode of engineering rock masses.

Specifically, structural planes can cause additional deformations of the surrounding rock, such as the closure and sliding of the structural planes. The existence of structural planes can seriously deteriorate the integrity and bearing capacity of the rock mass, and the existing structural planes are the source of new cracks, making the surrounding rock more prone to deformation and damage in engineering. Furthermore, structural planes can also alter the stress distribution pattern in engineering rock masses (such as changes in the magnitude and direction of the circumferential stress) [3,38,39], resulting in an abnormal stress distribution in local areas and thus causing different failure modes of the surrounding rock. In particular, the failure mode is directly related to the occurrence of structural planes. In addition, the mechanical properties of rock masses exhibit remarkable anisotropic characteristics, and the interaction between structural planes and between structural planes and engineering structures (such as tunnels) increases the complexity and unpredictability of the mechanical response and disaster behavior of engineering rock masses.

The stability analysis of rock masses often relies on a thorough understanding of the characteristics of structural planes. In practical engineering, the structural planes of rock masses may have complex geometric features (such as different shapes, orientations, dip angles, and distribution densities), which have a decisive impact on the strength, deformation, and failure modes of the rock mass. By using automated methods to identify and analyze these structural planes, the spatial distribution information of the planes can be accurately obtained, providing reliable data for rock mass stability assessment. Moreover, the characteristics of rock mass structural planes directly influence the design and construction of rock engineering projects (such as underground tunnels, mining, and underground storage facilities). Engineering design needs to comprehensively consider the geometric characteristics, mechanical properties, and distribution of structural planes within the rock mass. Thus, accurately mastering this information is crucial for optimizing design and improving construction safety. Additionally, during underground engineering and mining operations, the structural planes of rock masses are often the source of disasters such as rock bursts, spalling, and collapses. Changes in the distribution and characteristics of structural planes directly affect the failure behavior of the rock mass under external loading. Therefore, the automatic recognition technology of structural planes provides new ideas and methods for disaster warning and risk management in rock engineering. More accurate data on structural planes could enhance the scientific and effective prevention of disasters, ensuring the safety of the project. In particular, as rock engineering moves towards intelligence and digitization, the application of automatic recognition technology provides fundamental data for the intelligent management of rock engineering. Combined with big data analysis and machine learning, the technology of automatically recognizing structural planes in rock masses not only provides real-time monitoring information but also allows for long-term dynamic analysis, offering real-time support for engineering decision making. In summary, automatically identifying structural planes and accurately obtaining their geometric information can provide a data foundation for engineering rock mass quality grading, rock mass stability assessment and analysis, 3D geomechanical model construction, rock support design, and disaster monitoring and early warning involved in rock engineering disaster prevention and control. This is conducive to solving the dual scientific challenges of the prevention and precise control of rock engineering disasters and intelligent emergency decision making.

It should be noted that although considerable progress has been made in the theory and technology of the automatic identification of rock discontinuities, due to the complexity and uncertainty of geological bodies and the urgent need for accurate prevention and control of rock engineering disasters, relevant research is still needed in the future to strengthen the refined and intelligent identification, extraction, analysis, and characterization of discontinuities, so as to obtain complete and true information, such as the occurrence and spatial distribution of structural planes, to improve the theoretical and technical level of disaster prevention and control.

6. Conclusions

The traditional classification methods for rock mass structural planes have many limitations when dealing with complex rock mass structures and large-scale data. To overcome the drawback of the FCM clustering algorithm easily falling into local optima, in this paper, an improved ABC algorithm is proposed and combined with FCM clustering, and a new FCM-ABC method is developed. Using the measured structural plane data, the specific calculation process, the selection of parameters, the effectiveness of grouping, and the dominant orientation of the proposed method for structural plane occurrence classification are analyzed and discussed, and satisfactory clustering results are achieved, verifying the effectiveness of the method. Also, multiple aspects of the excellent performance of this method for the identification of structural plane sets compared to traditional clustering methods are demonstrated. The approach can not only solve the problem of the sensitivity of traditional clustering algorithms to the initial clustering center but also accelerate the convergence speed of this algorithm and improve the clustering accuracy and stability of the algorithm, showing great potential in solving complex optimization problems. In particular, this method can avoid the subjectivity of artificially defining classification boundaries when the grouping boundary of dominant structural planes is not clear and prior knowledge is insufficient. Additionally, it can give objective and reliable results without relying on engineering experience, providing a reliable basis for the identification and classification of rock discontinuity sets, the optimization design and stability analysis of rock engineering, and the accurate prevention and control of engineering disasters. The method theoretically expands the technology of rock mass structural plane identification and has important application value in practical engineering. Note that to further verify the applicability and reliability of this method, the sample data containing at least three groups of dominant structural planes should be used for clustering analysis based on more parameters in the future. Also, corresponding software platforms to achieve the intelligent and refined grouping of structural planes should be developed to guide engineering practice.

Author Contributions: Conceptualization, P.L. and M.C.; methodology, P.L., T.C. and M.C.; software, T.C.; validation, P.L., T.C. and L.S.; formal analysis, P.L., Y.L. and T.C.; investigation, P.L., Y.L. and P.W.; resources, P.L. and X.Z.; data curation, P.L. and T.C.; writing—original draft preparation, P.L. and T.C.; writing—review and editing, P.L. and X.Z.; visualization, Y.L. and Y.W.; supervision, M.C.; project administration, P.L. and M.C.; funding acquisition, P.L. and M.C. All authors have read and agreed to the published version of the manuscript.

Funding: This research was funded by the Science, Technology & Innovation Project of Xiongan New Area (Grant No. 2023XAGG0061), the Open Research Fund of the State Key Laboratory of Strata Intelligent Control and Green Mining co-founded by Shandong Province and the Ministry of Science and Technology, Shandong University of Science and Technology (Grant No. SICGM202303), the National Natural Science Foundation of China (Grant Nos. 52204084 and 52474091), the Interdisciplinary Research Project for Young Teachers of USTB (Fundamental Research Funds for the Central Universities) (Grant No. FRF-IDRY-GD22-002), and the National Key R&D Program of China (Grant Nos. 2022YFC2905600 and 2022YFC3004601).

Institutional Review Board Statement: Not applicable.

Informed Consent Statement: Not applicable.

Data Availability Statement: The datasets used during the current study are available from the corresponding author upon reasonable request.

Acknowledgments: The authors are deeply indebted to the financial supporters.

Conflicts of Interest: The authors declare no conflicts of interest.

References

1. Kulatilake, P.H.S.W.; Wathugala, D.N.; Poulton, M.; Stephansson, O. Analysis of Structural Homogeneity of Rock Masses. *Eng. Geol.* **1990**, *29*, 195–211. [CrossRef]
2. Martin, M.W.; Tannant, D.D. A Technique for Identifying Structural Domain Boundaries at the EKATI Diamond Mine. *Eng. Geol.* **2004**, *74*, 247–264. [CrossRef]
3. Sagong, M.; Park, D.; Yoo, J.; Lee, J.S. Experimental and Numerical Analyses of an Opening in a Jointed Rock Mass under Biaxial Compression. *Int. J. Rock Mech. Min. Sci.* **2011**, *48*, 1055–1067. [CrossRef]
4. Li, P.; Cai, M.F. Energy Evolution Mechanism and Failure Criteria of Jointed Surrounding Rock under Uniaxial Compression. *J. Cent. South Univ.* **2021**, *28*, 1857–1874. [CrossRef]

5. Wang, Y.; Yi, X.F.; Li, P.; Cai, M.F.; Sun, T. Macro-Meso Damage Cracking and Volumetric Dilatancy of Fault Block-in-Matrix Rocks Induced by Freeze–Thaw-Multistage Constant Amplitude Cyclic (F-T-MSCAC) Loads. *Fatigue Fract. Eng. Mater. Struct.* **2022**, *45*, 2990–3008. [CrossRef]
6. Li, P.; Cai, M.F.; Guo, Q.F.; Ren, F.H. Investigation on Acoustic Emission Characteristics of Hole-Joint Contained Granite under a Compressive Disturbance: Experimental Insights. *Lithosphere* **2022**, *2022*, 3594940. [CrossRef]
7. Cai, M.F.; Wang, P.; Zhao, K.; Zhang, D.K. Fuzzy C-Means Cluster Analysis Based on Genetic Algorithm for Automatic Identification of Joint Sets. *Chin. J. Rock Mech. Eng.* **2005**, *24*, 371–376.
8. Li, P.; Cai, M.F.; Gao, Y.B.; Wang, P.T.; Miao, S.J.; Wang, Y. Fracture Evolution and Failure Behavior around an Opening in Brittle Jointed Rocks Subjected to Uniaxial Compression. *Theor. Appl. Fract. Mech.* **2022**, *122*, 103651. [CrossRef]
9. Li, P.; Wu, Y.Q.; Cai, M.F. Failure Behavior of the Surrounding Rock of Jointed Rock Masses in a Gold Mine under Blasting Impact Disturbance. *Environ. Earth Sci.* **2022**, *81*, 106. [CrossRef]
10. Harrison, J.P. Fuzzy Objective Functions Applied to the Analysis of Discontinuity Orientation Data. In Proceedings of the Rock Characterization: ISRM Symposium, Eurock '92, Chester, UK, 14–17 September 1992; pp. 25–30.
11. Zhao, K.; Cai, M.F. Application of Fuzzy C-Means Cluster Algorithm in Identification of Structural Plane Group. *Met. Mine* **2002**, 13–15. [CrossRef]
12. Song, T.J.; Chen, J.P.; Zhang, W.; Xiang, L.J.; Yang, J.H. A Method for Multivariate Parameter Dominant Partitioning of Discontinuities of Rock Mass Based on Artificial Bee Colony Algorithm. *Rock Soil Mech.* **2015**, *36*, 861–868.
13. Terzaghi, R.D. Sources of Error in Joint Surveys. *Géotechnique* **1965**, *15*, 287–304. [CrossRef]
14. Zhang, L.; Einstein, H.H. Estimating the Mean Trace Length of Rock Discontinuities. *Rock Mech. Rock Eng.* **1998**, *31*, 217–235. [CrossRef]
15. Shanley, R.J.; Mahtab, M.A. Delineation and Analysis of Clusters in Orientation Data. *J. Int. Assoc. Math. Geol.* **1976**, *8*, 9–23. [CrossRef]
16. Mahtab, M.A.; Yegulalp, T.M. A Rejection Criterion for Definition of Clusters in Orientation Data. In Proceedings of the The 23rd U.S. Symposium on Rock Mechanics (USRMS), Berkeley, CA, USA, 25–27 August 1982; Volume All Days, p. ARMA-82-116.
17. Hammah, R.E.; Curran, J.H. Fuzzy Cluster Algorithm for the Automatic Identification of Joint Sets. *Int. J. Rock Mech. Min. Sci.* **1998**, *35*, 889–905. [CrossRef]
18. Dershowitz, W.; Busse, R.; Geier, J.; Uchida, M. A Stochastic Approach for Fracture Set Definition. In Proceedings of the 2nd North America Rock Mechanics Symposium, Montreal, QC, Canada, 19–21 June 1996; Volume All Days, p. ARMA-96-1809.
19. Zhou, Y.X.; Zhou, Z.F.; Sun, Q.G. Synthetical Fuzzy Clustering Analysis for Joints Occurrence of Rock Mass. *Chin. J. Rock Mech. Eng.* **2005**, *24*, 2283–2287.
20. Jimenez-Rodriguez, R.; Sitar, N. A Spectral Method for Clustering of Rock Discontinuity Sets. *Int. J. Rock Mech. Min. Sci.* **2006**, *43*, 1052–1061. [CrossRef]
21. Tokhmechi, B.; Memarian, H.; Moshiri, B.; Rasouli, V.; Noubari, H.A. Investigating the Validity of Conventional Joint Set Clustering Methods. *Eng. Geol.* **2011**, *118*, 75–81. [CrossRef]
22. Sirat, M.; Talbot, C.J. Application of Artificial Neural Networks to Fracture Analysis at the Äspö HRL, Sweden: Fracture Sets Classification. *Int. J. Rock Mech. Min. Sci.* **2001**, *38*, 621–639. [CrossRef]
23. Dunn†, J.C. Well-Separated Clusters and Optimal Fuzzy Partitions. *J. Cybern.* **1974**, *4*, 95–104. [CrossRef]
24. Bezdek, J.C. Pattern Recognition with Fuzzy Objective Function Algorithms. In *Advanced Applications in Pattern Recognition*; Springer: Berlin/Heidelberg, Germany, 1981.
25. Gao, X.B.; Li, J.; Xie, W.X. Optimal Choice of Weighting Exponent in a Fuzzy C Means Clustering Algorithm. *Pattern Recognit. Artif. Intell.* **2000**, *13*, 7–12.
26. Karaboğa, D. *An Idea Based on Honey Bee Swarm for Numerical Optimization*; Technical Report-TR06; Erciyes University: Kayseri, Türkiye, 2005.
27. Karaboga, D.; Basturk, B. A Powerful and Efficient Algorithm for Numerical Function Optimization: Artificial Bee Colony (ABC) Algorithm. *J. Glob. Optim.* **2007**, *39*, 459–471. [CrossRef]
28. Karaboga, D.; Basturk, B. On the Performance of Artificial Bee Colony (ABC) Algorithm. *Appl. Soft Comput.* **2008**, *8*, 687–697. [CrossRef]
29. Karaboga, D.; Ozturk, C. A Novel Clustering Approach: Artificial Bee Colony (ABC) Algorithm. *Appl. Soft Comput.* **2011**, *11*, 652–657. [CrossRef]
30. Akay, B.; Karaboga, D. Artificial Bee Colony Algorithm for Large-Scale Problems and Engineering Design Optimization. *J. Intell. Manuf.* **2012**, *23*, 1001–1014. [CrossRef]
31. Brajevic, I.; Tuba, M. An Upgraded Artificial Bee Colony (ABC) Algorithm for Constrained Optimization Problems. *J. Intell. Manuf.* **2013**, *24*, 729–740. [CrossRef]
32. Qin, Q.D.; Cheng, S.; Li, L.; Shi, Y.H. Artificial Bee Colony Algorithm: A Survey. *CAAI Trans. Intell. Syst.* **2014**, *9*, 127–135.
33. Gath, I.; Geva, A.B. Unsupervised Optimal Fuzzy Clustering. *IEEE Trans. Pattern Anal. Mach. Intell.* **1989**, *11*, 773–780. [CrossRef]

34. Dong, F.R.; Wang, S.H.; Hou, Q.K. Multi-Parameter Dominant Grouping Method of Rock Mass Discontinuity Based on Principal Component Analysis. *Rock Soil Mech.* **2022**, *43*, 2457–2470.
35. Park, C.H.; Bobet, A. Crack Coalescence in Specimens with Open and Closed Flaws: A Comparison. *Int. J. Rock Mech. Min. Sci.* **2009**, *46*, 819–829. [CrossRef]
36. Zhang, X.P.; Wong, L.N.Y. Crack Initiation, Propagation and Coalescence in Rock-like Material Containing Two Flaws: A Numerical Study Based on Bonded-Particle Model Approach. *Rock Mech. Rock Eng.* **2013**, *46*, 1001–1021. [CrossRef]
37. Li, P.; Cai, M.F.; Gao, Y.B.; Guo, Q.F.; Miao, S.J.; Ren, F.H.; Wang, Y. Mechanical Responses and Fracturing Behavior of Jointed Rock Masses with a Cavity under Different Dynamic Loads. *Int. J. Impact Eng.* **2023**, *178*, 104608. [CrossRef]
38. Afolagboye, L.O.; He, J.M.; Wang, S.J. Crack Initiation and Coalescence Behavior of Two Non-Parallel Flaws. *Geotech. Geol. Eng.* **2018**, *36*, 105–133. [CrossRef]
39. Li, P.; Cai, M.F.; Miao, S.J.; Li, Y.; Wang, Y. Correlation between the Rock Mass Properties and Maximum Horizontal Stress: A Case Study of Overcoring Stress Measurements. *Int. J. Miner. Metall. Mater.* **2025**, *32*, 39–48. [CrossRef]

geosciences

MDPI

Article

The High-Precision Monitoring of Mining-Induced Overburden Fractures Based on the Full-Space Inversion of the Borehole Resistivity Method: A Case Study

Zhongzhong Xu [1,2], Jiulong Cheng [1,2,*] and Hongpeng Zhao [1,2]

1 College of Earth Science and Surveying Engineering, China University of Mining and Technology, Beijing 100083, China; xzzcumtb@163.com (Z.X.); z5900955@163.com (H.Z.)
2 State Key Laboratory for Fine Exploration and Intelligent Development of Coal Resources, China University of Mining and Technology, Beijing 100083, China
* Correspondence: jlcheng@cumtb.edu.cn

Abstract

The evolution of mining-induced overburden fractures (MIOFs) and their dynamic monitoring are critical for preventing roof water hazards and gas disasters in coal mines. Conventional methods often fail to provide sufficient accuracy under the thin soft–hard interbedded roof strata, necessitating advanced alternatives. Here, we address this challenge by proposing a borehole resistivity method (BRM) based on Back-Propagation Neural Network full-space inversion (BPNN-FSI). Based on the Carboniferous Taiyuan Formation in the North China Coalfield, geoelectric models of MIOFs were established for different mining stages. Finite element simulations generated apparent resistivity responses to train and validate the BPNN-FSI model. At the 9-204 working face of Dianping Coal Mine (Shanxi Province), we compared the proposed BRM based on BPNN-FSI with an empirical formula, numerical simulation, similarity physical simulation, and underground inclined drilling water-loss observations (UIDWLOs). Results demonstrate that the BRM based on BPNN-FSI achieves sub-1% error in height of MIOF (HMIOF) monitoring, with a maximum detected fracture height of 52 m—significantly outperforming conventional methods. This study validates the accuracy and robustness of BRM based on BPNN-FSI for MIOF monitoring in thin soft–hard interbedded roof strata, offering a reliable tool for roof hazard prevention and sustainable mining practices.

Keywords: coal seam; mining-induced overburden fractures; borehole resistivity method; monitoring; Back-Propagation Neural Network full-space inversion

1. Introduction

Coal serves as a critical pillar for China's energy security [1,2]. The development patterns of mining-induced overburden fractures (MIOFs) are significantly influenced by the complex interplay between soft rock plasticity and hard rock brittleness, particularly under the challenging conditions of deep high ground stress and the dynamic evolution of fracture networks [3–5]. Under the coupled effects of high-intensity mining and deep high ground stress, the dynamic evolution mechanisms of MIOFs remain poorly understood [6,7]. Given that MIOFs serve as primary pathways for water inrush and gas migration, they pose direct threats to mine safety. Although monitoring methods for MIOFs have advanced in recent years, existing methods still lack high-precision dynamic monitoring capabilities under the thin soft–hard interbedded roof strata of central and western China, especially for fracture evolution in soft–hard interbedded roofs. High-precision dynamic monitoring of MIOFs can effectively elucidate these mechanisms, providing a scientific foundation for the proactive prevention of roof water hazards and gas disasters.

Current research methodologies for MIOF evolution can be classified into four primary categories. Empirical formula methods, based on statistical analysis of extensive field measurements, provide rapid predictive models for engineering applications. Mechanical analysis methods employ theoretical frameworks to develop analytical models, elucidating the mechanisms of MIOF evolution under varying geological conditions. Numerical simulations and similarity physical simulations visually reconstruct the dynamic development of MIOF through computational modeling and scaled physical testing. In situ monitoring techniques, integrating borehole television, distributed optical fiber sensing, hydrogeological testing, and geophysical prospecting, enable real-time monitoring of MIOF evolution. These approaches, each with distinct advantages, collectively form a comprehensive methodological framework for MIOF research. Liu Tianquan [8] pioneered empirical models for predicting MIOF height by systematically analyzing mining data from eastern China, establishing a quantitative foundation for MIOF studies. From a mechanical perspective, Zhang et al. [9] derived a plastic zone boundary equation using elastoplastic theory and the Mohr–Coulomb criterion, clarifying fracture initiation and propagation in thick–hard roofs. Meanwhile, Xiao et al. [10] applied key stratum theory to reveal how dominant strata control fracture orientation and extent. Numerical studies by Guo et al. [11] uncovered the dynamic evolution of annular fracture zones in longwall panels and their correlation with gas migration. In contrast, Han et al. [12] identified saddle-shaped plastic zones in gobs, linking their height to shear stress distribution through orthogonal numerical experiments. Sun et al. [13] combined physical and numerical simulations to demonstrate the inverse relationship between coal pillar width and overburden damage in twin-seam mining. For in situ monitoring advancement, Yang et al. [14] observed the development of MIOFs in a fully mechanized mining face using underground borehole television, obtaining the height of MIOF development. Meanwhile, Zhang et al. [15–17] monitored the evolution of MIOFs during coal mining using distributed optical fiber sensing and combined this with lithological analysis to determine the height of MIOF development. Further advancing the field, Zhang et al. [18] used theoretical analysis, numerical simulation, and underground inclined drilling water-loss observations (UIDWLOs) to clarify the failure mode of weak overburden under thick loose strata and the evolution of MIOFs. Wang et al. [19] studied the temporal effect of MIOF development under shallow loose strata by applying borehole flushing fluid consumption, borehole television, and rock core RQD indicators to investigate the post-mining closure effect of MIOFs. Beyond fracture characterization, Cheng et al. [20] proposed an algorithm for optimizing the layout of microseismic monitoring points, achieving high-precision source location in underground confined-space microseismic monitoring. Additionally, the borehole resistivity method (BRM) has emerged as a dynamic, full-space monitoring tool. For instance, Hou et al. [21] applied this technique to delineate fracture zones, guiding optimal gas drainage borehole placement.

Current methodologies for investigating MIOF evolution exhibit significant limitations in addressing the thin soft–hard interbedded roof strata prevalent in central and western China. Traditional approaches—including empirical formulas, mechanical analyses, numerical simulation, similarity physical simulation, and static monitoring techniques—are constrained by theoretical simplifications and static observation paradigms, resulting in inadequate accuracy for dynamic fracture characterization. While the UIDWLO method demonstrates high precision in HMIOF monitoring in overlying strata during coal seam roof failure and has been widely adopted in field applications, its effectiveness is constrained by its limited detection range (restricted to borehole vicinity), construction complexity, and susceptibility to borehole collapse [22]. While BRM offers dynamic monitoring capabilities, its dependence on conventional half-space inversion theory compromises inversion precision, failing to meet practical engineering requirements for accurate fracture identification. This coexistence of multiple methods with their respective limitations severely restricts the accurate revelation of the evolution of MIOFs under thin soft–hard interbedded roof strata.

In this study, we address the technical challenge of high-precision dynamic monitoring of MIOFs under thin soft–hard interbedded roof strata in the central and western regions of China. Taking the typical strata of the Taiyuan Formation in the Carboniferous System of the North China Coalfield as the research object, we constructed geoelectric models of the MIOFs of overburden at different mining stages. The apparent resistivity responses of these models were calculated using the finite element method to build the training and testing sets for the Back-Propagation Neural Network full-space inversion (BPNN-FSI) model. We then established the BPNN-FSI model and proposed a BRM based on BPNN-FSI. Using the 9-204 working face of Dianping Coal Mine in Shanxi Province as an engineering case, we comprehensively applied the empirical formula, numerical simulation, similarity physical simulation, UIDWLO, and the BRM based on BPNN-FSI to investigate the evolution of MIOFs. By comparing and analyzing the results obtained from these four methods, we validated the reliability of the BRM based on BPNN-FSI, which achieved high-precision dynamic monitoring of MIOFs. This method can effectively monitor the evolution of MIOFs and provides a new technical means for safe coal mining under thin soft–hard interbedded roof strata in the central and western regions.

2. Case Study Area and Methods

2.1. Geological Setting of the Case Study

The Dianping Coal Mine is located in Dianping Village, Dawu Town, Fangshan County, Lüliang City, Shanxi Province. The coal-bearing strata in the mining area primarily consist of the Upper Carboniferous Taiyuan Formation (C_3t) and the Lower Permian Shanxi Formation (P_1s). The recoverable coal seams include the No. 3, No. 5, and No. 9 seams, with burial depths of 248.9 m, 268.45 m, and 322.03 m, respectively. The interlayer spacing between the No. 5 and No. 9 seams ranges from 63 m to 67.1 m. According to the adjusted mining sequence plan, priority is given to extracting the deeper No. 9 seam. However, MIOFs due to the extraction of the No. 9 seam may compromise the structural integrity of the overlying No. 5 seam. To ensure the safe subsequent mining of the No. 5 seam, it is of theoretical significance to investigate the development height and evolution patterns of MIOFs. This study focuses on the 9-204 working face, with Figure 1 illustrating its roof structure and Figure 2 presenting the layout schematic. Based on the stratigraphic occurrence characteristics illustrated in Figure 1 and the comprehensive rock physical–mechanical parameters presented in Table 1 for the roof and floor strata of the No. 9 coal seam, this study conducted a systematic investigation of the mechanical properties of surrounding rocks. To ensure parameter accuracy, the research team performed rigorous rock mechanics tests on samples collected from the vicinity of the working face. The resultant mechanical properties of both roof and floor strata are systematically summarized in Table 1. The evolution patterns and development height of MIOFs are analyzed, providing a scientific basis for safe coal seam extraction.

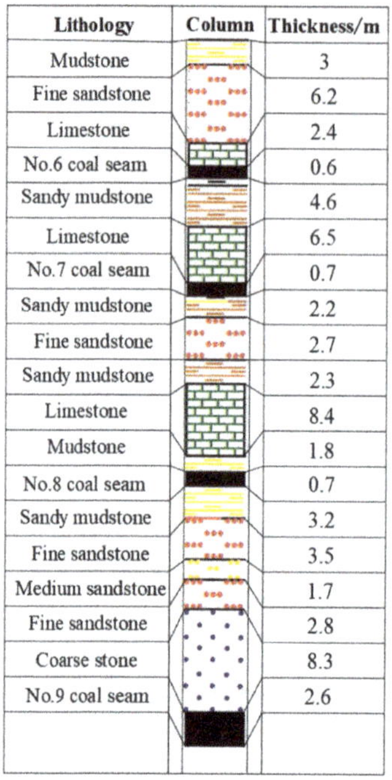

Figure 1. The stratigraphic column of the No. 9 seam.

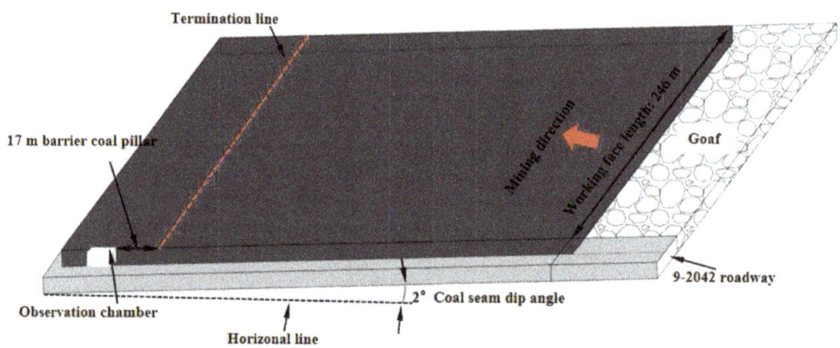

Figure 2. A schematic diagram of the 9-204 working face layout.

Table 1. Mechanical parameters of rock strata.

Lithology	Average Tensile Strength/MPa	Young's Modulus /GPa	Cohesion /MPa	Friction Angle/(°)	Poisson Ratio	Density/ (kg/m^3)	Compressive Strength/MPa
Limestone	2.70	8.69	11.4	38	0.18	2800	34.18
Mudstone	2.48	0.81	2.8	30	0.29	2699	17.8
Sandy mudstone	3.65	1.99	7.8	32	0.28	2600	18.15
Coarse sandstone	4.34	7.92	3.0	40	0.15	2700	49.35
Coal seam	0.20	1.12	0.2	31	0.30	1410	4.7
Fine sandstone	8.50	3.58	9.2	33	0.26	2570	59.96
Medium sandstone	3.50	3.92	5.0	34	0.22	2560	39.4

2.2. Numerical Simulation

Numerical simulation is an effective method for studying the evolution laws of mining-induced fractures in coal seam roofs during coal mining processes. This study employs the COMSOL Multiphysics 6.1 finite element analysis software to conduct numerical simulations investigating the evolution patterns and development height of MIOFs at the 9-204 working face. In this study, COMSOL Multiphysics 6.1 finite element analysis software was employed, utilizing the steady-state study in the Solid Mechanics module of the Structural Mechanics branch. The modeling procedure essentially follows the same workflow as conventional finite difference software (FLAC3D). The software utilizes an automatic remeshing technology for large-deformation geometries, effectively addressing the mesh adaptation challenges encountered by conventional numerical methods when simulating large roof deformations. Through its advanced visualization capabilities, the spatial distribution characteristics of MIOFs are quantitatively characterized, while the dynamic evolution of fracture development height is visually demonstrated.

2.2.1. Numerical Simulation Model

Based on the geological conditions of the 9-204 working face in the No. 9 coal seam, this study established a three-dimensional numerical model to simulate the evolution characteristics of MIOFs. The model dimensions are 349 m × 349 m × 100 m, incorporating two roadways and one working face (Figure 3). The simulated coal seam has a burial depth of 322 m, with a working face length of 246 m and an average dip angle of 2°. To accurately represent the mining process, a progressive excavation method was adopted to simulate the complete 246 m face advance. For boundary effect mitigation, 50 m wide boundary coal pillars were set on both sides along the x and y directions. This configuration ensures computational reliability by reducing interference from boundary constraints on the analysis of MIOF evolution, thereby enhancing the accuracy of numerical simulations.

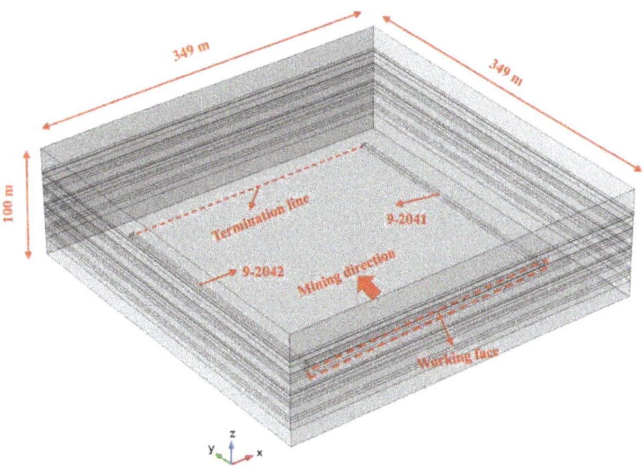

Figure 3. A schematic diagram of the numerical simulation model.

2.2.2. Numerical Simulation Scheme

This study employs numerical simulation methods to analyze the in situ stress environment in 250 m deep underground rock strata. The initial in situ stress field (Figure 4) was established by applying 6 MPa vertical stress (Z-axis) and equivalent horizontal stresses (X and Y axes), consistent with the regional in situ stress characteristic of a horizontal-to-vertical stress ratio of 1. The model boundary conditions were configured with fixed constraints at the bottom and normal constraints in horizontal directions. Based on the borehole columnar diagram of the 9-204 working face in Shanxi Dianping Coal Mine, the rock strata were appropriately consolidated and simplified to establish the computational model. The numerical simulation was performed using the Mohr–Coulomb failure criterion [23,24].

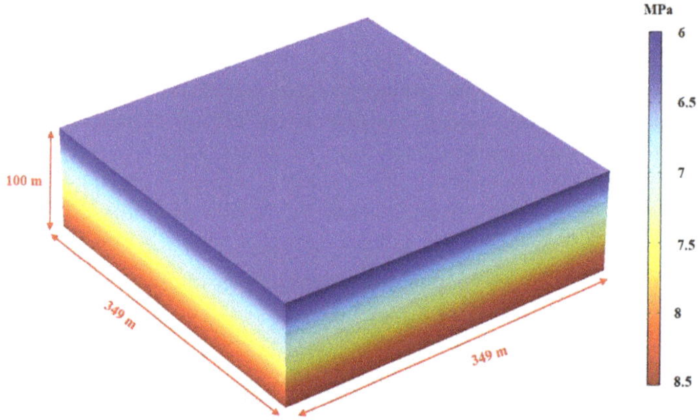

Figure 4. A schematic of the initial in situ stress field.

2.3. Similarity Physical Simulation

Similar physical simulation of coal mining can visually demonstrate the development process of MIOFs in the coal seam roof. Based on similarity theory, a physical simulation experiment was designed to replicate the mining process of the No. 9 coal seam in Dianping Coal Mine, considering the geological conditions and mining parameters. The model utilized artificially proportioned materials to simulate the mechanical properties of coal and rock strata, with scaling ratios applied to geometry, time, and strength (including density, elastic modulus, uniaxial compressive strength, and Poisson's ratio). Displacement sensors and stress sensors were employed to monitor real-time fracture development and evolution in the overburden strata during simulated mining. The experimental data were then converted into full-scale strata movement parameters using similarity principles for practical mining applications.

The experiment employed a similarity material model with dimensions of 2.0 m (length) × 0.2 m (width) × 1.15 m (height). The scaling ratios were set as follows: geometric similarity ratio (1:150), time similarity ratio (1:12), density similarity ratio (1:1.5), stress similarity ratio (1:225), and Poisson's ratio (1:1). The material composition of the model was determined based on the actual physico-mechanical parameters and layer thicknesses of the coal seam roof and floor in the 9-204 working face. These parameters were converted using the aforementioned similarity ratios to ensure that the experimental conditions strictly adhered to the in situ geological conditions.

In the similarity physical simulation, the completed model was first subjected to 72 h of shade drying before removing the channel steels and formworks. Based on the geometric similarity ratio, horizontal and vertical displacement measurement lines were arranged on the No. 9 coal seam roof surface at 10 cm intervals, with densified monitoring points in critical areas (roof strata of both No. 9 and No. 5 coal seams). Reflective targets (2 cm × 2 cm crosses) were fixed using thumbtacks and monitored via total station to track overburden movement (Figure 5). During the mining simulation, a 10 cm protective coal pillar (equivalent to 15 m in situ) was maintained outside the starter cut. The No. 9 coal seam was excavated for 180 cm, with each 10 cm advance followed by a 30 min interval to ensure complete strata movement. The experiment was conducted by two operators working cooperatively to promptly remove excavated materials. High-resolution cameras were employed after each mining increment to document overburden deformation characteristics.

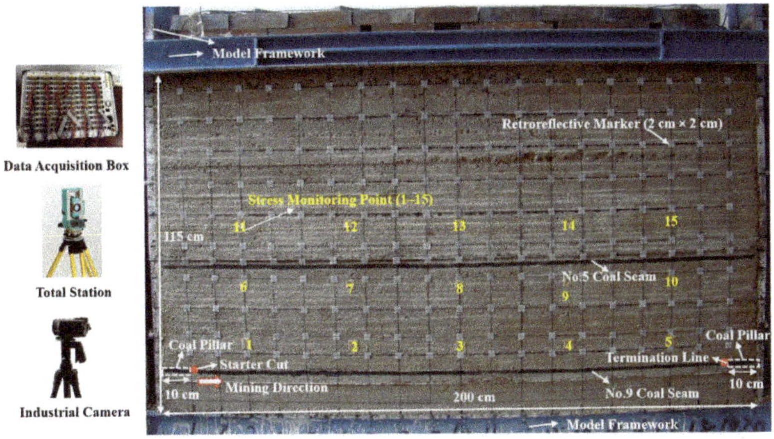

Figure 5. Layout of the similarity physical simulation.

2.4. UIDWLO Monitoring Method

The UIDWLO monitoring method is an effective approach for determining the development height of MIOFs during coal seam roof failure. The UIDWLO monitoring method (Figure 6) represents a high-precision technique for assessing MIOFs in overburden strata. This approach involves establishing an observation chamber at the longwall panel's stopping line and drilling upward-inclined boreholes to enable accurate monitoring of fracture height development in the overlying strata. The implementation procedure was conducted as follows: A monitoring station was positioned 17 m west of the stopping line in Gateway 9-2042, where a 5 m long drilling chamber was excavated to install the No. 2 hole (detailed drilling parameters are presented in Table 2). The borehole was specifically designed with the following technical specifications, ensuring comprehensive coverage of the fracture development zone within the roof strata: a diameter ranging from 73 to 89 mm, an azimuth angle of 56°, an inclination angle of 45°, and a total depth of 94 m (controlling a vertical height of 66 m), the arrangement diagram of UIDWLO is presented in Figure 7. The monitoring system employed a "Double-Packer Leakage Detection Device" to conduct segmented water injection tests, which precisely measured leakage flow rates in each borehole segment (with positioning errors maintained below 5%). This methodology provides reliable quantitative data for characterizing fracture development patterns and determining fracture heights in the overburden strata, offering crucial technical support for stability assessment in mining operations.

Table 2. The parameter of the observation borehole.

No.	Diameter (mm)	Orientation Angle (°)	Elevation Angle (°)	Drilling Depth (m)	Measured Height (m)
No. 2 hole	73~89	56	45	94	66

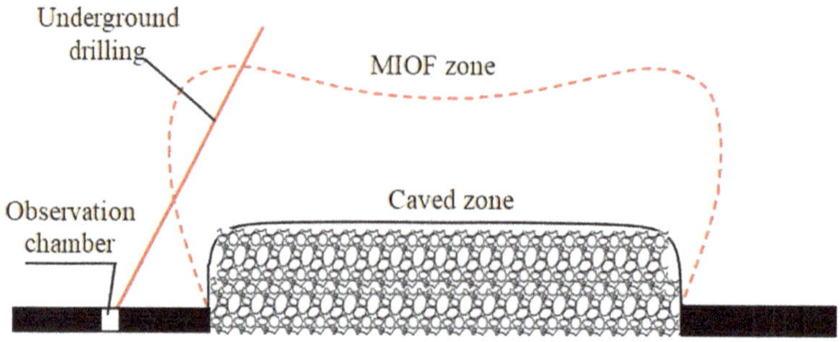

Figure 6. Schematic diagram of the UIDWLO.

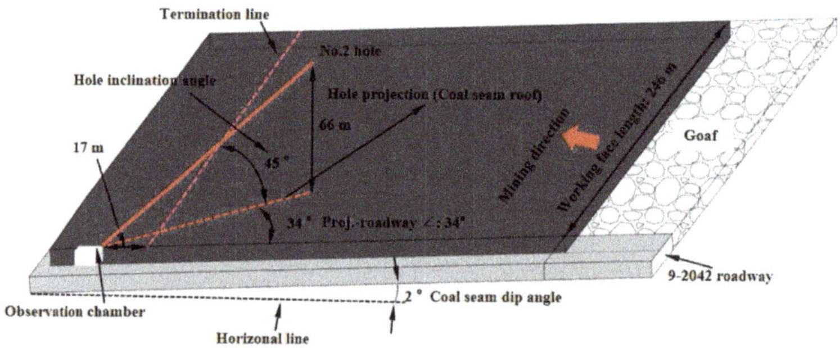

Figure 7. Layout diagram of the UIDWLO.

2.5. BRM Based on BPNN-FSI Monitoring Method

2.5.1. Working Principle and Advantages of BRM

The BRM is a full-space resistivity method based on the electrical property differences in rock masses. This technique integrates electrical sounding, electrical profiling systems, and drilling monitoring technology to achieve dynamic monitoring of MIOF development in overlying strata during coal seam extraction [25–27]. The fundamental principle lies in the fact that stress redistribution in the roof strata caused by coal mining alters the pore structure, mechanical state, and water-bearing characteristics of rock masses, consequently inducing significant changes in their electrical parameters [28–31]. In practical implementation, electrode arrays identical to those used in conventional high-density resistivity methods are installed in underground boreholes (Figure 8). Current is injected into surrounding rocks through current electrodes while simultaneously measuring potential differences between monitoring electrodes. The apparent resistivity (ρ_S) at each measurement point is then calculated using the apparent resistivity formula [32] (Equation (1)). Through programmed control of electrode combinations and spacings, potential data at different depths can be automatically acquired, enabling real-time monitoring of electrical property evolution from initial strata deformation to the complete process of MIOF development. This approach provides critical data for investigating the evolution patterns of MIOFs and serves as an important foundation for hazard monitoring.

$$\rho_S = K\frac{\nabla U}{I} \tag{1}$$

In the formula, K represents the full-space array coefficient, ∇U denotes the potential difference between measuring electrodes M and N, and I stands for the current intensity.

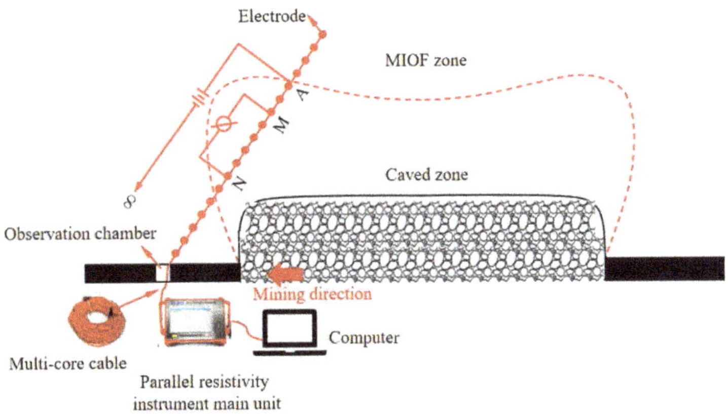

Figure 8. Schematic diagram of the BRM Wenner tripole configuration.

2.5.2. Engineering Design of BRM

This study established a dynamic monitoring method for MIOFs at the 9-204 working face using the BRM. The implementation procedure comprised (1) installation of a monitoring station 17 m west of the stopping line in the 9-2042 Roadway, where a 5 m drilling chamber was excavated to deploy the No. 1 hole (parameters detailed in Table 3); (2) configuration of the borehole with a 73–89 mm diameter, 56° azimuth angle, 50° inclination angle, and 105 m depth (controlling a vertical height of 72 m) to cover the roof fracture development zone; (3) deployment of a Wenner tripole array featuring 50 electrodes at 2 m intervals through integrated push-in installation technology. The system incorporated borehole cleaning for fractured strata and cement grouting to achieve optimal electrode–wall coupling, forming a complete technical protocol encompassing "borehole design–electrode arrangement–device installation–data acquisition" (layout schematic shown in Figure 9). This solution enabled real-time monitoring of MIOFs within 72 m of the roof strata, providing reliable technical support for safe mining operations. The methodology demonstrated three key innovations: (a) enhanced spatial resolution through optimized electrode spacing, (b) improved data quality via advanced coupling techniques, (c) operational efficiency gains through automated measurement sequences.

Table 3. Parameters of BRM observation borehole.

No.	Diameter (mm)	Orientation Angle (°)	Elevation Angle (°)	Drilling Depth (m)	Measured Height (m)
No. 1 hole	73~89	56	50	105	72

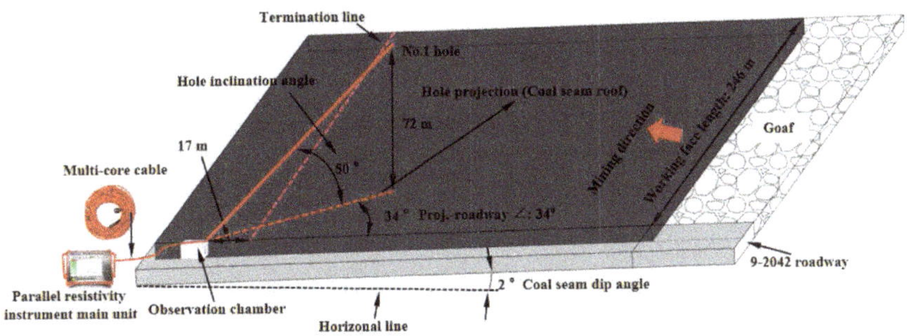

Figure 9. Layout diagram of the BRM.

2.5.3. BRM Based on BPNN-FSI

(1) BPNN-FSI model structure

This study conducts research into a full-space inversion borehole resistivity method based on a BPNN algorithm, with particular focus on solving the full-space inversion problem of the geoelectric model for overburden in "two zones". The BPNN-FSI process essentially represents an iterative optimization of network weights and thresholds, where the training terminates when the mean square error between the network output (inverted resistivity) and desired output (true resistivity of geoelectric model) reaches the preset accuracy. Figure 10 illustrates the network structure of the BPNN-FSI model. In this structure, the input layer consists of apparent resistivity observation data from the geoelectric model, while the output layer produces corresponding inverted resistivity values.

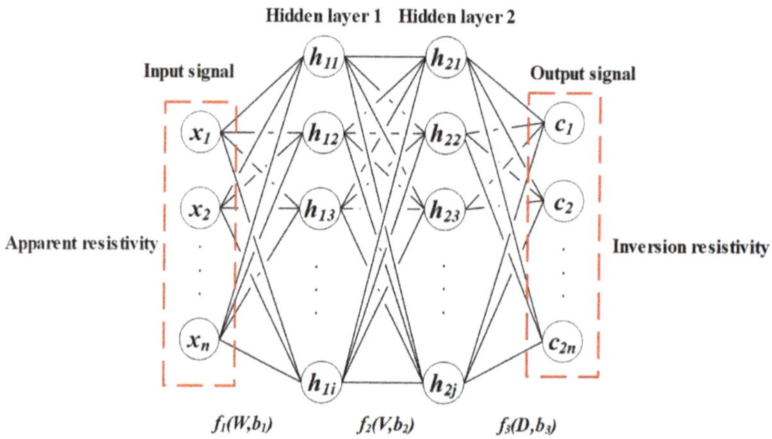

Figure 10. The BPNN-FSI model structure.

The BPNN-FSI model employs a dual-hidden-layer structure, mathematically expressed as follows: The input layer $X = (x_1, x_2, \ldots, x_n)$ is mapped to the first hidden layer H_1 through weight matrix W and bias vector b_1, with its output computed by activation function f_1:

$$H_1 = f_1(WX + b_1) \tag{2}$$

where the i node of H_1 is denoted as h_{1i}.

The second hidden layer H_2 is computed via weight matrix V and bias vector b_2, using activation function f_2:

$$H_2 = f_2(VX + b_2) \tag{3}$$

where the j node of H_2 is denoted as h_{2j}.

The final output layer $C = (c_1, c_2, \ldots, c_{2n})$ is determined by weight matrix D, bias vector b_3, and activation function f_3:

$$H_3 = f_3(DX + b_3) \tag{4}$$

This model achieves the mapping from apparent resistivity data to inverted resistivity values through layer-wise nonlinear transformations.

(2) Model Training and Testing

Taking the typical stratigraphy of the Carboniferous Taiyuan Formation in North China coalfield as a case study, this research established geoelectric models of MIOFs for different mining stages, based on the developmental patterns of MIOFs during longwall retreating [33,34]. The COMSOL Multiphysics 6.1 finite element method was employed to calculate the apparent resistivity responses of these models, generating the training and testing datasets for BPNN inversion [35]. Field data collection utilized a tripole electrode configuration with 51 measuring electrodes deployed at 2 m intervals. The study systematically constructed 810 representative geoelectric models [35], producing 442,260 sample data points through models. The dataset was partitioned into 730 models (398,580 data points) for training and 80 models (43,680 data points) for independent testing, with the test set rigorously excluded from all training phases to ensure unbiased model evaluation.

To ensure high prediction accuracy while maintaining reasonable training duration, a 4-layer BPNN-FSI network structure was selected [36], as illustrated in Figure 10. The number of neurons in the hidden layers critically determines model performance: insufficient neurons compromise prediction accuracy, whereas excessive neurons not only prolong training time but also degrade the model's generalization capability. To optimize this parameter, the study first constrained the neuron count range using empirical formulas [36], then systematically evaluated different configurations under fixed hyperparameters (learning rate = 0.01, maximum iterations = 3000, convergence error = 10^{-6}). The final neuron count was determined through this rigorous trial process, balancing computational efficiency with inversion accuracy.

$$m = \sqrt{n+l} + \alpha \tag{5}$$

In the formula, m represents the number of neurons in the hidden layer, n denotes the number of neurons in the input layer, and l indicates the number of neurons in the output layer. The constant $\alpha \in [1, 10]$. For the BPNN-FSI prediction, the number of input features is 3, that is, $n = 3$, and the number of outputs is 2, that is, $l = 2$. Based on the empirical Formula (5), substituting the feature numbers into the formula yields the range of the hidden layer neurons m as 3–12.

Figure 11 presents comparative results after 3000 training iterations, demonstrating that both prediction accuracy and training duration exhibit positive correlations with neuron quantity. While increasing the number of neurons prolongs training time, the correlation coefficient peaks at 12 neurons. Although this configuration requires marginally longer training, the selected 12-neuron structure optimally balances computational efficiency with predictive precision.

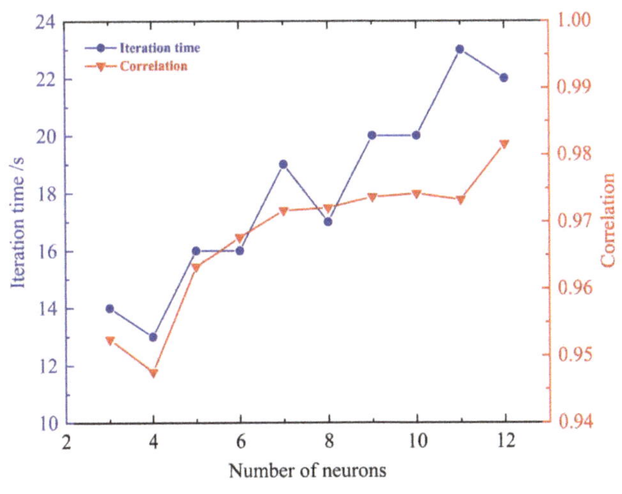

Figure 11. Relationship between hidden layer neuron quantity and prediction accuracy.

To validate the generalization capability of the BPNN-FSI model, an independent test set (80 model samples excluded from training) was employed for evaluation. The results indicate that inversion accuracy ranges between 0.81 and 0.96 across test samples, with all cases exceeding the 0.8 threshold. This confirms the model's robust training effectiveness and reliable predictive performance for geophysical inversion tasks.

3. Results

3.1. Numerical Simulation Results

Numerical simulation can effectively reproduce the evolution of MIOFs during coal seam extraction. Under the combined action of tensile and shear stresses, the MIOFs exhibit a stage-wise development characteristic. These fractures include vertical tensile fractures and horizontal shear fractures, which interconnect with each other, ultimately forming a complex three-dimensional spatial network of MIOFs. The failure modes of the rock strata are primarily manifested as follows: tensile failure occurs when the tensile stress exceeds the tensile strength of the rock mass, while shear failure occurs when the shear stress surpasses the peak strength of the rock mass [37,38].

The study analyzed four typical stages of working face advancement at 50 m, 100 m, 150 m, and 246 m (Figure 12). The results demonstrate that the development of MIOFs exhibits distinct stage-wise characteristics. When the 9-204 working face advanced to 50 m, the maximum fracture height reached 8.3 m; this increased to 12.7 m at 100 m and rose significantly to 24.7 m at 150 m. Upon reaching 246 m, the fracture height peaked at 50 m. Notably, when the working face advanced beyond 246 m, fracture development was primarily characterized by horizontal propagation, with the height stabilizing, indicating that the MIOFs entered a stable phase. Further analysis revealed that the fully developed fracture zones were predominantly concentrated directly above the surrounding rock of the stope. Based on these findings, pre-reinforcement measures can be implemented in the coal mining face and roadways to ensure safe extraction.

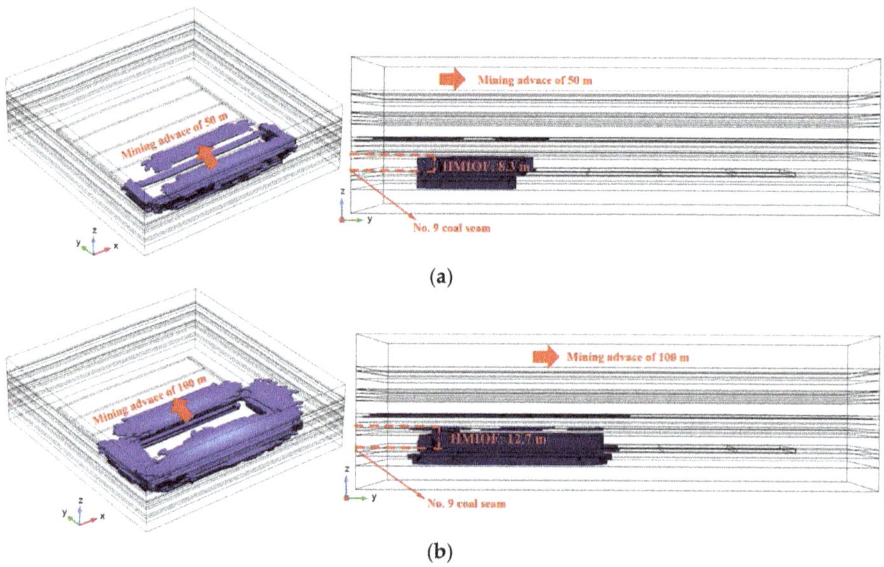

Figure 12. *Cont.*

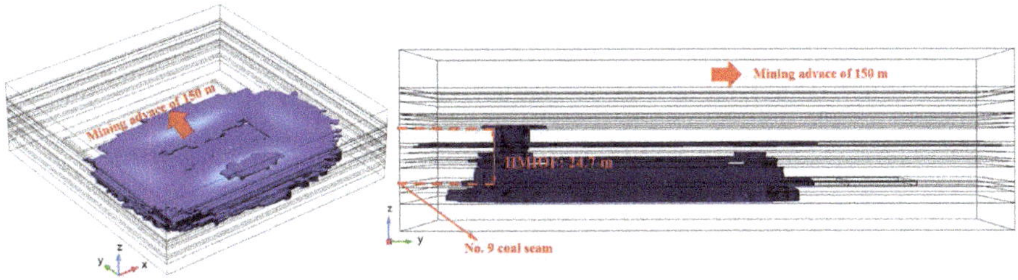

(c)

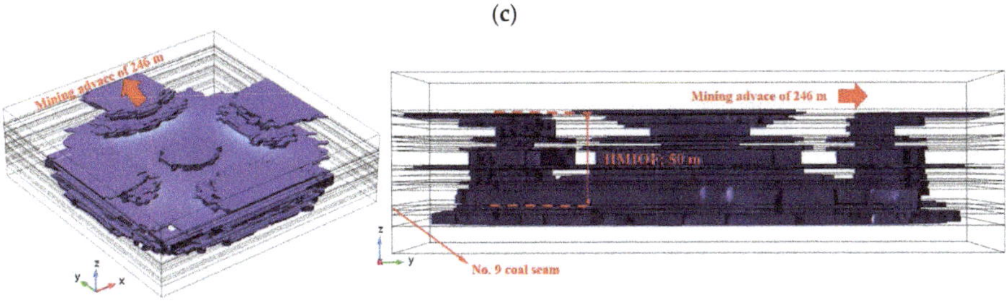

(d)

Figure 12. Fractures evolution in coal seam roof and floor due to mining. (**a**) Mining advance of 50 m; (**b**) mining advance of 100 m; (**c**) mining advance of 150 m; (**d**) mining advance of 246 m.

3.2. Similarity Physical Simulation Results

The interburden strata, serving as the key rock layers between the No. 9 coal seam and No. 5 coal seam, function both as the roof of the No. 9 coal seam and the floor of the No. 5 coal seam. Their structural stability and fracture development characteristics directly influence the minability of the No. 5 coal seam. Similarity physical simulation tests reveal that MIOFs exhibit two dominant patterns: alternating development of bed-separation fractures and newly generated fractures, with fractures primarily inclined from the starter cut toward the goaf direction. These rock layers not only act as potential pathways connecting the goafs of the upper and lower coal seams but also critically affect mine safety through their fracture and movement behavior. Research indicates that as mining depth increases, the fracture patterns of overlying strata become more regular, with damage severity decreasing progressively from the lower to upper sections. This phenomenon provides theoretical support for safe extraction from the No. 5 coal seam.

The development of MIOFs exhibits distinct stage-wise characteristics (Figure 13). The initial fracture stage demonstrates intense rock breakage, followed by periodic fracturing that results in regularly arranged rock blocks with reduced bulking factors, effectively constraining the movement of upper strata. As the goaf expands, the strata transfer fracture forces through voussoir beam articulated structures, while the suspended roof effect promotes upward propagation of bed separation. Notably, fracture development near the coal seam is significantly more pronounced than in distant strata. Interlayer deformation analysis reveals that the fracture of thick, hard strata can induce grouped movement of overlying formations, forming enclosed bed-separation structures near the No. 5 coal seam. Physical simulation tests confirm that after mining the No. 9 coal seam, the HMIOF reaches 30 cm (equivalent to 45 m in prototype scale).

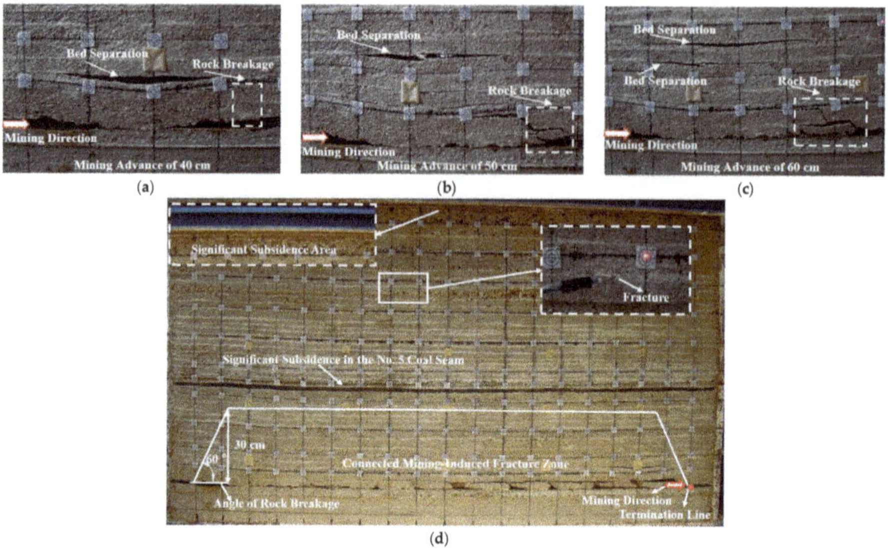

Figure 13. Evolutionary stages of MIOF above No. 9 coal seam. (**a**) Mining advance of 40 cm; (**b**) mining advance of 50 cm; (**c**) mining advance of 60 cm; (**d**) terminal mining configuration (180 cm).

3.3. UIDWLO Results

The results of UIDWLO (Figure 14) show that the mining-affected zone can be divided into two typical areas: Zone I (the zone of MIOFs) exhibits water leakage rates of 4.8–30.2 L/min, reflecting significant fracture development, while Zone II (the bending subsidence zone) shows reduced leakage rates of 1.2–3.2 L/min, notably, within the borehole interval of 79–83 m, localized increases in drilling fluid loss were observed. However, analysis of the borehole log indicates that this elevated loss is attributable to bed separation in the coal seam roof. Based on the geometric relationship of boreholes, the measured data indicate that the maximum HMIOF above the 9-204 working face reaches 51.62 m, with the fracture development apex located at a borehole depth of 73 m. These quantitative results provide important evidence for accurately evaluating the degree of overburden damage.

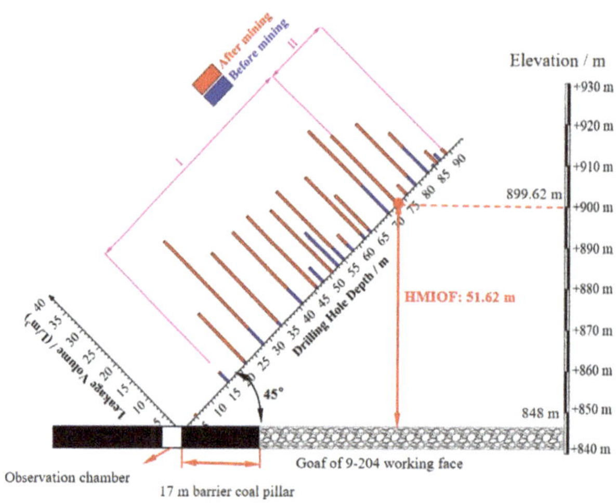

Figure 14. The water leakage before and after mining.

3.4. BRM Based on BPNN-FSI Results

The monitoring results obtained from the BRM based on BPNN-FSI reveal the dynamic evolution characteristics of MIOFs (Figure 15a–d). When the working face was 40 m from the termination line, the fracture development height reached 44 m (from the roof elevation of 848 m to 892 m in the No. 9 coal seam). The resistivity profile showed a high-resistivity fracture zone at borehole depths of 5–28 m, while micro-fracture development features appeared in surrounding rocks at 30–55 m. As the face advanced to 20.5 m, the fracture height increased to 49 m (elevation 897 m), with a significant high-resistivity response observed in the 30–54 m range. The maximum development height of 52 m (elevation 900 m) was achieved when approaching 3.5 m from the stopping line, where a distinct high-resistivity zone formed between 5 and 71 m due to moisture loss. After 67 days of stopping, the height stabilized at 51 m (elevation 899 m). This study demonstrates that the dynamic evolution process of MIOFs can be accurately characterized through resistivity anomalies, providing crucial empirical evidence for understanding fracture development patterns in overlying strata.

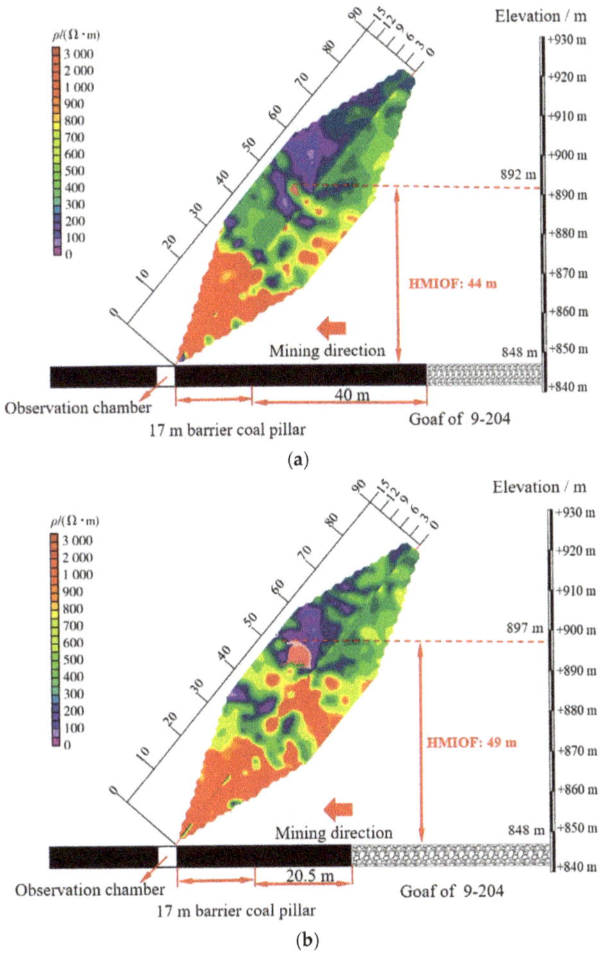

Figure 15. *Cont.*

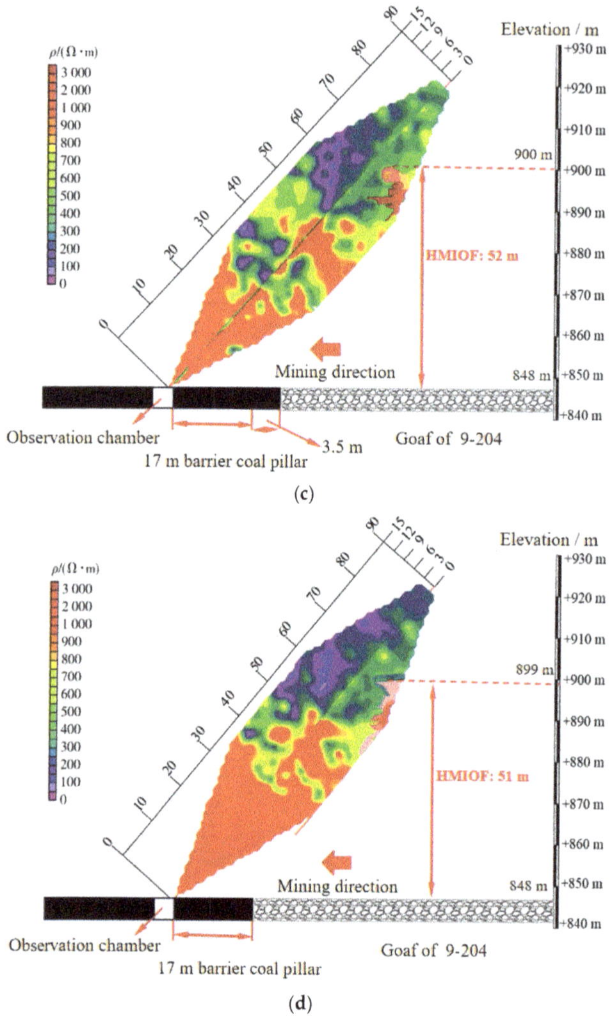

Figure 15. Resistivity profile monitoring results of MIOF evolution. (**a**) Working face 40 m from termination line; (**b**) working face 20.5 m from termination line; (**c**) working face 3.5 m from termination line; (**d**) working face reaches termination line.

The BRM based on BPNN-FSI successfully revealed the evolutionary characteristics of MIOFs during the extraction of the No. 9 coal seam. Through detailed analysis of resistivity profiles at various monitoring stages, the maximum fracture development height in the overlying strata of the 9-204 working face was determined to be 52 m.

4. Discussion

Considering the geological characteristics of thin interbedded soft–hard strata in the Taiyuan Formation coal measures, this study systematically investigated fracture development patterns in MIOFs at the 9-204 working face through four complementary methodologies: (1) numerical simulation, (2) physical similarity modeling, (3) upward-inclined borehole water leakage observation (UIDWLO), (4) a BRM based on BPNN-FSI. The measured HMIOF showed strong methodological consistency, yielding 50 m (numerical simulation), 45 m (physical simulation), 51.62 m (UIDWLO), and 52 m (BRM based on BPNN-FSI), respectively. While UIDWLO demonstrated superior reliability in HMIOF accuracy monitoring compared to conventional methods, three inherent limitations were identified: (i) restricted spatial resolution limited to borehole-adjacent fractures, (ii) operational complexity due to sophisticated construction requirements, (iii) susceptibility to borehole collapse, impairing field applicability. Notably, the developed BRM based on BPNN-FSI exhibited exceptional agreement with UIDWLO results (relative error < 1%), validating its effectiveness as an alternative monitoring approach. All methods conclusively demonstrated that MIOF propagation from the No. 9 coal seam terminated below the 63–67.1 m interburden zone, preserving the structural integrity of the No. 5 coal seam's floor. This multi-methodological verification provides robust evidence for (1) absence of hydraulic connectivity between seams, (2) maintenance of load-bearing capacity in underlying strata, (3) feasibility of safe extraction in the No. 5 coal seam under current mining conditions.

Comparative analysis reveals that traditional empirical formulas [8] exhibit significant calculation errors ranging from 18.18% to 24.26% (Table 4), while the novel method proposed in this study controls the error within 1%. Regarding monitoring techniques, microseismic monitoring—which captures the microseismic signals released by rock mass fracturing and inversely deduces the location, intensity, and evolution patterns of the fractures [39,40]—while suitable for high-energy rock fracturing, still requires improvement in locating weak fracturing signals within the thin interbedded strata of the Taiyuan Formation. This study innovatively proposes a BRM based on BPNN-FSI, which not only achieves high-precision monitoring of MIOF height (relative error < 1%) but also accurately characterizes the spatial distribution of fractures. It should be noted that the method's accuracy heavily depends on the initial geoelectric model construction, suggesting that future research should focus on optimizing neural network algorithms to enhance model adaptability. The research outcomes provide a reliable technical approach for safe mining in Shanxi's complex coal measures, particularly demonstrating that the No. 5 coal seam maintains structural integrity after No. 9 coal seam extraction and can be safely mined without water inrush or strata instability risks.

Table 4. Comparison of fracture height.

	Equation	HMIOF (m)		Relative Error (%)	
		H_1	H_2	H_1	H_2
Empirical formula [8]	$H_1 = \frac{100\sum M}{1.5M+3.6} \pm 5.6$ $H_2 = 20\sqrt{\sum M} + 10$	39.11	42.25	24.26	18.18
Numerical simulation		50		3.18	
Similarity physical simulation		45		12.84	
BRM based on BPNN-FSI		52		0.7	
UIDWLO		51.62			

However, the current research still faces several limitations: the BRM remains at the qualitative analysis stage and has yet to achieve quantitative characterization of fracture development intensity; additionally, the inversion algorithm's precision requires further improvement. Addressing these challenges will constitute critical future research directions to advance methodological refinement and practical applications.

5. Conclusions

This study systematically investigated the evolution law of MIOFs under thin soft–hard interbedded roof strata in the Taiyuan Formation in Shanxi Province through integrated approaches, including numerical simulation, similarity physical modeling, UIDWLO, and a BRM based on BPNN-FSI. The main conclusions are as follows:

(1) Comprehensive results from numerical simulation, similarity physical modeling, UIDWLO, and the BRM based on BPNN-FSI consistently demonstrate that MIOFs from the No. 9 coal seam did not propagate to the No. 5 coal seam, confirming the structural integrity of the No. 5 coal seam's floor without rupture occurrence.
(2) BPNN-FSI achieved high-precision dynamic monitoring of fracture development height (relative error < 1%), showing significant advantages over traditional empirical formulas, numerical simulation, and similarity physical modeling approaches.
(3) The monitoring accuracy of the proposed BRM based on BPNN-FSI depends on initial geoelectric model construction, suggesting that future research should incorporate artificial intelligence algorithms for model optimization.
(4) The research outcomes not only provide technical support for dynamic monitoring of MIOFs, but also offer methodological references for similar open-pit slope engineering. Under thin soft–hard interbedded roof strata, these findings significantly guide safe and efficient coal mining operations by effectively preventing roof water inrush and gas disasters, while providing a scientific basis for sustainable ecological development in mining areas.

Author Contributions: All authors contributed to the study's conception and design. Material preparation, data collection, and analysis were performed by Z.X., J.C. and H.Z. The first draft of the manuscript was written by Z.X. and all authors commented on previous versions of the manuscript. All authors have read and agreed to the published version of the manuscript.

Funding: This research was funded by National Natural Science Foundation of China (42274194).

Data Availability Statement: Data and materials can be made available upon request, if suitable.

Conflicts of Interest: The authors declare no conflicts of interest.

Abbreviations

The following abbreviations are used in this manuscript:

MIOF	Mining-induced overburden fracture
BPNN-FSI	Back-Propagation Neural Network full-space inversion
BRM	Borehole resistivity method
HMIOF	Height of mining-induced overburden fractures
UIDWLO	Underground inclined drilling water-loss observations

References

1. Ren, B.; Ding, K.; Wang, L.; Wang, S.; Jiang, C.; Guo, J. Research on an Intelligent Mining Complete System of a Fully Mechanized Mining Face in Thin Coal Seam. *Sensors* **2023**, *23*, 9034. [CrossRef]
2. Wang, J.L.; Jiang, L.J.; Cang, T.C.; Zhou, X.Z.; Wang, B.C. Simulation of a Multi-Stage Stress Field and Regional Prediction of Structural Fractures in the Tucheng Syncline, Western Guizhou, China. *Geosciences* **2025**, *15*, 132. [CrossRef]
3. Nejati, H.R.; Ghazvinian, A. Brittleness Effect on Rock Fatigue Damage Evolution. *Rock Mech. Rock Eng.* **2014**, *47*, 1839–1848. [CrossRef]
4. Liu, Y.B.; Cheng, J.L.; Jiao, J.J.; Gao, Z.; Cheng, P. Influences of the Hard Rock Proportion Coefficient on the Evolution Pattern and Fractal Characteristics of Mining Fractures in a Composite Roof. *Int. J. Geomech.* **2024**, *24*, 04024038. [CrossRef]
5. Zhang, T.; Chen, Q.Z.; Zhang, J.Z.; Zhou, X.P. Influences of Mechanical Contrast on Failure Characteristics of Layered Composite Rocks Under True-Triaxial Stresses. *Rock Mech. Rock Eng.* **2023**, *56*, 5363–5381. [CrossRef]
6. Huang, W.P.; Li, C.; Zhang, L.W.; Yuan, Q.; Zheng, Y.S.; Liu, Y. In situ identification of water–permeable fractured zone in overlying composite stratum. *Int. J. Rock Mech. Min. Sci.* **2018**, *105*, 85–97. [CrossRef]
7. Ju, J.F.; Xu, J.L.; Zhao, F.Q.; Wang, Y.Z. Surface Subsidence Observations and Strata Breaking Activity Inversion from Underground Coal Mining: A Case Study in Western China. *Rock Mech. Rock Eng.* **2024**, *57*, 10935–10952. [CrossRef]

8. Liu, T.Q. Current Status and Prospects of Coal Mining Technology Under Buildings, Water Bodies, Railways and Above Confined Water. *Coal Sci. Technol.* **1995**, *1*, 5–7+62. [CrossRef]
9. Zhang, C.W.; Jin, Z.X.; Song, X.M.; Feng, G.R.; Li, Z.; Gao, R.; Zhu, D.F.; Li, C. Failure mechanism and fracture aperture characteristics of hard thick main roof based on voussoir beam structure in longwall coal mining. *Energy Sci. Eng.* **2020**, *8*, 340–352. [CrossRef]
10. Xiao, P.; Han, K.; Shuang, H.Q.; Ding, Y.; Kong, X.G.; Lin, H.F. The effects of key rock layer fracturing on gas extraction during coal mining over a large height. *Energy Sci. Eng.* **2021**, *4*, 520–534. [CrossRef]
11. Guo, H.; Yuan, L.; Shen, B.T.; Qu, Q.D.; Xue, J.H. Mining-induced strata stress changes, fractures and gas flow dynamics in multi-seam longwall mining. *Int. J. Rock Mech. Min. Sci.* **2012**, *54*, 129–139. [CrossRef]
12. Han, Y.C.; Cheng, J.L.; Huang, Q.S.; Zou, D.H.S.; Zhou, J.; Huang, S.H.; Long, Y. Prediction of the height of overburden fractured zone in deep coal mining: Case study. *Arch. Min. Sci.* **2018**, *63*, 617–631. [CrossRef]
13. Sun, X.Y.; Zhang, Q.; Li, C.; Zhang, L. Comparative simulation study on the influence of double-seam mining on overburden strata in a northern Shaanxi mine. *Coal Geol. Explor.* **2020**, *48*, 183–189. [CrossRef]
14. Yang, D.M.; Guo, W.B.; Zhao, G.B.; Tan, Y.; Yang, W.Q. Height of water-conducting zone in longwall top-coal caving mining under thick alluvium and soft overburden. *J. China Coal Soc.* **2019**, *44*, 3308–3316. [CrossRef]
15. Zhang, D.; Wang, J.C.; Zhang, P.S.; Shi, B. Internal strain monitoring for coal mining similarity model based on distributed fiber optical sensing. *Measurement* **2017**, *97*, 234–241. [CrossRef]
16. Du, W.G.; Chai, J.; Zhang, D.D.; Lei, W.L. The study of water-resistant key strata stability detected by optic fiber sensing in shallow–buried coal seam. *Int. J. Rock Mech. Min. Sci.* **2021**, *141*, 104604. [CrossRef]
17. Zhang, D.D.; Chen, Q.; Wang, Z.S.; Yang, J.F.; Chai, J. Optical Fiber Frequency Shift Characterization of Overburden Deformation in Short-Distance Coal Seam Mining. *Geofluids* **2021**, *2021*, 1751256. [CrossRef]
18. Zhang, G.C.; Tao, G.Z.; Meng, X.J.; Li, Y.; Qu, Z.; Xu, R.H.; Yu, S.C.; Chen, M.; Zhou, G.L.; Luan, H.J. Failure law of weak overburden stratum under lying extra-thick alluvium. *J. China Coal Soc.* **2022**, *47*, 3998–4010. [CrossRef]
19. Wang, W.X.; Sui, W.H.; Dong, Q.H.; Hu, W.W.; Gu, S.X. Post-mining closure effect of overburden fractures under unconsolidated layers and prediction of overburden failure in repeated mining. *J. China Coal Soc.* **2013**, *38*, 1728–1734. [CrossRef]
20. Cheng, J.L.; Song, G.D.; Sun, X.Y.; Wen, L.F.; Li, F. Research Developments and Prospects on Microseismic Source Location in Mines. *Engineering* **2018**, *4*, 653–660. [CrossRef]
21. Hou, W.G.; Cheng, J.L.; Li, D.; Zhang, P.; Qin, J.H.; Chen, T. Determination of gas drainage layer in overburden rock of coal seam based on dynamic monitoring by borehole resistivity method: A case study of Liyazhuang coal mine. *Sci. Technol. Eng.* **2021**, *21*, 7046–7052. [CrossRef]
22. Zhang, P.S.; Xu, S.A.; Guo, L.Q.; Wu, R.X. Research progress and prospects of monitoring technology for deformation and failure of stope surrounding rock. *Coal Sci. Technol.* **2020**, *48*, 14–48. [CrossRef]
23. Li, S.; Fan, C.J.; Luo, M.K.; Yang, Z.H.; Lan, T.W.; Zhang, H.F. Structure and deformation measurements of shallow overburden during top coal caving longwall mining. *Int. J. Min. Sci. Technol.* **2017**, *27*, 1081–1085. [CrossRef]
24. He, X.; Zhao, Y.X.; Zhang, C.; Han, P.H. A model to estimate the height of the water-conducting fracture zone for longwall panels in Western China. *Mine Water Environ.* **2020**, *39*, 823–838. [CrossRef]
25. Yu, J.H.; Liu, J.J.; Zhang, H.H.; Lu, H.T. Research and application of wavelet neural network in electrical resistivity imaging inversion. *J. Appl. Geophys.* **2023**, *215*, 105114. [CrossRef]
26. Yue, J.H.; Zhang, H.R.; Yang, H.Y. Electrical prospecting methods for advance detection: Progress, problems, and prospects in Chinese coal mines. IEEE Geosci. *Remote Sens. Mag.* **2019**, *7*, 94–106. [CrossRef]
27. Cheng, J.L.; Yu, S.J. Simulation experiment on the response of resistivity to deformation and failure of overburden. *Chin J. Geophys.* **2000**, *43*, 699–706. [CrossRef]
28. Brace, W.F.; Orange, A.S. Electrical Resistivity Changes in Saturated Rock under Stress. *Science* **1966**, *153*, 1525–1526. [CrossRef] [PubMed]
29. Kaselow, A.; Shapiro, S.A. Stress sensitivity of elastic moduli and electrical resistivity in porous rocks. *J. Geophys. Eng.* **2004**, *1*, 1–11. [CrossRef]
30. Kahraman, S.; Yeken, T. Electrical resistivity measurement to predict uniaxial compressive and tensile strength of igneous rocks. *Bull. Mater. Sci.* **2010**, *33*, 731–735. [CrossRef]
31. Wu, R.X.; Hu, Z.A.; Hu, X.W. Principle of using borehole electrode current method to monitor the overburden stratum failure after coal seam mining and its application. *J. Appl. Geophys.* **2020**, *179*, 104111. [CrossRef]
32. Tao, T.; Han, P.; Ma, H.; Tan, H.D. 3D Time–lapse resistivity inversion. *Chin. J. Geophys.* **2024**, *67*, 3973–3988. [CrossRef]
33. Liu, Y.B.; Cheng, J.L.; Jiao, J.J.; Meng, X.X. Feasibility study on multi-seam upward mining of multi-layer soft–hard alternate complex roof. *Environ. Earth Sci.* **2022**, *81*, 424. [CrossRef]

34. Bai, X.X.; Cao, A.Y.; Wang, C.B.; Liu, Y.Q.; Xue, C.C.; Yang, X.; Yang, Y.; Wang, S.W.; Hao, Q. The focal mechanism and field investigations of mining-induced earthquake by super-thick and weak cementation overburden strata fracturing. *Geomech. Geophys. Geo-Energy Geo-Resour.* **2025**, *11*, 7. [CrossRef]
35. Cheng, J.L.; Chen, T.; Cheng, P.; Liu, Y.B.; Zhang, Y.Q.; Xu, Z.Z.; Cheng, Q. A Borehole Resistivity Full-Space Imaging Method for Monitoring the Evolution of Overburden Fractures in Coal Seams. CN 115680612 A, 13 December 2024.
36. Lang, J.; He, Q.Q.; Fan, X.M.; Huang, P.F.; Zhang, X.X. Prediction of airflow classification effect of wet coal based on BP neural network. *J. China Coal Soc.* **2021**, *46*, 1001–1010. [CrossRef]
37. Wen, Z.J.; Xu, C.L.; Gong, F.Q.; Zuo, Y.J.; Song, Z.Q. Mechanical response and impact tendency index correction of gangue-coal combined structure. *J. Cent. South Univ.* **2025**, *32*, 2288–2306. [CrossRef]
38. Mahetaji, M.; Brahma, J. A Critical Review of Rock Failure Criteria: A Scope of Machine Learning Approach. *Eng. Fail. Anal.* **2024**, *159*, 107998. [CrossRef]
39. Barthwal, H.; Mirko, V.D.B. Microseismicity observed in an underground mine: Source mechanisms and possible causes. *Geomech. Energy Envir.* **2020**, *22*, 100167. [CrossRef]
40. Himanshu, B.; Frank, J.C.; Mirko, V.D.B. 3-D attenuation tomography from microseismicity in a mine. *Geophys. J. Int.* **2019**, *219*, 1805–1817. [CrossRef]

materials

MDPI

Article

Experimental Study on Damage Evolution Characteristics of Granite Under Short-Term Freeze–Thaw Cycles

Xianda Yang [1], Peng Zeng [1,2,*], Kui Zhao [1,2,*], Dong Zhang [1], Hepeng Zhang [1], Nan Liang [1] and Lihui Sun [3]

1 School of Mining Engineering, Jiangxi University of Science and Technology, Ganzhou 341000, China; xdy141516@gmail.com (X.Y.); zhangdong6344@163.com (D.Z.); zhanghepengmail@163.com (H.Z.); ln961110@163.com (N.L.)

2 Jiangxi Province Key Laboratory of Safe and Efficient Mining of Rare Metal Resources, Ganzhou 341000, China

3 School of Mining and Geomatics Engineering, Hebei University of Engineering, Handan 056038, China; slh2002789@sina.com

* Correspondence: zengpeng@jxust.edu.cn (P.Z.); yglmf_zk@163.com (K.Z.)

Abstract

Rock engineering problems in short-term freeze–thaw zones and short-term freeze–thaw cycles can lead to a reduction in rock strength, thereby inducing engineering disasters. Granite in short-term freeze–thaw zones was selected as the research object. Taking the freezing time (1 h) required for the internal temperature of the rock to reach the target freezing temperature as the reference, freeze–thaw cycle tests with durations of 1 h, 2 h, and 3 h were carried out in sequence. Combined with uniaxial compression acoustic emission (AE) tests, the effects of freeze–thaw duration and freeze–thaw cycle number on the physical and mechanical properties and AE characteristics of the rock were systematically investigated. A multi-index damage characterization system was established. Results show that: (1) Both freeze–thaw duration and cycle number show a positive correlation with the attenuation of peak strength and elastic modulus. (2) With the increase in freeze–thaw duration and cycle number, the AE cumulative ringing count rate and cumulative energy rate show an exponential decay trend. (3) Their proportion shows an increasing trend with the increase in freeze–thaw duration and cycle number. (4) Establishing a multi-index coupled damage variable to replace the traditional single-index characterization can improve the objectivity and reliability of freeze–thaw damage assessment.

Keywords: granite; short-term freeze–thaw; uniaxial compression; acoustic emission; damage variable

1. Introduction

Although some southern regions of China are neither permafrost regions nor seasonal frozen ground regions, they are frequently affected by abrupt temperature changes and intermittent precipitation in winter, and driven by the coupling of repeated alternating short-term freeze–thaw cycles, resulting in a typical short-term freeze–thaw cycle effect. Although the single duration of such freeze–thaw processes is short, the superposition of their frequent occurrence characteristics and intense phase transition effects poses a significant threat to the structural integrity and long-term stability of rock masses. The severe low-temperature rain, snow and freezing disaster that struck southern China in 2008 fully demonstrated the severity of this risk: it triggered more than 3100 geological disasters directly across over 20 provinces (municipalities and autonomous regions) including

Jiangxi and Hunan, among which there were over 800 collapse disasters. Unlike the long-term stable freezing that characterizes the seasonal or permafrost regions in northern China, short-term freeze–thaw events in southern China exhibit unique features of short cycles, complex processes and repeated frost heave effects, causing the interior of rocks to withstand repeated disturbances induced by water-ice phase transitions in a short period of time [1]. Therefore, focusing on the research of physical and mechanical properties and damage evolution characteristics of rocks in short-term freeze–thaw zones not only makes up for the limitation that traditional freeze–thaw research mostly focuses on long-cycle frozen ground regions, but also provides a fundamental scientific basis for disaster early warning and stability analysis of geotechnical engineering in short-term freeze–thaw zones.

Rock freeze–thaw damage evolution has long been a key research focus in the field of geotechnical engineering, where the scientific setting of freeze–thaw duration parameters is a core prerequisite for experimental research. Table 1 [2–13] summarizes the key parameters of granite freeze–thaw tests conducted over the past three years. The results show that the freezing duration in existing studies is mostly set to more than 4 h; even though a small number of studies set the freezing duration to 3 h, they do not clearly specify the basis for such parameter setting. This indicates that the setting of freeze–thaw duration parameters for granite in short-term freeze–thaw zones lacks scientific support, and the relationship between these parameters and damage evolution still needs to be systematically investigated.

Table 1. Summary of the granite freeze–thaw cycle scheme.

Researcher	Lithology	Freeze–Thaw Time	Number of Cycles	Region	Time
Zhang	Granite	4 h, 4 h	0, 9, 18, 27	Permafrost	2025 [2]
Gong	Granite	4 h, 4 h	0, 25, 50, 100, 150, 200	Seasonally frozen ground	2025 [3]
Jia	Granite	4 h, 4 h	0, 5, 10, 20, 40, 80	Seasonally frozen ground	2024 [4]
Cao	Granite	4 h, 4 h	0, 10, 20, 30, 40	Seasonally frozen ground	2024 [5]
Song	Granite	4 h, 4 h	0, 25, 50, 100, 150, 200	Seasonally frozen ground	2024 [6]
Gong	Granite	4 h, 4 h	0, 50, 100, 200	Seasonally frozen ground	2024 [7]
Qi	Granite	4 h, 4 h	0, 20, 40, 60, 80, 100, 120	Seasonally frozen ground	2024 [8]
Liu	Granite	4 h, 4 h	0, 10, 20, 30, 40	Seasonally frozen ground	2024 [9]
Dun	Granite	3 h, 3 h	0, 20, 40, 70, 100	Permafrost	2023 [10]
Ullah	Granite	4 h, 4 h	0, 25	Short-term frozen ground	2023 [11]
Zhang	Granite	4 h, 4 h	0, 15, 30, 45, 60, 80	Seasonally frozen ground	2023 [12]
Yu	Granite	6 h, 6 h	0, 20, 40, 60	Seasonally frozen ground	2024 [13]

In recent years, AE technology has been widely applied to the real-time monitoring and mechanism investigation of rock damage evolution under freeze–thaw cycles, thanks to its high sensitivity to the microcrack propagation process. Existing studies have achieved a series of results focusing on the application of AE technology in freeze-thawed rocks.

At the level of basic parameters, indicators such as ringing count, cumulative energy, event rate, and energy rate can effectively characterize the evolutionary laws of rock failure stages under different freeze–thaw cycles. With the increase in the number of freeze–thaw cycles, some of these parameters generally decrease during the peak failure stage, indicating that the internal structural deterioration of rocks leads to a reduction in energy release capacity [14–18]. In terms of derivative parameter analysis, researchers can further reveal the microscopic mechanisms of freeze–thaw damage by means of the b-value, r-value, and RA–AF (rise angle–average frequency) method. The results show that with the accumulation of freeze–thaw cycles, both the b-value and r-value during rock failure exhibit regular variations; the distribution characteristics of RA-AF indicate that the proportion of tensile cracks gradually increases, and the rock failure mode gradually transforms from shear-dominated to tensile-dominated [19–23]. In the aspect of waveform

https://doi.org/10.3390/ma19050853

characteristic analysis, researchers clarify the intrinsic waveform properties of AE signals through methods such as time-frequency analysis and spectral decomposition. For instance, Shi et al. adopted the hyper-element wavelet transform method proposed by Moca, which achieves time-frequency super-resolution while maintaining favorable temporal locality, and successfully elucidated the AE amplitude evolution laws of freeze–thawed granite within three frequency domains [24,25]. In the field of novel signal processing, Ullah realized continuous identification of the initiation point of volume expansion and the peak point of instability failure during granite deformation by fusing the AE energy curve with the AEEMR (AE energy moment rate) [11]. In terms of damage quantification, multifractal dimension and Mel-frequency cepstral coefficients (MFCCs) have been applied to the characterization of freeze–thaw damage degree and precursor early warning [6,26]. However, existing AE studies mostly focus on long-term freeze–thaw scenarios. Targeting the characteristics of high frequency and short duration of short-term freeze–thaw in southern China, the influence law of the coupling between freeze–thaw duration and cycle number on AE response characteristics has not yet been clarified; there is also a lack of a damage quantification method based on the collaboration of AE multi-parameters and physical-mechanical indicators. Therefore, the research on AE characteristics of rocks in short-term freeze–thaw zones still needs to be further advanced.

In view of this, this study takes granite in short-term freeze–thaw zones as the research object. Combining with the characteristics of meteorological data in short-term freeze–thaw zones, the freezing temperature was set at -10 °C and the thawing temperature at 20 °C. Through preliminary freeze–thaw tests, the critical duration required for specimen cores to reach the target freezing temperature (1 h) was determined, and then freeze–thaw cycle tests with durations of 1 h, 2 h, and 3 h were designed. Subsequently, uniaxial compression AE tests were conducted on the freeze–thawed specimens. This study focuses on investigating the variation laws of porosity, peak strength and elastic modulus of granite specimens after freeze–thaw cycles, explores the AE ringing count, cumulative ringing count rate, energy rate, cumulative energy rate and RA-AF characteristics during the rock failure process, and finally, based on the elastic modulus, peak strength, porosity, cumulative ringing counts, and cumulative energy rate, the coupled damage variables of these five parameters were calculated using the Entropy Weight Method. Subsequently, the damage evolution characteristics of the rock under different freeze–thaw durations and cycle numbers were systematically analyzed. The research findings aim to clarify the damage evolution laws of granite under short-term freeze–thaw conditions and provide a theoretical basis for the stability analysis of geotechnical engineering in short-term freeze–thaw environments.

2. Materials and Methods

2.1. Specimen Preparation

The granite specimens used in the test were collected from short-term freeze–thaw zones and processed into standard cylindrical specimens with a diameter of 50 mm and a height of 100 mm. Specimens with obvious surface defects were discarded, and an ultrasonic velocity tester was employed to select samples with consistent wave velocities for subsequent experiments. XRD analysis revealed that the main mineral components of the granite were as follows: quartz (46.8%), albite (27.6%), potassium feldspar (15.8%), mica (6.3%), and calcite (3.5%), as shown in Figure 1.

https://doi.org/10.3390/ma19050853

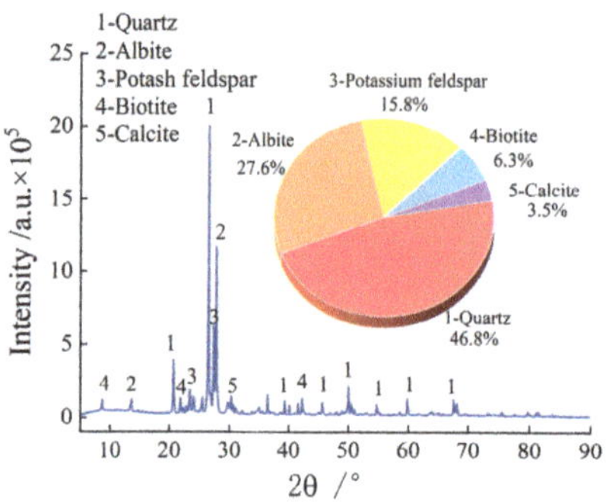

Figure 1. Granitic composition.

2.2. Parameters Related to Freeze–Thaw Tests

2.2.1. Target Freeze–Thaw Temperatures

Based on the Frozen Soil Zoning and Classification Map of China (Figure 2), an investigation of the winter temperature data of the area where the granite was collected shows that the daily maximum temperature can exceed 15 °C. Meanwhile, the superposition effect of topographic relief in mountainous areas and surrounding environmental factors results in significant diurnal temperature variation characteristics in some regions, with the minimum temperature dropping to the range of −5 °C to −10 °C. The duration of sub-zero temperatures in winter lasts from several hours to several days, which is basically consistent with the working conditions of short-term freeze–thaw [1]. Considering the freeze–thaw test temperatures reported in relevant domestic and international studies, the target temperatures designed for this test were set as −10 °C for freezing and 20 °C for thawing.

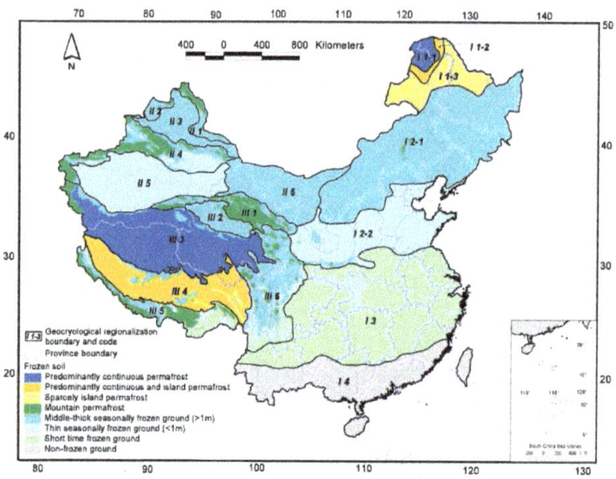

Figure 2. Geocryological regionalization and classification map of the frozen soil in China [27].

2.2.2. Number of Freeze–Thaw Cycles

Based on previous research experience and the relevant parameters of granite freeze–thaw tests summarized in Table 1, the number of freeze–thaw cycles in the test was determined as 25, 50, 75, and 100 cycles, respectively. In addition, a separate non-freeze–thaw cycle group was set up as the control group.

https://doi.org/10.3390/ma19050853

2.2.3. Freeze–Thaw Cycle Duration

Targeting the typical characteristics of short duration and high frequency of freeze–thaw cycles in short-term freeze–thaw zones, the core objective of this test design is to determine the period from the start of cooling to the moment when the rock reaches the preset target freezing temperature. For frozen soil and rocks in seasonal permafrost regions, the freeze–thaw duration in relevant test schemes ranges from 4 to 6 h. Based on a comprehensive review of freeze–thaw protocols and findings from both domestic and international studies, researchers recommend a freeze–thaw protocol of 2 h freezing and 4 h thawing for rocks with a porosity greater than 10%. For dense and hard rocks with a porosity less than 10%, a duration of 1 h for both freezing and thawing is suggested [28]. Therefore, in this study, the specific freeze–thaw durations were determined through subsequent preliminary experiments.

The specimens were subjected to forced saturation for 48 h using a ZK-270 rock vacuum saturation device; after calculation, the degree of saturation is 99.11%. A hole with a diameter of 2 mm and a depth of 26 mm was drilled at the center of the axial curved surface of each specimen, and a PT100 platinum resistance temperature sensor was inserted deep into the hole. The opening of the hole on the end face was coated with 704 adhesive to isolate the hole from external water and air. After the adhesive solidified, the specimens were immersed in water for inspection; the absence of air bubbles indicated a qualified seal. Subsequently, the specimens were wrapped with polyethylene film to maintain a saturated state. The data sampling rate of the thermometer was set at 1 sample per minute.

With reference to previous research findings, and in distinction from the freeze–thaw test schemes applied to permafrost and seasonal frozen ground regions, the test was designed with a 3 h freezing stage and a 3 h thawing stage, where the durations of the heating and cooling processes were not included in the cycle time. The core temperature variation in the granite specimens is presented in Figure 3. Specifically, the start of cooling was recorded as point a; the moment when the temperature approached 0 °C was marked as point b, corresponding to a freezing duration of 5 ± 2 min; the end of the temperature plateau was denoted as point c, with a freezing duration of 12 ± 2 min; and the time when the target temperature of −10 °C was reached was defined as point d1, representing a stable freezing period of 60 ± 2 min. Notably, this duration exhibited no significant variation with the increase in the number of freeze–thaw cycles.

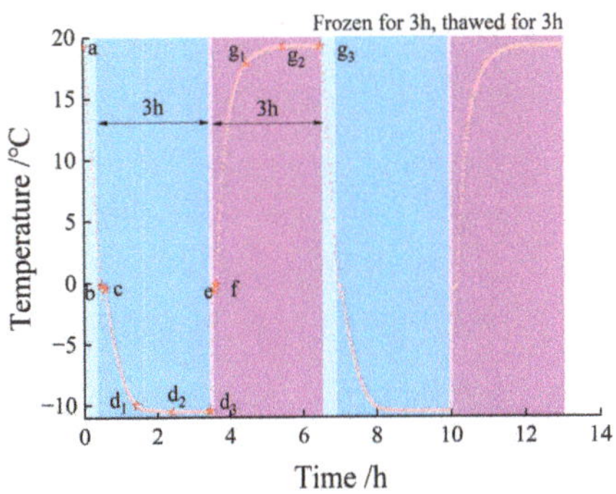

Figure 3. Temperature with 2 cycles of frozen and thawed for 3 h.

Based on the results of the preliminary freeze–thaw tests, the granite achieves stable freezing after approximately 1 h, with the core temperature reaching the target temperature at point d1. Therefore, the freeze–thaw test protocol was finally determined as follows: freezing for 1 h and thawing for 1 h, freezing for 2 h and thawing for 2 h, and freezing for

https://doi.org/10.3390/ma19050853

3 h and thawing for 3 h. Time points d1, d2, and d3 correspond to the endpoints of the 1 h, 2 h, and 3 h freezing stages, respectively, while time points g1, g2, and g3 correspond to the endpoints of the 1 h, 2 h, and 3 h thawing stages, respectively.

2.3. Freeze–Thaw Cycle Tests

For each testing condition, three identical granite specimens were prepared and tested. The specimens were placed in a drying oven at 105 °C for 24 h. After the specimens were allowed to cool naturally to room temperature, they were subjected to forced saturation for 48 h using a ZK-270 rock vacuum saturation device. According to the calculation, the saturation degree of all samples is above 99%.

The entire specimens were hermetically wrapped with polyethylene film to maintain a saturated state during freeze–thaw cycles. Freeze–thaw cycle tests were conducted in a DB-TH-22-D temperature and humidity test chamber (Danbo Co., Ltd., Danyang, Jiangsu, China). After the preset number of freeze–thaw cycles was completed, porosity tests were performed on the specimens using an NM-60 nuclear magnetic resonance (NMR) analyzer (Suzhou Niumag Analytical Instrument Corporation, Suzhou, Jiangsu, China), followed by a 24 h drying treatment of the specimens.

Uniaxial compression tests were carried out on an RMT-150C testing system (Institute of Rock and Soil Mechanics, Wuhan, Hubei, China) under a displacement-controlled mode with a loading rate of 0.002 mm/s. During the uniaxial compression tests, AE signal monitoring was performed synchronously. The AE monitoring system used was the Micro-II system (Physical Acoustics Corporation, Princeton Junction, NJ, USA), with a preamplifier gain of 40 dB, a sampling rate of 1 MSPS, a sampling length of 1024 points, and a pre-trigger time of 256 μs. The threshold level was set at 40 dB. The precise definition of an AE hit was controlled by three timing parameters: a peak definition time (PDT) of 50 μs, a hit definition time (HDT) of 100 μs, and a hit lockout time (HLT) of 300 μs. Processed through a band-pass filter of 20–500 kHz. Regarding the AE parameters used in this study, ringing count is defined as the number of times the signal amplitude exceeds the threshold, and the AE energy is defined as the 'absolute energy,' calculated as the time integral of the squared pre-amplified sensor signal voltage (V(t)) divided by a reference impedance of 10 kΩ. This parameter is quantified in units of attojoules (aJ, 10^{-18} J).

Figure 4 shows the experimental testing system. The Chinese meaning of "UC Device" in Figure 4 is "No operation without proper training".

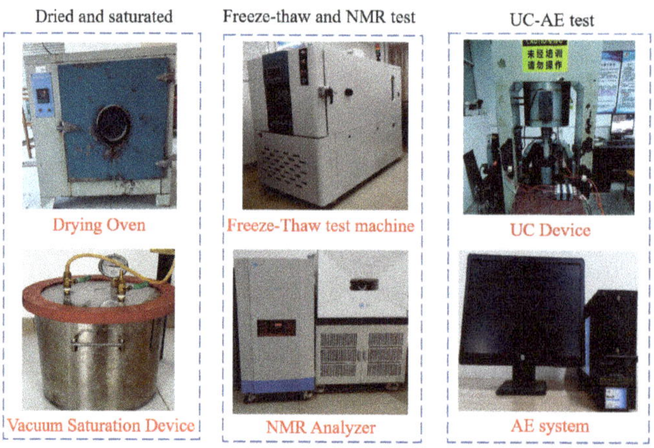

Figure 4. Diagram of the test system.

https://doi.org/10.3390/ma19050853

3. Results

3.1. Variation Law of Porosity

Figure 5 shows the variation law of specimen porosity with the number of freeze–thaw cycles. Error bars represent the standard deviation (SD) of the mean (n = 3).

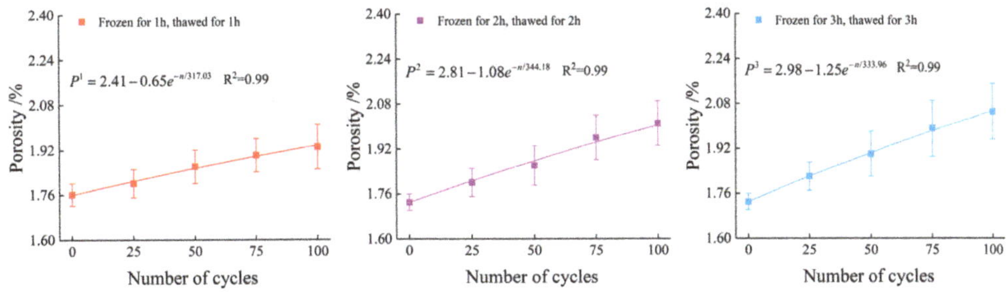

Figure 5. The variation trend of porosity.

To explicitly quantify the expansion of the pore structure, we introduced a normalized indicator, defined as the ratio of the porosity of the freeze–thaw-treated specimens to that of the unfrozen reference:

$$P_{norm} = \frac{P_N}{P_0} \tag{1}$$

where P_{norm} is the normalized porosity indicator; P_N represents the porosity for the specimen subjected to N freeze–thaw cycles; and P_0 denotes the initial porosity for the unfrozen reference specimen.

The evolutionary trends of the porosity indicator across 25, 50, 75, and 100 cycles for the 1 h, 2 h, and 3 h durations are presented as follows:

For the 1 h duration, the normalized values increased gradually to 1.02 (25 cycles), 1.06 (50 cycles), 1.08 (75 cycles), and 1.10 (100 cycles).

For the 2 h duration, a more pronounced expansion was observed, with the values rising to 1.04, 1.08, 1.13, and 1.16 at the corresponding cycle intervals.

For the 3 h duration, the most significant proliferation occurred, as the indicator surged to 1.05, 1.10, 1.15, and 1.18.

This detailed quantitative comparison confirms that increasing both the number of freeze–thaw cycles and the freeze–thaw duration leads to an exponential growth of the rock porosity.

It can be concluded that freeze–thaw duration and the number of freeze–thaw cycles exhibit a positive correlation with specimen porosity. A comparison of the specimens subjected to 100 freeze–thaw cycles reveals that the porosities of the specimens under 1 h, 2 h, and 3 h freeze–thaw conditions have increased by 1.10, 1.16, and 1.18, respectively. This demonstrates that at the same number of freeze–thaw cycles, the porosity of the specimens exhibits a nonlinear trend of accelerated growth followed by decelerated growth as the freeze–thaw duration increases.

At a fixed number of freeze–thaw cycles, the growth rate of porosity gradually diminished and tended to plateau as the freeze–thaw duration increased. This phenomenon is primarily attributed to the saturation of freeze–thaw damage within the rock. Specifically, pre-existing micro-cracks continuously propagated and coalesced to form macroscopic fissures. Simultaneously, the closure of certain micropores occurred under the action of local stress, leading to the gradual stabilization of the pore structure evolution.

Figure 6 presents the nuclear magnetic resonance (NMR) T2 spectra of the specimens under three freeze–thaw durations. According to the pore classification criteria for saturated

https://doi.org/10.3390/ma19050853

rocks [7], the pores can be categorized into three types, namely micropores, mesopores, and macropores.

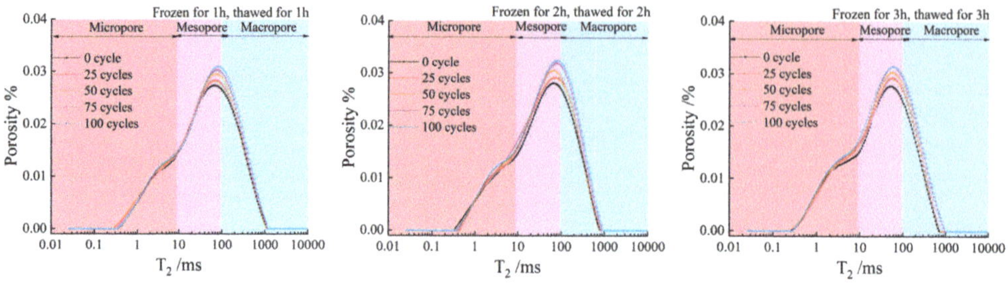

Figure 6. T2 spectrum distribution evolution of the three freeze–thaw durations.

Figure 7 presents the area percentages and growth rates of micropores, mesopores, and macropores inside the specimens after different freeze–thaw durations and cycle numbers. It can be seen from Figure 7 that the proportions of micropores and mesopores in all specimens decrease with the increase in the number of freeze–thaw cycles. Among these, the variation range of micropores is within 1%, that of mesopores is within 20%, and the proportion of macropores is greater than 80%.

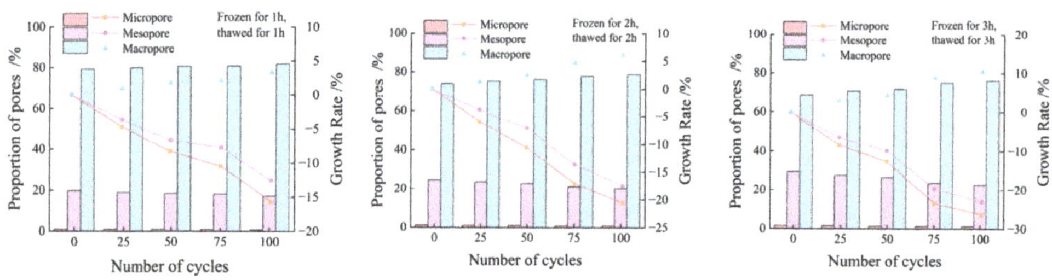

Figure 7. Proportion of each spectrum area and growth rate.

The longer the freeze–thaw duration is, the more fully the water in the internal fractures of the rock freezes during the freezing stage; the volume expansion of ice consequently intensifies the development of internal fractures. After thawing, the uneven stress inside the rock drives the seepage and migration of water. During the subsequent freezing cycle, this water redistribution further exacerbates crack propagation. After repeated freeze–thaw cycles, micropores and mesopores merge and expand into macropores. Thus, the proportion of micropores and mesopores decreases while that of macropores increases accordingly.

3.2. Variation Laws of Peak Strength and Elastic Modulus

Figure 8 illustrates the relationships between the peak strength, elastic modulus, and number of freeze–thaw cycles of the specimens under three freeze–thaw durations. Error bars represent the standard deviation (SD) of the mean (n = 3). As can be seen from Figure 8, with the increase in the number of freeze–thaw cycles, the peak strength and elastic modulus of the specimens generally exhibit a nonlinear attenuation trend.

To explicitly quantify the mechanical degradation, we introduced a normalized indicator, defined as the ratio of the peak strength of the freeze–thaw-treated specimens to that of the unfrozen reference:

$$\sigma_{norm} = \frac{\sigma_N}{\sigma_0} \quad (2)$$

 https://doi.org/10.3390/ma19050853

where σ_{norm} is the normalized peak strength indicator; σ_N represents the peak strength for the specimen subjected to N freeze–thaw cycles; and σ_0 denotes the peak strength for the unfrozen reference specimen.

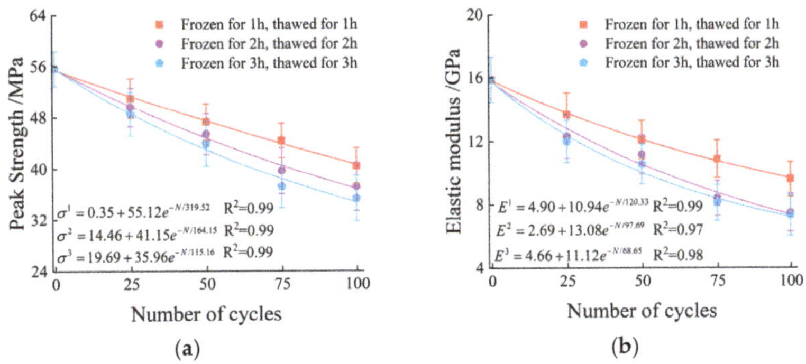

Figure 8. The variation trend of peak strength and elastic modulus: (**a**) characteristics of peak strength; (**b**) characteristics of elastic modulus.

The evolutionary trends of the peak strength indicator across 25, 50, 75, and 100 cycles for the 1 h, 2 h, and 3 h durations are presented as follows:

For the 1 h duration: The normalized values decreased gradually to 0.91 (25 cycles), 0.85 (50 cycles), 0.79 (75 cycles), and 0.72 (100 cycles).

For the 2 h duration, a more pronounced weakening was observed, with the values dropping to 0.89, 0.81, 0.73, and 0.66 at the corresponding cycle intervals.

For the 3 h duration, the most severe deterioration occurred, as the indicator plummeted sharply to 0.87, 0.78, 0.66, and 0.63.

This detailed quantitative comparison confirms that increasing both the number of freeze–thaw cycles and the freeze–thaw duration leads to an exponential degradation of the rock peak strength.

To explicitly quantify the degradation of rock stiffness, we introduced a normalized indicator, defined as the ratio of the elastic modulus of the freeze–thaw-treated specimens to that of the unfrozen reference:

$$E_{norm} = \frac{E_N}{E_0} \tag{3}$$

where E_{norm} is the normalized elastic modulus indicator; E_N represents the elastic modulus for the specimen subjected to N freeze–thaw cycles; and E_0 denotes the elastic modulus for the unfrozen reference specimen.

The evolutionary trends of the elastic modulus indicator across 25, 50, 75, and 100 cycles for the 1 h, 2 h, and 3 h durations are presented as follows:

For the 1 h duration, the normalized values decreased gradually to 0.86 (25 cycles), 0.76 (50 cycles), 0.68 (75 cycles), and 0.60 (100 cycles).

For the 2 h duration, a more pronounced reduction was observed, with the values dropping to 0.77, 0.70, 0.52, and 0.47 at the corresponding cycle intervals.

For the 3 h duration, the most severe deterioration occurred, as the indicator plummeted sharply to 0.75, 0.66, 0.51, and 0.46.

This detailed quantitative comparison confirms that increasing both the number of freeze–thaw cycles and the freeze–thaw duration leads to an exponential degradation of the rock elastic modulus.

The exponential attenuation of elastic modulus and peak strength can be intrinsically attributed to the pore structure evolution discussed in Section 3.1. Specifically, the domi-

https://doi.org/10.3390/ma19050853

nance of macropores (>80%) significantly weakens the rock, leading to the loss of bearing capacity and resistance to deformation.

3.3. AE Response Characteristics

3.3.1. Ringing Count Rate and Cumulative Ringing Count Rate

Figure 9 depicts the relationships between the AE ringing count rate, cumulative ringing count rate, and stress–time curve. During the initial loading stage, the ringing count rate remains at a low level, corresponding to the elastic deformation stage of the rock. At this stage, primary microcracks inside the rock undergo slight initiation, with a few microfracture events occurring. When the stress approaches the peak strength, the ringing count rate surges to the peak value, while the cumulative ringing count rate increases rapidly. Newly generated cracks propagate and coalesce, marking a period of intensive microfracture activity. The burst of AE signals corresponds well to the macroscopic failure of the rock.

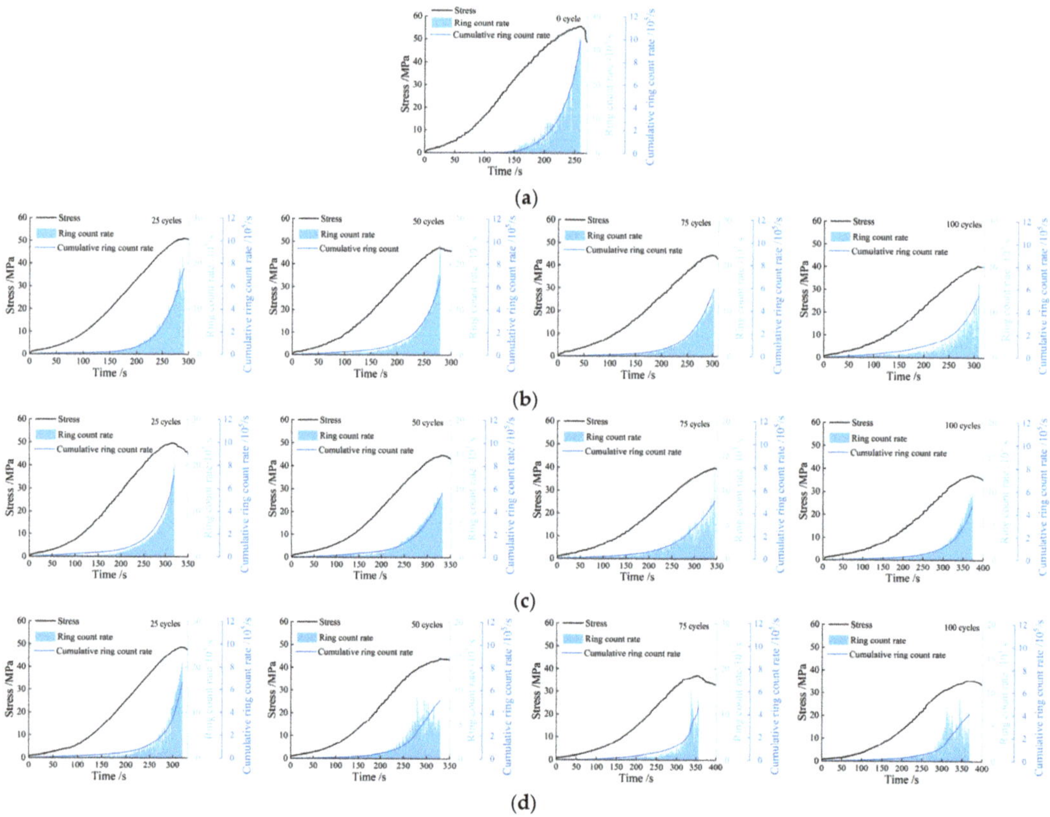

Figure 9. The relationship between the AE ringing count rate, cumulative ringing count rate, and stress–time under different cycle numbers. (**a**) Unfrozen-thawed; (**b**) 1 h freeze-thawed; (**c**) 2 h freeze-thawed; (**d**) 3 h freeze-thawed.

Taking the AE characteristics of specimen failure under the 3 h freeze–thaw condition as an example, after 25 freeze–thaw cycles, the peak stress is approximately 48 MPa, the peak ringing count rate exceeds 2×10^4, and the cumulative ringing count rate is about 6.6×10^5. In contrast, after 100 freeze–thaw cycles, the peak stress decreases to around 35 MPa, the peak ringing count rate is only about 1×10^4, and the cumulative value drops

https://doi.org/10.3390/ma19050853

significantly to approximately 4.1×10^5. With the increase in the number of freeze–thaw cycles, the ringing count rate generated during the specimen failure process decreases progressively. Figure 10 shows the relationship between the number of freeze–thaw cycles and the cumulative ringing count rate. It can be seen from the figure that the variation laws of the AE ringing count rate and cumulative ringing count rate during the failure of specimens under 1 h and 2 h freeze–thaw conditions are generally consistent with those of the specimens under 3 h freeze–thaw conditions.

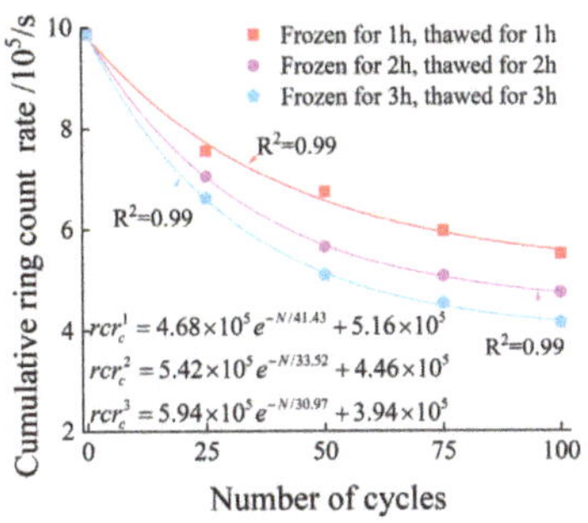

Figure 10. The relationship between freeze–thaw cycle numbers and cumulative ringing count rate.

To explicitly quantify the attenuation of AE activity, we introduced a normalized indicator, defined as the ratio of the cumulative ringing count rate at peak stress for the freeze–thaw-treated specimens to that of the unfrozen reference:

$$I_{norm} = \frac{crcr_N}{crcr_0} \quad (4)$$

where I_{norm} is the normalized cumulative ringing count indicator; $crcr_N$ represents the cumulative ringing counts at peak stress for the specimen subjected to N freeze–thaw cycles; and $crcr_0$ denotes the cumulative ringing counts at peak stress for the unfrozen reference specimen.

The evolutionary trends of the cumulative ringing counts indicator across 25, 50, 75, and 100 cycles for the 1 h, 2 h, and 3 h durations are presented as follows:

For the 1 h duration, the normalized values decreased gradually to 0.76 (25 cycles), 0.68 (50 cycles), 0.60 (75 cycles), and 0.55 (100 cycles).

For the 2 h duration, a more pronounced attenuation was observed, with the values dropping to 0.71, 0.57, 0.51, and 0.47 at the corresponding cycle intervals.

For the 3 h duration, the most severe deterioration occurred, as the indicator plummeted sharply to 0.66, 0.51, 0.45, and 0.42.

This detailed quantitative comparison confirms that increasing both the number of freeze–thaw cycles and the freeze–thaw duration leads to an exponential degradation of the cumulative ringing count rate.

3.3.2. Energy Rate and Cumulative Energy Rate

Figure 11 depicts the relationships between the AE energy rate, cumulative energy rate, and the stress–time curve. During the initial loading stage, the energy rate remains at a low level, corresponding to the elastic deformation stage of the rock. At this stage, primary microcracks inside the rock undergo slight initiation, accompanied by low energy release from a small number of microfracture events. When the stress approaches the peak strength, the energy rate surges to its peak value, while the cumulative energy rate increases rapidly.

 https://doi.org/10.3390/ma19050853

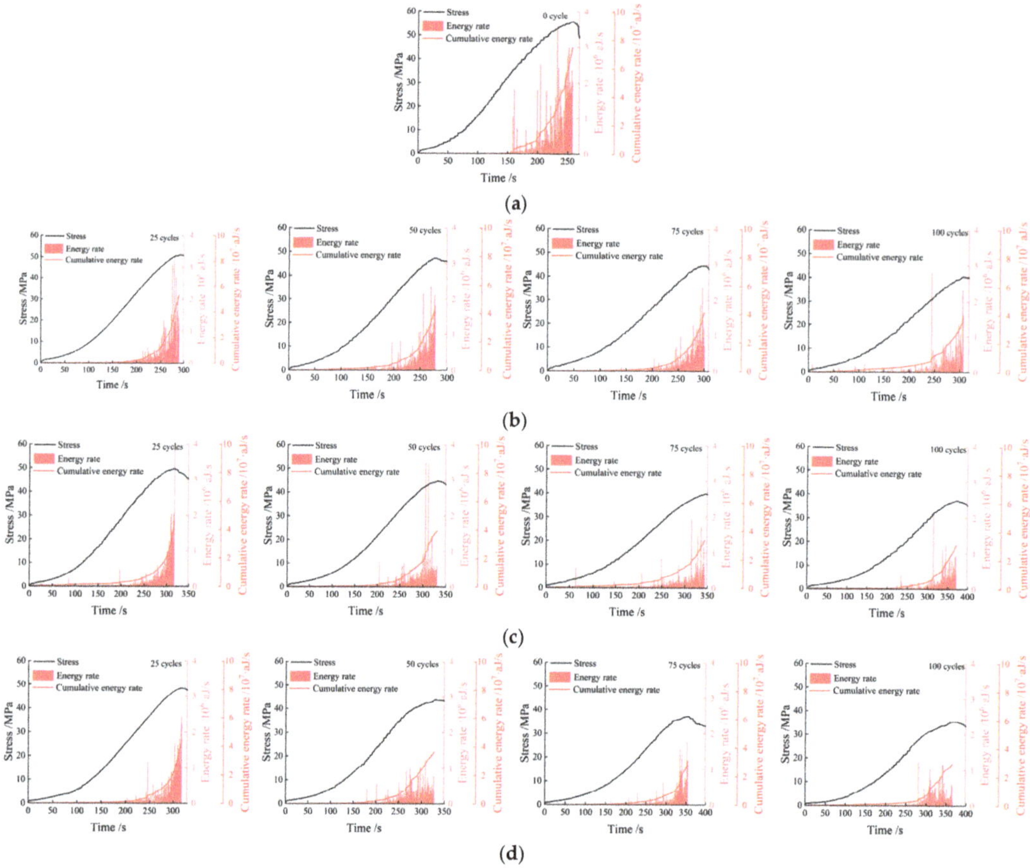

Figure 11. The relationship between the AE ringing energy rate, cumulative energy rate, and stress–time under different cycle numbers. (**a**) Unfrozen-thawed; (**b**) 1 h freeze-thawed; (**c**) 2 h freeze-thawed; (**d**) 3 h freeze-thawed.

Taking the specimens under the 3 h freeze–thaw condition as an example, after 25 freeze–thaw cycles, the peak stress of the specimens is approximately 48 MPa, the peak energy rate reaches 2.4×10^6 aJ/s, and the cumulative energy rate is about 4.4×10^7 aJ/s. In contrast, after 100 freeze–thaw cycles, the peak stress decreases to around 35 MPa, the peak energy rate drops to below 1×10^6 aJ/s, and the cumulative energy rate decreases synchronously to 2.9×10^7 aJ/s. With the increase in the number of freeze–thaw cycles, the energy generated during the failure process of the specimens decreases progressively. Figure 12 shows the relationship between the number of freeze–thaw cycles and the cumulative energy rate. It can be seen from the figure that the variation laws of the AE energy rate and cumulative energy rate during the failure of specimens under 1 h and 2 h freeze–thaw conditions are generally consistent with those of the specimens under 3 h freeze–thaw conditions.

To explicitly quantify the attenuation of AE energy release, we introduced a normalized indicator, defined as the ratio of the cumulative energy rate at peak stress for the freeze–thaw-treated specimens to that of the unfrozen reference:

$$J_{norm} = \frac{cer_N}{cer_0} \tag{5}$$

https://doi.org/10.3390/ma19050853

where J_{norm} is the normalized cumulative energy rate indicator; cer_N represents the cumulative energy at peak stress for the specimen subjected to N freeze–thaw cycles; and cer_0 denotes the cumulative energy at peak stress for the unfrozen reference specimen.

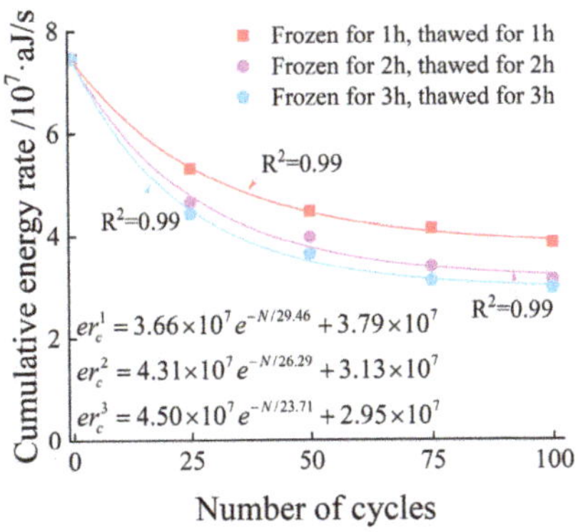

Figure 12. The relationship between freeze–thaw cycle numbers and cumulative energy rate.

The evolutionary trends of the cumulative energy rate indicator across 25, 50, 75, and 100 cycles for the 1 h, 2 h, and 3 h durations are presented as follows:

For the 1 h duration, the normalized values decreased gradually to 0.71 (25 cycles), 0.60 (50 cycles), 0.55 (75 cycles), and 0.51 (100 cycles).

For the 2 h duration, a more pronounced attenuation was observed, with the values dropping to 0.62, 0.53, 0.45, and 0.41 at the corresponding cycle intervals.

For the 3 h duration, the most severe deterioration occurred, as the indicator plummeted sharply to 0.59, 0.48, 0.44, and 0.39.

This detailed quantitative comparison confirms that increasing both the number of freeze–thaw cycles and the freeze–thaw duration leads to an exponential degradation of the cumulative energy rate.

3.3.3. RA-AF Distribution Characteristics

The classification of crack modes using AE parameters RA (the ratio of rise time to amplitude) and AF (average frequency) was initially standardized by the Japanese Construction Materials Standard (JCMS-III B5706) in 2003 [29]. By defining a diagonal transition zone in the RA-AF coordinate system, this method effectively differentiates tensile from shear cracking, as shown in Figure 13. Given its robustness in waveform analysis, this approach has been extensively implemented in rock mechanics to track the transition of micro-cracking mechanisms under various loading conditions. The calculation procedures are as follows [30]:

$$RA = \frac{Rise\ time}{Amplitude} \tag{6}$$

$$AF = \frac{Count}{Duration\ \text{time}} \tag{7}$$

Figure 14 shows the microcrack type diagrams of the specimens during failure after different freeze–thaw durations and cycle numbers. Among them, the internal microcracks of the unfrozen-thawed specimens are dominated by the tensile mode, accounting for 83.96%. For the specimens subjected to 1 h freeze–thaw, after 25, 50, 75, and 100 freeze–thaw cycles, the proportions of tensile microcracks are 85.39%, 87.49%, 88.56%, and 90.2%, respectively. As for the specimens with 2 h freeze–thaw, the proportions of tensile microcracks are 87.00%, 88.95%, 91.35%, and 92.89%, respectively, after 25, 50, 75, and 100 freeze–thaw

 https://doi.org/10.3390/ma19050853

cycles. For the specimens under 3 h freeze–thaw conditions, after 25, 50, 75, and 100 freeze–thaw cycles, the proportions of tensile microcracks reach 89.37%, 92.32%, 93.69%, and 94.39%, respectively.

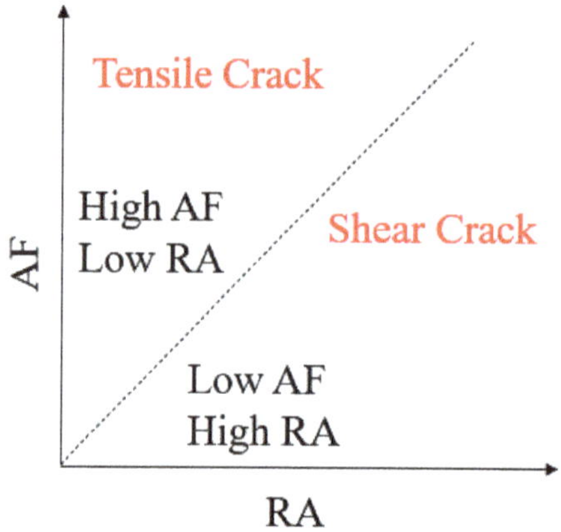

Figure 13. Cracking pattern resolution based on RA-AF analysis.

With an increase in the number of freeze–thaw cycles, the proportion of tensile cracks in specimens subjected to 1 h, 2 h, and 3 h freeze–thaw durations increased from 83.96% for the unfrozen state to 90.2%, 92.89%, and 94.39% after 100 freeze–thaw cycles, respectively, and the failure mode of the specimens was tensile-dominated.

It can be seen from Figure 15 that under the same freeze–thaw duration, the proportion of tensile microcracks inside the specimens shows an increasing trend with the increase in the number of freeze–thaw cycles. At the same number of freeze–thaw cycles, the proportion of tensile microcracks inside the specimens also increases gradually with the extension of freeze–thaw duration. These results indicate that the number of freeze–thaw cycles and freeze–thaw duration exhibit a positive correlation with the proportion of tensile microcracks inside the specimens.

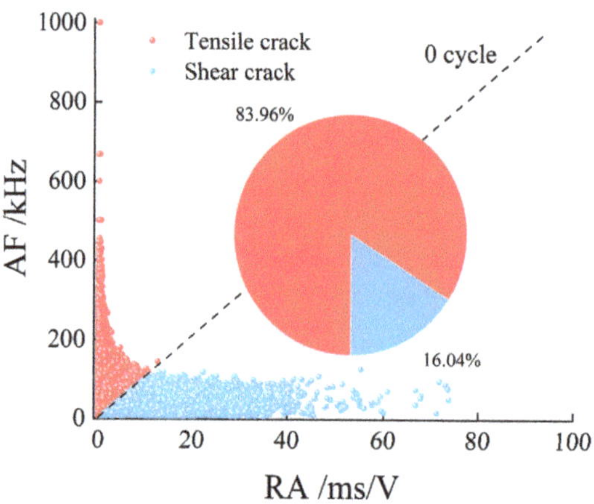

Figure 14. *Cont.*

 https://doi.org/10.3390/ma19050853

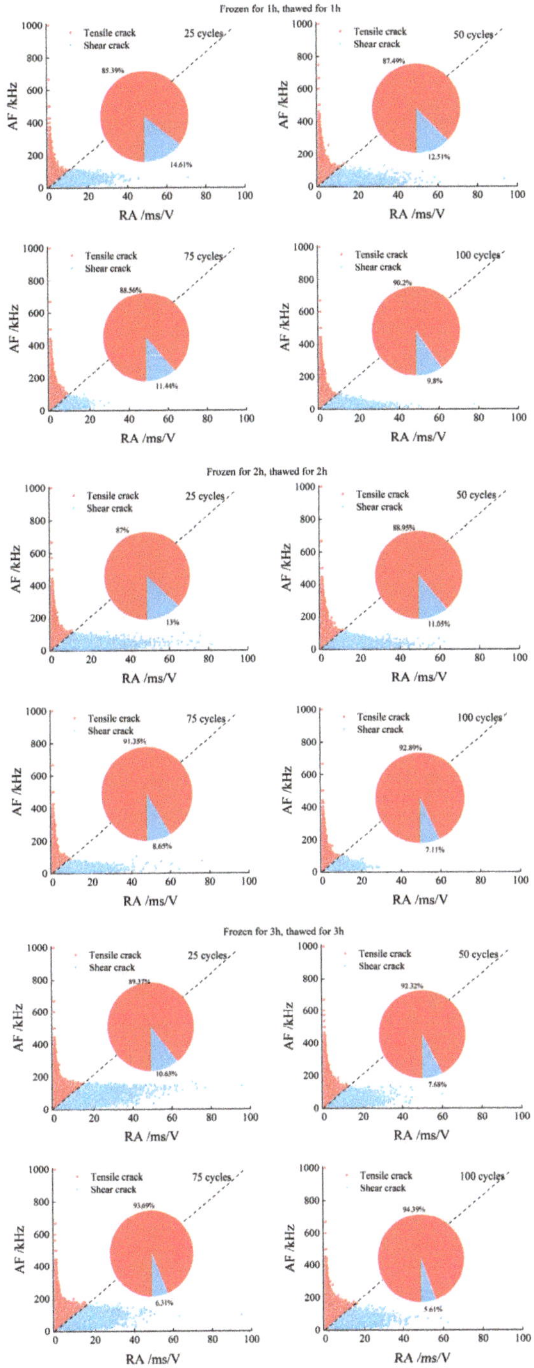

Figure 14. The AF/RA distribution of AE signals from rock failure.

https://doi.org/10.3390/ma19050853

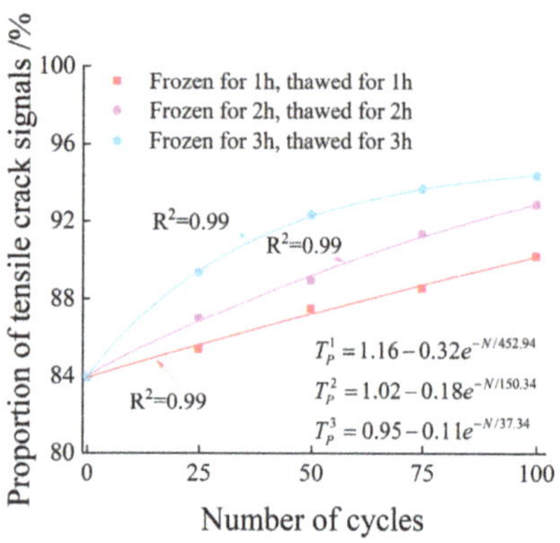

Figure 15. Changes in the proportion of tensile crack signals.

3.4. Rock Damage Evolution Characteristics

3.4.1. Damage Characteristics Based on Peak Strength

The frost heaving force induced by the freezing expansion of pore water during the freeze–thaw cycles of rock drives the initiation, propagation, and coalescence of internal microcracks, resulting in a decline in rock structural integrity. The accumulation of such micro-damage impairs the load-bearing capacity of the rock, which is manifested as the attenuation of peak strength. According to the principles of damage mechanics, the damage variable (D_σ) based on the peak strength of rock after freeze–thaw cycles can be expressed as follows:

$$D_\sigma = 1 - \sigma_n/\sigma_0 \tag{8}$$

where σ_n is the peak strength of the rock after N freeze–thaw cycles, and σ_0 is the peak strength of the unfrozen-thawed rock. The relationship between the number of freeze–thaw cycles and the damage value (D_σ) is presented in Figure 16.

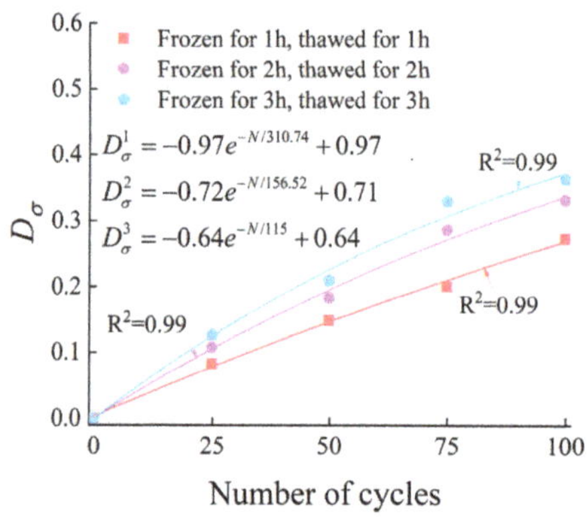

Figure 16. Relationship between the D_σ and the number of cycles.

3.4.2. Damage Characteristics Based on Elastic Modulus

The variation in elastic modulus can reflect the internal damage state of rock. Therefore, the damage variable based on the elastic modulus of rock after freeze–thaw cycles can be expressed as follows:

$$D_E = 1 - E_n/E_0 \tag{9}$$

https://doi.org/10.3390/ma19050853

where E_0 is the elastic modulus of the unfrozen-thawed rock, and E_n is the elastic modulus of the rock after N freeze–thaw cycles.

The relationship between the number of freeze–thaw cycles and the damage value (D_E) is presented in Figure 17.

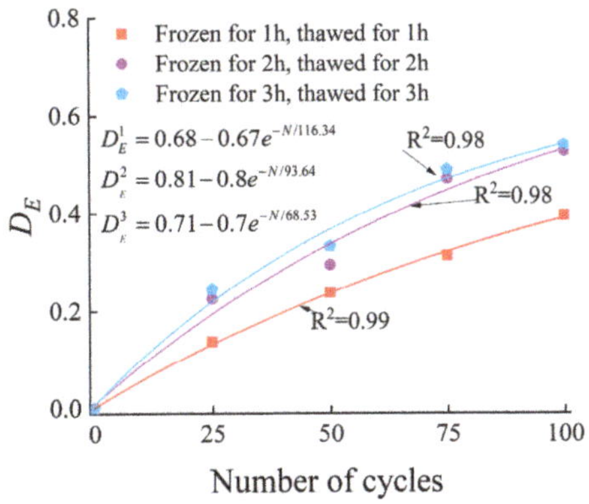

Figure 17. Relationship between the D_E and the number of cycles.

3.4.3. Damage Characteristics Based on Porosity

By establishing a quantitative relationship between the elastic modulus and porosity of rock, the influence mechanism of pore structure evolution on the elastic modulus of rock after freeze–thaw cycles was investigated [10]. Based on the calculation method proposed in Reference [10], the porosity variation (ΔP) of rock after freeze–thaw cycles was obtained, with the detailed results presented in Figure 18 showing the relationship between porosity variation and elastic modulus. It can be concluded from the analysis that the elastic modulus decreases with the increase in porosity variation, and the two parameters exhibit a linear relationship.

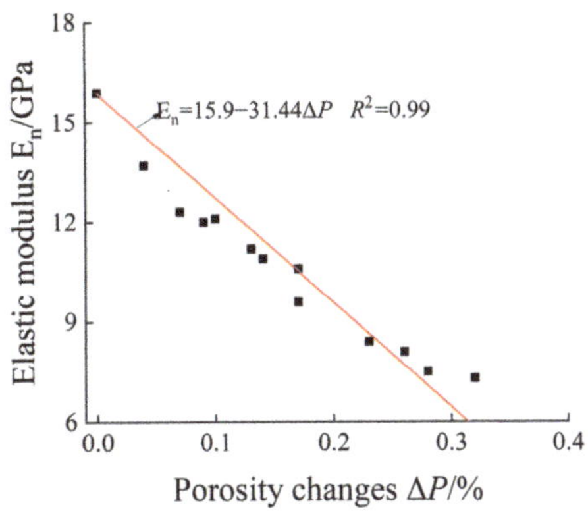

Figure 18. The relationship between elastic modulus and ΔP.

Therefore, by combining Formula (2) (the damage variable defined by elastic modulus) with the fitting formula in Figure 18, Formula (3) is derived, the damage variable based on porosity variation (D_P) is obtained, with the detailed results presented in Figure 19.

$$D_P = 1.97 \cdot \Delta P \quad (10)$$

https://doi.org/10.3390/ma19050853

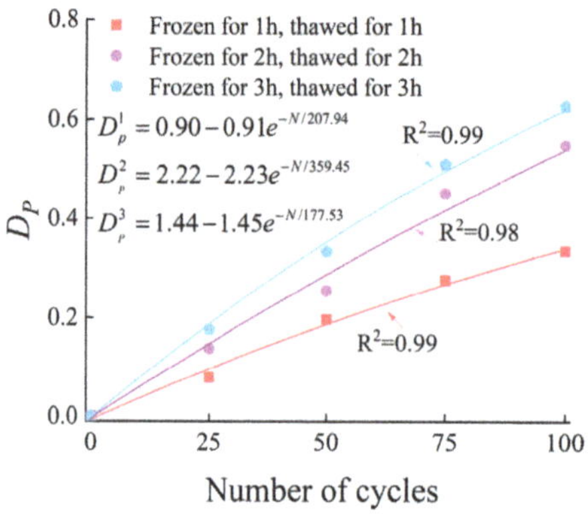

Figure 19. The relationship between D_P and the number of cycles.

3.4.4. Damage Characteristics Based on Cumulative Ringing Count Rate

According to the characteristics of the AE cumulative ringing count rate during the rock failure process, the damage variable based on the cumulative ringing count rate (D_{rcr}) can be expressed as follows:

$$D_{crcr} = 1 - \frac{crcr_N}{crcr_0} \tag{11}$$

where $crcr_N$ is the cumulative ringing count rate during the failure process of rock after N freeze–thaw cycles, and $crcr_0$ is the cumulative ringing count rate during the failure process of unfrozen-thawed rock.

The relationship between the number of freeze–thaw cycles and the damage value (D_{crcr}) is presented in Figure 20.

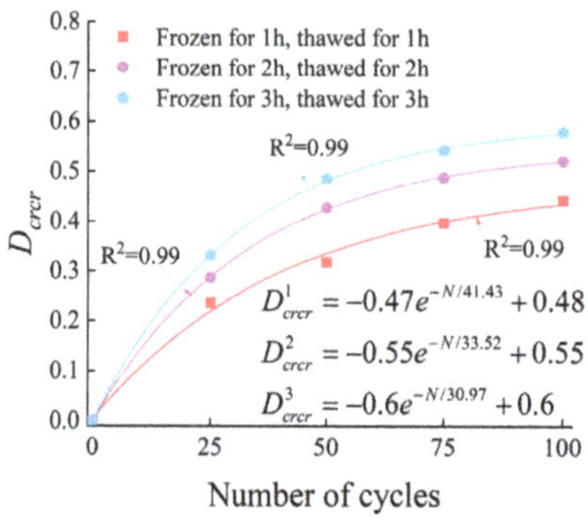

Figure 20. The relationship between D_{crcr} the number of cycles.

3.4.5. Damage Characteristics Based on Cumulative Energy Rate

Similarly, the damage variable based on cumulative energy rate (D_{cer}) can be expressed as follows:

$$D_{cer} = 1 - \frac{cer_N}{cer_0} \tag{12}$$

where cer_N is the cumulative energy rate during the failure process of rock after N freeze–thaw cycles, and cer_0 is the cumulative energy rate during the failure process of unfrozen-thawed rock.

The relationship between the number of freeze–thaw cycles and the damage value (D_{cer}) is presented in Figure 21.

https://doi.org/10.3390/ma19050853

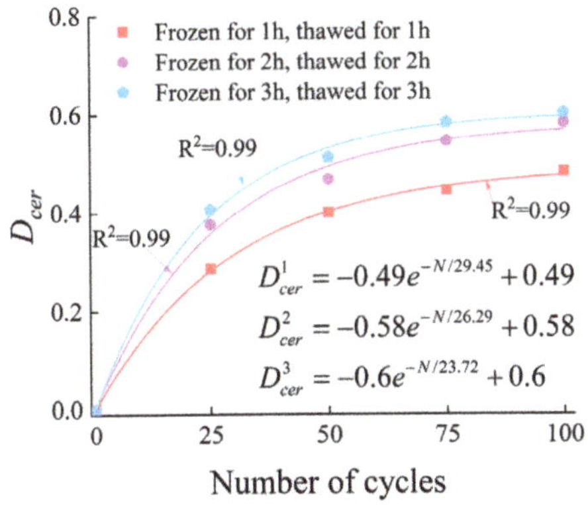

Figure 21. The relationship between D_{cer} and the number of cycles.

3.4.6. Multi-Index Coupling Damage Variable

The damage variables of granite calculated based on different characterization parameters exhibit a consistent evolution trend, while there are significant differences in the quantitative characteristics among various parameters (Tables 2–4). Such differences are likely to lead to discrepancies in the evaluation conclusions of freeze–thaw cycle damage effects, ultimately affecting the accurate characterization of rock damage degree.

Table 2. Different damage parameters under the 1 h freeze–thaw duration.

Freeze–Thaw Cycles	D_σ	D_E	D_P	D_{crcr}	D_{cer}
25	0.08	0.14	0.08	0.24	0.29
50	0.15	0.24	0.2	0.32	0.40
75	0.20	0.31	0.3	0.40	0.45
100	0.27	0.40	0.33	0.44	0.48

Table 3. Different damage parameters under the 2 h freeze–thaw duration.

Freeze–Thaw Cycles	D_σ	D_E	D_P	D_{crcr}	D_{cer}
25	0.11	0.23	0.14	0.29	0.38
50	0.18	0.30	0.25	0.43	0.47
75	0.29	0.47	0.45	0.49	0.55
100	0.33	0.53	0.55	0.52	0.58

Table 4. Different damage parameters under the 3 h freeze–thaw duration.

Freeze–Thaw Cycles	D_σ	D_E	D_P	D_{crcr}	D_{cer}
25	0.13	0.25	0.18	0.33	0.41
50	0.21	0.33	0.33	0.49	0.51
75	0.33	0.49	0.51	0.54	0.58
100	0.36	0.54	0.63	0.58	0.60

In view of this, the Entropy Weight Method (EWM) was utilized to determine the entropy coefficients for five damage indices—peak strength, elastic modulus, porosity, cumulative ringing count, and cumulative energy—to quantitatively evaluate their respective sensitivities in characterizing freeze–thaw damage.

https://doi.org/10.3390/ma19050853

Based on the damage variables presented in Tables 2–4, the proportion P_{ij} of the j-th damage parameter in the i-th freeze–thaw cycle is calculated as follows:

$$P_{ij} = \frac{D_{ij}}{\sum_{i=1}^{m} D_{ij}} \quad (13)$$

where D_{ij} is the damage value of the j-th damage parameter in the i-th freeze–thaw cycle.

Entropy reflects the degree of data dispersion. A higher degree of dispersion corresponds to a smaller entropy value, indicating that the indicator provides more valid information and thus carries a greater weight. The entropy value e_j for each damage indicator is calculated as follows:

$$e_j = -\frac{1}{ln(m)} \sum_{i=1}^{m} P_{ij} \ln(P_{ij}) \quad (14)$$

The coefficient of difference d_j characterizes the discriminating ability of each indicator. A larger coefficient implies that the damage parameter has a greater influence on the comprehensive evaluation. The formula is as follows:

$$d_j = 1 - e_j \quad (15)$$

Based on the coefficient of difference, the entropy weight w_j for each damage parameter is calculated, ensuring the sum of weights equals 1. The formula is as follows:

$$w_j = \frac{d_j}{\sum_{j=1}^{n} d_j} \quad (16)$$

where $\sum_{j=1}^{n} d_j$ is the sum of the difference coefficients for the n damage parameters.

The entropy weight results (w_j) are presented in Table 5. The analysis indicates that porosity consistently holds the maximum weight across different freeze–thaw durations, exhibiting a 'rise-then-fall' trend (0.360 to 0.410 to 0.391) with increasing duration. This reveals the phased mechanism of pore water phase transition and crack propagation: The increase in weight from 0.360 to 0.410 is primarily controlled by the degree of water-ice phase transition. At 1 h of freeze–thaw, pore water in the deep rock may not be fully frozen, resulting in insufficient phase transition and limited frost heaving. When the duration increases to 2 h, the phase transition becomes more complete, inducing rapid evolution from micro-crack initiation to coalescence. Consequently, the internal pore structure deteriorates significantly, leading to the strongest sensitivity for damage evaluation. As the duration further extends to 3 h, the weight decreases slightly. This is attributed to the fact that after prolonged frost heaving, the internal micro-crack network has largely coalesced. The deterioration of the pore structure enters a relatively stable period, and the variation amplitude of porosity tends to level off.

Table 5. Entropy weight of the damage variable.

Freeze–Thaw Time	D_σ	D_E	D_P	D_{crcr}	D_{cer}
Frozen for 1 h, thawed for 1 h	0.282	0.217	0.360	0.085	0.057
Frozen for 2 h, thawed for 2 h	0.279	0.186	0.410	0.080	0.045
Frozen for 3 h, thawed for 3 h	0.292	0.186	0.391	0.087	0.044

 https://doi.org/10.3390/ma19050853

The weights of macroscopic mechanical parameters (peak strength and elastic modulus) exhibit distinct evolutionary patterns. The weight of peak strength remains remarkably stable across different freeze–thaw durations (0.282, 0.279, and 0.292), indicating a consistent response to freeze–thaw damage. As a core mechanical parameter, peak strength effectively reflects the degradation of mechanical performance caused by frost action throughout the process. In contrast, the weight of the elastic modulus decreases from 0.217 at 1 h to 0.186 at 2 h and 3 h, where it stabilizes. At 1 h of freeze–thaw, the internal micro-cracks are primarily in the stages of compaction and initial propagation, contributing significantly to the comprehensive damage and thus resulting in a relatively higher weight. As the duration extends to 2–3 h, the internal micro-crack network gradually becomes interconnected and homogenized, leading to a weakened contribution and a subsequent decline in weight until stabilization.

The weights of AE indicators (cumulative ringing count and cumulative energy) consistently remain at a low level (total sum < 0.15). Specifically, the weight of cumulative energy exhibits a slight but continuous decline with increasing freeze–thaw duration (0.057, 0.045, 0.044), while the weight of the cumulative ringing count fluctuates marginally (0.085, 0.080, 0.087). During the initial stages of fewer cycles or shorter durations, the damage to the rock's dense structure leads to a decrease in both cumulative ringing count and energy. As the cycles increase and the duration extends, the internal micro-crack network becomes substantially interconnected. Consequently, the rock enters a deteriorated state characterized by low-energy release, where AE data tends to stabilize, and the numerical discrepancies between different experimental conditions diminish.

By integrating the five damage parameters (peak strength, elastic modulus, porosity, cumulative ringing count, and cumulative energy), a multi-indicator coupled damage variable ($\overline{D}$) based on the Entropy Weight Method is constructed as follows:

$$\overline{D} = w_\sigma D_\sigma + w_E D_E + w_P D_P + w_{crcr} D_{crcr} + w_{cer} D_{cer} \tag{17}$$

where w_σ, w_E, w_P, w_{crcr}, and w_{cer} are the entropy weight coefficients for the damage variables of peak strength, elastic modulus, porosity, cumulative ringing count, and cumulative energy, respectively.

It can be seen from Figure 22 that after 25, 50, 75, and 100 freeze–thaw cycles, the coupling damage variables of the specimens in the 1 h freeze–thaw group are 0.12, 0.21, 0.29, and 0.34, respectively; those of the 2 h freeze–thaw group are 0.17, 0.26, 0.41, and 0.48, respectively; and those of the 3 h freeze–thaw group are 0.20, 0.31%, 0.45, and 0.52, respectively.

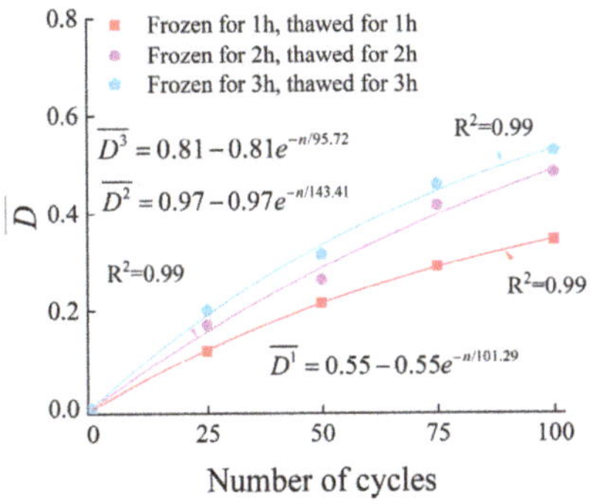

Figure 22. The relationship between $\overline{D}$ and the number of cycles.

The above results indicate that the coupling damage variables of specimens with different freeze–thaw durations all show a monotonically increasing trend with the increase

 https://doi.org/10.3390/ma19050853

in the number of freeze–thaw cycles. Specifically, for the specimens subjected to 1 h freeze–thaw, the damage increases from 0.12 (after 25 freeze–thaw cycles) to 0.34 (after 100 freeze–thaw cycles); for the 2 h freeze–thaw specimens, the damage rises from 0.17 to 0.48; and for the 3 h freeze–thaw specimens, the damage increases from 0.20 to 0.52. The total increase amplitudes of the damage variables are 0.22, 0.31, and 0.32, respectively, which further verifies the law that the extension of freeze–thaw duration aggravates damage accumulation while the increase amplitude gradually slows down.

Different types of parameters exhibit distinct limitations in characterizing the freeze–thaw damage of granite.

Macroscopic mechanical parameters (peak strength and elastic modulus) represent the macroscopic bearing capacity of the rock against deformation and failure. As the number of freeze–thaw cycles and duration increase, the propagation and coalescence of internal micro-cracks lead to the deterioration of mechanical properties, resulting in a non-linear growth of the damage variable. However, these parameters primarily reflect the macroscopic attenuation characteristics after damage has accumulated to a certain threshold, characterizing the final state of damage accumulation rather than fully revealing the micro-mesoscopic process of damage evolution.

Microstructural parameters (porosity) quantify the microstructure deterioration induced by frost action. Their values increase monotonically with the number of cycles, effectively characterizing the process from crack initiation and propagation to coalescence and demonstrating high sensitivity to damage accumulation throughout the entire freeze–thaw cycle. Nevertheless, as a static physical index, porosity only reflects the damage state at a specific moment and fails to characterize the dynamic mechanical behavior of crack propagation under loading.

AE parameters (cumulative ringing count and cumulative energy), serving as dynamic monitoring indices, can record real-time damage information during the loading process. However, in the late stages of freeze–thaw damage, the destruction of the rock medium's integrity causes significant attenuation and scattering of acoustic wave propagation. This leads to a reduction in cumulative AE values and a decline in resolution for distinguishing different damage levels, while also being susceptible to environmental interference.

By integrating the characterization information from macroscopic mechanics, microstructure, and AE dynamic response, the multi-indicator coupled damage variable effectively overcomes the limitations of using single parameters. Comparative analysis demonstrates that the evolution trend of the coupled damage variable is highly consistent with the patterns observed in individual parameters. It incorporates the high sensitivity of porosity to damage accumulation, accounts for the macroscopic characterization provided by mechanical parameters, and integrates the dynamic response features of AE parameters. Ultimately, this approach ensures that the damage assessment results possess higher objectivity and reliability.

4. Conclusions

(1) The increase in freeze–thaw duration and number of freeze–thaw cycles both aggravate the deterioration of peak strength and elastic modulus of granite, and the attenuation amplitude of strength and modulus decreases with the extension of freeze–thaw duration under the same number of freeze–thaw cycles. Compared with unfrozen-thawed specimens, after 100 freeze–thaw cycles, the normalized values of peak strength in the 1 h, 2 h, and 3 h freeze–thaw groups are 0.72, 0.66, and 0.63, respectively, and the normalized values of elastic modulus were 0.60, 0.47, and 0.46, respectively. This phenomenon is attributed to the fact that only the surface layer of the rock in the 1 h freeze–thaw group is fully frozen, and more freeze–thaw cycles are

 https://doi.org/10.3390/ma19050853

required for the accumulation of internal damage, whereas the 2 h and 3 h groups undergo more sufficient freezing, and the main damage evolution is completed in the early cycles.

(2) During the loading and failure process of granite, the AE cumulative ringing count rate and cumulative energy rate exhibit exponential attenuation with the increase in freeze–thaw duration and number of freeze–thaw cycles, and the attenuation amplitude decreases with the extension of freeze–thaw duration. Compared with unfrozen-thawed specimens, after 100 freeze–thaw cycles, the normalized values of the AE cumulative ringing count rate of specimens in the 1 h, 2 h, and 3 h freeze–thaw groups are 0.55, 0.47, and 0.42, respectively, and the normalized values of the cumulative energy rate are 0.51, 0.41, and 0.39, respectively.

(3) The failure process of all specimens is dominated by tensile cracks, and the proportion of tensile cracks shows an increasing trend with the extension of freeze–thaw duration and the accumulation of freeze–thaw cycles.

(4) A single damage parameter is insufficient to comprehensively characterize the extent of freeze–thaw damage. In this study, the Entropy Weight Method (EWM) was employed to determine the weight of each parameter, integrating peak strength, elastic modulus, porosity, cumulative AE ringing count, and cumulative energy to construct a multi-indicator coupled damage variable. Compared with traditional single-indicator methods, this approach effectively accounts for structural damage sensitivity, macroscopic characterization precision, and dynamic evolution features. Consequently, it enhances the objectivity and reliability of freeze–thaw damage assessment in granite.

(5) Despite the insightful findings, several limitations of this study should be acknowledged. First, the experiments were conducted on a single lithology (granite) from a specific site; thus, the generalizability to other rock types requires further validation. Second, this study focused on fully saturated specimens under uniaxial compression, which may not fully represent the complex triaxial stress states and partially saturated conditions found in deep-seated engineering projects. Additionally, only a single pair of freezing and thawing temperatures was tested. Future research will aim to incorporate triaxial loading, varying degrees of saturation, and different thermal amplitudes to enhance the broader applicability of the proposed damage model.

Author Contributions: Conceptualization, P.Z. and X.Y.; methodology, P.Z.; software, P.Z.; validation, P.Z., H.Z. and D.Z.; formal analysis, P.Z.; investigation, P.Z.; resources, P.Z.; data curation, P.Z.; writing—original draft preparation, X.Y.; writing—review and editing, P.Z.; visualization, D.Z., H.Z., L.S. and N.L.; supervision, P.Z.; project administration, P.Z.; funding acquisition, P.Z., K.Z. and L.S. All authors have read and agreed to the published version of the manuscript.

Funding: This research was funded by the Deep Earth Probe and Mineral Resources Exploration—National Science and Technology Major Project (2025ZD1010901), the National Natural Science Foundation of China (52464007, 52164004, 52474128), and the project supported by the Jiangxi Provincial Natural Science Foundation (20252BAC240059).

Institutional Review Board Statement: Not applicable.

Informed Consent Statement: Not applicable.

Data Availability Statement: The original contributions presented in this study are included in the article. Further inquiries can be directed to the corresponding authors.

Acknowledgments: The authors would like to acknowledge the Jiangxi University of Science and Technology.

Conflicts of Interest: The authors declare no conflict of interest.

 https://doi.org/10.3390/ma19050853

References

1. Wang, Y.Z.; Liu, X.J.; Ai, C.J.; Peng, Y. Experimental investigation on shear strength parameters C and φ for a temporarily frozen soil in South China. *Eng. J. Wuhan Univ.* **2010**, *43*, 198–202.
2. Zhang, Y.; Zhang, Y.; Jiang, X. Study of the degradation and microstructural characteristics of granite porphyry with freeze–thaw cycles. *Sci. Rep.* **2025**, *15*, 5091. [CrossRef]
3. Gong, Y.F.; Wu, S.Z.; Song, J.X. Pore characteristics and constitutive model of constrained granite under freeze-thaw cycles. *Constr. Build. Mater.* **2025**, *491*, 142706. [CrossRef]
4. Jia, S.; Yu, Q.; Yin, H.; Dai, Z.; Yin, S.; Kong, Y.; Thanh, H.V.; Soltanian, M.R. Analysis of damage evolution and study on mesoscopic damage constitutive model of granite under freeze–thaw cycling. *Bull. Eng. Geol. Environ.* **2024**, *83*, 236. [CrossRef]
5. Cao, J.; Hu, J.; Wang, X.; Yang, B.; Xia, Z.; Wang, H.; Zhang, L. Mechanical properties and acoustic emission characteristics of mixed granite after different numbers of freeze–thaw cycles. *Sci. Rep.* **2024**, *14*, 14074. [CrossRef] [PubMed]
6. Song, J.X.; Gong, Y.F.; Wu, S.Z.; Zhang, Y. Damage deterioration behavior of granite subjected to freeze-thaw cycles: Experiments and theories. *Constr. Build. Mater.* **2024**, *450*, 138540. [CrossRef]
7. Gong, Y.F.; Song, J.X.; Wu, S.Z.; Zhang, Y. Evolution of pore structure and analysis of freeze damage in granite during cyclic freeze-thaw using NMR technique. *Eng. Geol.* **2024**, *335*, 107545. [CrossRef]
8. Qi, L.; Wang, J.; Zhang, D.; Zhang, Y.; Ma, J. Evaluation of the influence of freeze–thaw cycles on the joint strength of granite in the Eastern Tibetan Plateau, China. *Sci. Rep.* **2024**, *14*, 24085. [CrossRef]
9. Liu, J.X.; Guo, C.B.; Deng, T.Y.; Ren, S. Investigation of the deterioration of Basu granite mechanical properties caused by freeze–thaw cycles in high-altitude mountains in the eastern part of the Tibetan plateau China. *Sustainability* **2024**, *16*, 319. [CrossRef]
10. Chen, D.; Li, G.; Li, J.; Du, Q.; Zhou, Y.; Mao, Y.; Qi, S.; Tang, L.; Jia, H.; Peng, W. Mechanical characteristics and damage evolution of granite under freeze–thaw cycles. *Front. Energy Res.* **2023**, *10*, 983705. [CrossRef]
11. Ullah, B.; Zhou, Z.; Cai, X.; Lu, J.; Zhao, C.C. Failure prediction and microcracks development based on acoustic emission and energy evolution for different rocks treated with freeze–thaw weathering. *Bull. Eng. Geol. Environ.* **2023**, *82*, 471. [CrossRef]
12. Zhang, C.Y.; Tan, T.; Ke, B.; Ma, B.; Dai, B.; Fang, Z.; Zhao, E. CT measurement of damage characteristics of meso-structure of freeze-thawed granite in cold regions and preliminary exploration of its mechanical behavior during a single freeze-thaw process. *Appl. Rheol.* **2023**, *33*, 20230106. [CrossRef]
13. Zhao, Y.; Liu, B.; Wang, C.L.; Bi, J.; Yang, Z. Study on deterioration mechanism of granite with a natural single weak surface after freeze-thaw treatment. *Cold Reg. Sci. Technol.* **2024**, *218*, 104079. [CrossRef]
14. Lian, S.; Zhao, Y.; Wan, W.; Wu, Q.; Gao, R.; Wang, M. Study on the freezing-thawing damage mechanism and evolution model of saturated gray sandstone based on NMR and AE technology. *Bull. Eng. Geol. Environ.* **2025**, *84*, 560. [CrossRef]
15. Wang, C.Y.; You, R.; Lv, W.Y.; Sui, Q.; Yan, Y.; Zhu, H. Damage Evolution and Acoustic Emission Characteristics of Sandstone under Freeze–Thaw Cycles. *ACS Omega* **2024**, *9*, 4892–4904. [CrossRef]
16. Chen, G.Q.; Guo, T.Y.; Serati, M.; Pei, B. Microcracking mechanisms of cyclic freeze–thaw treated red sandstone: Insights from acoustic emission and thin-section analysis. *Constr. Build. Mater.* **2022**, *329*, 127097. [CrossRef]
17. Wang, Y.; Feng, W.K.; Wang, H.J.; Li, C.; Hou, Z. Rock bridge fracturing characteristics in granite induced by freeze-thaw and uniaxial deformation revealed by AE monitoring and post-test CT scanning. *Cold Reg. Sci. Technol.* **2020**, *177*, 103115. [CrossRef]
18. Wang, Y.; Feng, W.K.; Wang, H.J.; Han, J.; Li, C. Geomechanical and Acoustic Properties of Intact Granite Subjected to Freeze-Thaw Cycles during Water-Ice Phase Transformation in Beizhan's Open Pit Mine Slope, Xinjiang, China. *Water* **2019**, *11*, 2309. [CrossRef]
19. Yang, H.M.; Song, Y.J.; Ren, J.X.; Xie, L.; Chen, Y. Study of the evolution of characteristic parameters and damage mechanism of sandstone under the synergistic effects of F-T and stepped cyclic loading-unloading via real-time CT scanning and AE monitoring. *Measurement* **2025**, *256*, 118078. [CrossRef]
20. Yu, C.Y.; Huang, S.H.; Li, J.K.; Wu, X.; Tian, Y.; Bao, X. Study on the Failure Process and Acoustic Emission Characteristics of Freeze–Thawed Sandstone under Cyclic Loading and Unloading. *Buildings* **2024**, *14*, 1264. [CrossRef]
21. Zhang, S.W.; Hang, L.; Chen, Y.F.; Wang, Y.; Zhao, Y. Acoustic emission and failure characteristics of cracked rock under freezing-thawing and shearing. *Theor. Appl. Fract. Mech.* **2022**, *121*, 103537. [CrossRef]
22. Lian, S.L.; Zheng, K.; Zhao, Y.; Bi, J.; Wang, C.; Huang, Y.S. Investigation the effect of freeze–thaw cycle on fracture mode classification in concrete based on acoustic emission parameter analysis. *Constr. Build. Mater.* **2023**, *362*, 129789. [CrossRef]
23. Li, H.R.; Xu, Z.; Wang, P.; Gong, X.; Jiang, S.; Tian, Y.; Qian, K.; Liu, Q. Study on the failure characteristics and acoustic emission precursor parameters of concrete under uniaxial compression after freeze–thaw damage. *Eng. Fail. Anal.* **2025**, *181*, 109981. [CrossRef]
24. Moca, V.V.; Bârzan, H.; Nagy, D.A.; Mureșan, R.C. Time-frequency super-resolution with superlets. *Nat. Commun.* **2021**, *12*, 337. [CrossRef] [PubMed]
25. Shi, C.D.; Nie, W.; Ma, G.W.; He, M.; Chen, Y. Effects of freeze-thaw cycles on granite failure using acoustic emission test. *J. Rock Mech. Geotech. Eng.* **2025**, *17*, 4512–4526. [CrossRef]

 https://doi.org/10.3390/ma19050853

26. Zhang, X.; Liu, X.; Jin, J.; Wang, H.; Lv, Z.; Xiao, Y. Investigation of damage characteristics in freeze-thawed grey-white sandstone using AE multifractal analysis. *Geomech. Geophys. Geo-Energ. Geo-Resour.* **2025**, *11*, 93. [CrossRef]
27. Nan, Z.T. *China 1:10 Million Permafrost Zoning and Type Map*; National Glacier, Frozen Soil and Desert; Science Data Center, Chinese Academy of Sciences: Beijing, China, 2019.
28. Shen, Y.J.; Yang, G.S.; Rong, T.L.; Liu, H.; Lv, W.Y. Proposed scheme for freeze-thaw cycle tests on rock. *Chin. J. Geotech. Eng.* **2016**, *38*, 1775–1782.
29. *JCMS-III B5706*; Monitoring Method for Active Cracks in Concrete by Acoustic Emission. Federation of Construction Materials Industries: Tokyo, Japan, 2003.
30. Bi, J.; Zhao, Y.; Wu, Z.; Li, J.; Wang, C. Research on crack classification method and failure precursor index based on RA-AF value of brittle rock. *Theor. Appl. Fract. Mech.* **2024**, *129*, 104179. [CrossRef]

https://doi.org/10.3390/ma19050853

MDPI AG
Grosspeteranlage 5
4052 Basel
Switzerland
Tel.: +41 61 683 77 34

MDPI Books Editorial Office
E-mail: books@mdpi.com
www.mdpi.com/books

www.ingramcontent.com/pod-product-compliance
Lightning Source LLC
Chambersburg PA
CBHW061443210926
51850CB00008B/1598
9783725872589